The IMA Volumes in Mathematics and its Applications

Volume 41

Series Editors
Avner Friedman Willard Miller, Jr.

Institute for Mathematics and its Applications
IMA

The **Institute for Mathematics and its Applications** was established by a grant from the National Science Foundation to the University of Minnesota in 1982. The IMA seeks to encourage the development and study of fresh mathematical concepts and questions of concern to the other sciences by bringing together mathematicians and scientists from diverse fields in an atmosphere that will stimulate discussion and collaboration.

The IMA Volumes are intended to involve the broader scientific community in this process.

Avner Friedman, Director
Willard Miller, Jr., Associate Director

* * * * * * * * * *

IMA ANNUAL PROGRAMS

1982–1983	**Statistical and Continuum Approaches to Phase Transition**
1983–1984	**Mathematical Models for the Economics of Decentralized Resource Allocation**
1984–1985	**Continuum Physics and Partial Differential Equations**
1985–1986	**Stochastic Differential Equations and Their Applications**
1986–1987	**Scientific Computation**
1987–1988	**Applied Combinatorics**
1988–1989	**Nonlinear Waves**
1989–1990	**Dynamical Systems and Their Applications**
1990–1991	**Phase Transitions and Free Boundaries**
1991–1992	**Applied Linear Algebra**
1992–1993	**Control Theory and its Applications**
1993–1994	**Emerging Applications of Probability**

IMA SUMMER PROGRAMS

1987	**Robotics**
1988	**Signal Processing**
1989	**Robustness, Diagnostics, Computing and Graphics in Statistics**
1990	**Radar and Sonar**
1990	**Time Series**
1991	**Semiconductors**
1992	**Environmental Studies: Mathematical, Computational, and Statistical Analysis**

* * * * * * * * * *

SPRINGER LECTURE NOTES FROM THE IMA:

The Mathematics and Physics of Disordered Media
Editors: Barry Hughes and Barry Ninham
(Lecture Notes in Math., Volume 1035, 1983)

Orienting Polymers
Editor: J.L. Ericksen
(Lecture Notes in Math., Volume 1063, 1984)

New Perspectives in Thermodynamics
Editor: James Serrin
(Springer-Verlag, 1986)

Models of Economic Dynamics
Editor: Hugo Sonnenschein
(Lecture Notes in Econ., Volume 264, 1986)

David A. Yuen
Editor

Chaotic Processes in the Geological Sciences

With 146 Illustrations, 20 in Full Color

Springer-Verlag
New York Berlin Heidelberg London Paris
Tokyo Hong Kong Barcelona Budapest

David A. Yuen
Minnesota Supercomputer Institute
Army High Performance Computing
and Research Center
University of Minnesota
Minneapolis, MN 55415
USA

Series Editors:
Avner Friedman
Willard Miller, Jr.
Institute for Mathematics and its
Applications
University of Minnesota
Minneapolis, MN 55455
USA

Mathematics Subject Classifications: 86-08, 86A04, 86A17, 76D07, 76D10, 76E15, 76E20, 76E25, 76E30, 76F05, 76F20, 76F99, 76R10, 76R50, 65M70, 65M60

Library of Congress Cataloging-in-Publication Data
Chaotic processes in the geological sciences [edited by] David Yuen.
p. cm. – (The IMA volumes in mathematics and its applications ; v. 41)
Papers from the Workshop on Chaotic Processes in the Geological Sciences, held June 11–15, 1990, at the Institute for Mathematics and Its Applications and at the Minnesota Supercomputer Institute, University of Minnesota Twin Cities campus.
Includes bibliographical references.
ISBN 0-387-97789-9. – ISBN 3-540-97789-9
1. Chaotic behavior in systems–Congresses. 2. Geophysics–Mathematics–Congresses. 3. Geology–Mathematics–Congresses.
I. Yuen, David. II. Series.
QE33.2.C48C52 1992
550′.1′17–dc20 91-46552

Printed on acid-free paper.

ISBN-13: 978-1-4684-0645-0

Production managed by Hal Henglein; manufacturing supervised by Robert Paella.
Camera-ready copy prepared by the IMA.

9 8 7 6 5 4 3 2 1

ISBN-13: 978-1-4684-0645-0 e-ISBN-13: 978-1-4684-0643-6
DOI: 10.1007/978-1-4684-0643-6

The IMA Volumes
in Mathematics and its Applications

Current Volumes:

Volume 1: Homogenization and Effective Moduli of Materials and Media
Editors: Jerry Ericksen, David Kinderlehrer, Robert Kohn, J.-L. Lions

Volume 2: Oscillation Theory, Computation, and Methods of Compensated Compactness
Editors: Constantine Dafermos, Jerry Ericksen, David Kinderlehrer, Marshall Slemrod

Volume 3: Metastability and Incompletely Posed Problems
Editors: Stuart Antman, Jerry Ericksen, David Kinderlehrer, Ingo Muller

Volume 4: Dynamical Problems in Continuum Physics
Editors: Jerry Bona, Constantine Dafermos, Jerry Ericksen, David Kinderlehrer

Volume 5: Theory and Applications of Liquid Crystals
Editors: Jerry Ericksen and David Kinderlehrer

Volume 6: Amorphous Polymers and Non-Newtonian Fluids
Editors: Constantine Dafermos, Jerry Ericksen, David Kinderlehrer

Volume 7: Random Media
Editor: George Papanicolaou

Volume 8: Percolation Theory and Ergodic Theory of Infinite Particle Systems
Editor: Harry Kesten

Volume 9: Hydrodynamic Behavior and Interacting Particle Systems
Editor: George Papanicolaou

Volume 10: Stochastic Differential Systems, Stochastic Control Theory and Applications
Editors: Wendell Fleming and Pierre-Louis Lions

Volume 11: Numerical Simulation in Oil Recovery
Editor: Mary Fanett Wheeler

Volume 12: Computational Fluid Dynamics and Reacting Gas Flows
Editors: Bjorn Engquist, M. Luskin, Andrew Majda

Volume 13: Numerical Algorithms for Parallel Computer Architectures
Editor: Martin H. Schultz

Volume 14: Mathematical Aspects of Scientific Software
Editor: J.R. Rice

Volume 15: Mathematical Frontiers in Computational Chemical Physics
Editor: D. Truhlar

Volume 16: Mathematics in Industrial Problems
by Avner Friedman

Volume 17: Applications of Combinatorics and Graph Theory to the Biological and Social Sciences
Editor: Fred Roberts

Volume 18: q-Series and Partitions
Editor: Dennis Stanton

Volume 19: Invariant Theory and Tableaux
Editor: Dennis Stanton

Volume 20: Coding Theory and Design Theory Part I: Coding Theory
Editor: Dijen Ray-Chaudhuri

Volume 21: Coding Theory and Design Theory Part II: Design Theory
Editor: Dijen Ray-Chaudhuri

Volume 22: Signal Processing: Part I - Signal Processing Theory
Editors: L. Auslander, F.A. Grünbaum, J.W. Helton, T. Kailath, P. Khargonekar and S. Mitter

Volume 23: Signal Processing: Part II - Control Theory and Applications of Signal Processing
Editors: L. Auslander, F.A. Grünbaum, J.W. Helton, T. Kailath, P. Khargonekar and S. Mitter

Volume 24: Mathematics in Industrial Problems, Part 2
by Avner Friedman

Volume 25: Solitons in Physics, Mathematics, and Nonlinear Optics
Editors: Peter J. Olver and David H. Sattinger

Volume 26: Two Phase Flows and Waves
Editors: Daniel D. Joseph and David G. Schaeffer

Volume 27: Nonlinear Evolution Equations that Change Type
Editors: Barbara Lee Keyfitz and Michael Shearer

Volume 28: Computer Aided Proofs in Analysis
Editors: Kenneth Meyer and Dieter Schmidt

Volume 29: Multidimensional hyperbolic problems and computations
Editors Andrew Majda and Jim Glimm

Volume 30: Microlocal Analysis and Nonlinear Waves
Editors: Michael Beals, R. Melrose and J. Rauch

Volume 31: Mathematics in Industrial Problems, Part 3
by Avner Friedman

Volume 32: Radar and Sonar, Part 1
by Richard Blahut, Willard Miller, Jr. and Calvin Wilcox

Volume 33: Directions in Robust Statistics and Diagnostics: Part 1
Editors: Werner A. Stahel and Sanford Weisberg

Volume 34: Directions in Robust Statistics and Diagnostics: Part 2
Editors: Werner A. Stahel and Sanford Weisberg

Volume 35: Dynamical Issues in Combustion Theory
Editors: P. Fife, A. Liñán and F.A. Williams

Volume 36: Computing and Graphics in Statistics
Editors: Andreas Buja and Paul Tukey

Volume 37: Patterns and Dynamics in Reactive Media
Editors: Harry Swinney, Gus Aris and Don Aronson

Volume 38: Mathematics in Industrial Problems, Part 4
by Avner Friedman

Volume 39: Radar and Sonar, Part 2
Editors: F. Alberto Grünbaum, Martin Bernfeld and Richard E. Blahut

Volume 40: Nonlinear Phenomena in Atmospheric and Oceanic Sciences
Editors: G.F. Carnevale and R.T. Pierrehumbert

Volume 41: Chaotic Processes in the Geological Sciences
Editor: David Yuen

Forthcoming Volumes:

1989-1990: *Dynamical Systems and Their Applications*

Partial Differential Equations with Minimal Smoothness and Applications

Twist Mappings and Their Applications

Dynamical Theories of Turbulence in Fluid Flows

Summer Program 1990: *Time Series in Time Series Analysis*

Time Series (2 volumes)

1990-1991: *Phase Transitions and Free Boundaries*

On the Evolution of Phase Boundaries

Shock Induced Transitions and Phase Structures

Microstructure and Phase Transitions

Statistical Thermodynamics and Differential Geometry
of Microstructured Material

Free Boundaries in Viscous Flows

Summer Program 1991: *Semiconductors*

Semiconductors (2 volumes)

1991-1992: *Phase Transitions and Free Boundaries*

Sparse Matrix Computations: Graph Theory Issues and Algorithms

FOREWORD

This IMA Volume in Mathematics and its Applications

CHAOTIC PROCESSES IN THE GEOLOGICAL SCIENCES

is based on the proceedings of a workshop which was an integral part of the 1989-90 IMA program on "Dynamical Systems and their Applications". The workshop was intended to be an arena for scientific exchanges between earth scientists and mathematical researchers, especially with experts in dynamical systems.

We thank Shui-Nee Chow, Martin Golubitsky, Richard McGehee, George R. Sell and David Yuen for organizing the meeting. We especially thank David Yuen for editing the proceedings.

We also take this opportunity to thank those agencies whose financial support made the workshop possible: the Army Research Office, the Minnesota Supercomputer Institute, the National Science Foundation, and the Office of Naval Research.

Avner Friedman

Willard Miller, Jr.

PREFACE

The problems in geological sciences have many nonlinearities from the nature of the complicated physical laws which give rise to strongly chaotic behavior. Foremost and most visible are earthquakes and volcanic eruptions, more subtle are the time-dependent variations of the Earth's magnetic fields and motions of the surface plates. There are indeed many interesting mathematical issues dealing with geophysical problems which require the tools of modern mathematics and numerical analysis, as many of them involve processes with sharp transitions and a multiplicity of spatial and temporal scales. The physical laws governing them are in fact not well known and well constrained, such as the constitutive relationship used to describe mantle flow, earthquakes or volcanic eruptions. This aspect certainly distinguishes solid-earth geophysics from the atmospheric and oceanic sciences, where the equations describing the dynamics are well known and generally agreed upon.

The workshop on Chaotic Processes in the Geological Sciences was held June 11 to June 15, 1990 at the Institute for Mathematics and its Applications and at the Minnesota Supercomputer Institute on the University of Minnesota Twin Cities campus as part of the 1989-1990 I.M.A. program on Dynamical Systems and Their Applications. There were over 150 participants, including the fourteen invited lecturers whose contributions form the chapters of this volume. We have also included the abstracts of all of the speakers in order to bring across the interdisciplinary nature of this workshop. There was also a video session in which numerical simulations were shown on various workstations in the form of videotapes and online movies directly fed from the computers.

The chapters discuss a wide variety of topics in the general subject area of time-dependent nonlinear phenomena related to geophysical processes. We have focussed in particular on three areas: convection in the Earth's mantle, geodynamo processes and dynamics associated with volcanic phenomena. Many challenging problem areas are raised and in many cases the lecturers have emphasized unsolved aspects of their subject field, which provide fertile ground for future research. It is my hope that the collection of topics in this volume will help to stimulate further interaction for interested mathematicians and researchers from theoretical and applied mechanics to delve into some of the most promising areas of current research in geophysics and geophysical fluid dynamics of the Earth's interior.

I would like to express my thanks and that of all the participants in the workshop to Avner Friedman and Willard Miller, Jr. of IMA and George Sell for their help and hospitality in planning and hosting the workshop. It is a pleasure to thank Patricia Brick, Stephan Skogerboe, and Kaye Smith for their assistance in producing this volume.

David A. Yuen

CONTENTS

TURBULENT THERMAL CONVECTION

S. BALACHANDAR*

1. Introduction. Thermal convection has been the subject of extensive research for more than a century due to its fundamental role in geophysical, astrophysical and engineering applications. Recent experiments by the Chicago group (Heslog *et al*[1], and Casting *et al*[2]) have shown that at high Rayleigh numbers (Ra) the flow undergoes a transition in its structure from soft turbulence to hard turbulence. In particular, in the central turbulent core the probability distribution of temperature fluctuations change from a Gaussian to an exponential distribution. This observation has prompted new interest in the study of high Rayleigh number thermal convection.

At sufficiently high Rayleigh numbers the convective heat transfer is dominated both by large-scale convective rolls and by coherent thermal plumes. The complex nature of mutual interaction and competition of these flow structures makes this problem of turbulent thermal convection particularly interesting. Identification and characterization of these coherent flow structures is important in understanding their interaction process. A rational analytical approach for extracting the coherent structures of a flow field in terms of the eigenfunctions of the two-point velocity correlation tensor was first proposed by Lumley[3,4]. Though this approach provides an objective and rational tool to extract flow structures, structural correspondence between these eigenfunctions and actual coherent structures, often observed in experiments, is admittedly weak. This difference partly arises from the fact that, while the eigenfunctions are evaluated based on an ensemble averaged correlation tensor, the actual coherent structures are instantaneous and have a temporal dependence. Methodology for extracting the actual coherent structures from these eigenfunctions remain an open question. A number of other techniques (VITA[5,6] and stochastic estimation[7,8]) have also been proposed to identify coherent structures both from experimental and computational flow fields.

Tarman *et al*[9,10] have employed the eigenfunction technique to extract coherent structures from their simulation of weakly turbulent thermal convection. Here we extend this approach to a high Rayleigh number convection in the hard-turbulence regime. We will compare the results obtained with those of the low Rayleigh number simulation. Issues related to eigenfunctions as an optimal basis for efficient data compression and flow field reconstruction will also be addressed.

2. Problem description. The numerical simulation we now report on is a study of turbulent thermal convection in a box of square plan form with an aspect ratio of width (L) to height (H) equal to $2\sqrt{2}$. The temperatures are specified and the fluid is allowed to slip at the horizontal bounding planes. The flow field is taken

*Theoretical and Applied Mechanics Department, University of Illinois, Urbana, IL 61801.

to be periodic in both the horizontal x and y directions. The governing Boussinesq equations when normalized the standard[11] way reduce to

$$\frac{\partial \mathbf{u}}{\partial t} + \mathbf{u} \cdot \nabla \mathbf{u} = -\nabla p + \mathrm{RaPr} T \mathbf{e}_z + \mathrm{Pr} \nabla^2 \mathbf{u} \tag{1}$$

$$\frac{\partial T}{\partial t} + \mathbf{u} \cdot \nabla T = \mathbf{u} \cdot \mathbf{e}_z + \nabla^2 T \tag{2}$$

$$\nabla \cdot \mathbf{u} = 0 \tag{3}$$

where $\mathbf{u} = (u, v, w)$ is the velocity in the $\mathbf{x} = (x, y, z)$ directions, $\mathbf{e}_z$ is the unit vector in the z direction, T and p are temperature and pressure departures from the pure conduction and hydrostatic states, respectively. The two non-dimensional parameters, Rayleigh number (Ra) and Prandtl number (Pr) are given by

$$\mathrm{Ra} = \frac{g \alpha \Delta T H^3}{\kappa \nu} \quad \text{and} \quad \mathrm{Pr} = \frac{\nu}{\kappa} \tag{4}$$

where g is the acceleration due to gravity, α is the coefficient of thermal expansion, ΔT is the temperature difference between the top and the bottom horizontal planes, κ is the coefficient of thermal conductivity and ν is the kinematic viscosity. The numerical computation reported here is performed at a Rayleigh number of 6.5×10^6 and a Prandtl number of 0.72, corresponding to that of air. The isothermal, non-penetrative, no-stress vertical boundary condition used in the present simulation reduce to

$$w = T = \frac{\partial u}{\partial z} = \frac{\partial v}{\partial z} = 0 \quad \text{at} \quad z = \pm\frac{1}{2} \tag{5}$$

The above boundary condition along with the aspect ratio $2\sqrt{2}$, yield a critical Rayleigh number (Ra_c) of $27\pi^4/4$, which results in a ratio of actual to critical Rayleigh number ($\mathrm{Ra}/\mathrm{Ra}_c$) of 9800. According to the classification of the Chicago group, at this ratio of actual to critical Rayleigh number the computed flow field is in the hard turbulence regime.

3. Numerical method. The governing equations (1)-(3) are solved using Fourier collocation scheme, with 96 grid points resolving each of the two horizontal directions and 97 grid points spanning the vertical direction. The velocity and temperature are expanded by Fourier series in the two periodic horizontal directions. The particular choice of vertical boundary condition permits sine (for T and w) or cosine (for u and v) expansion without loss of spectral accuracy. During the first stage of the fractional time step procedure, a Crank-Nicholson scheme is used for the viscous terms and a second order Adams-Bashforth scheme for the nonlinear and buoyancy terms. In the final pressure step, the incompressibility condition is strictly enforced at the new time level $n+1$ by filtering the intermediate ($*$) level velocity field as follows

$$\hat{u}_i^{n+1} = \left(\boldsymbol{\delta}_{ij} - \frac{k_i k_j}{|\mathbf{k}|^2} \right) \hat{u}_j^* \tag{6}$$

where $\hat{u}_j^*$ is the fourier coefficient of the jth component of velocity before the pressure step, $\hat{u}_i^{n+1}$ is the Fourier velocity coefficient at the new time level, k_i is the ith component of the wave number and δ_{ij} is Kronecker delta. Since each expansion function individually satisfies the boundary conditions, the filtering process does not affect the overall boundary condition.

4. Proper scaling. [9,10] The structure of the two-point correlation matrix, to be used later in the eigenfunction calculation, depends on the normalization used for velocity and temperature. Proper scaling which will lead to comparable magnitudes for the three components of velocity and temperature is imperative in extracting meaningful eigenfunctions. Let us define the characteristic velocity (u_c) and temperature fluctuation (T_c) as follows

(7) $$u_c = \sqrt{g\alpha\delta\Delta T} \quad \text{and} \quad T_c = \frac{\kappa\Delta T}{2\delta u_c}$$

where δ is the sub-layer thickness defined as

(8) $$\delta = \frac{H}{2\text{Nu}}$$

and Nu is the Nusselt number. Then the proper scaling can be written as

(9) $$L_c = H, \quad u_c = \frac{\kappa}{H}\sqrt{\frac{\text{Ra Pr}}{2\text{Nu}}}, \quad T_c = \Delta T\sqrt{\frac{4\text{Nu}^3}{\text{Ra Pr}}}$$

where L_c, u_c and T_c are the proper length, velocity and temperature scales. Unlike classical scaling[11], proper scaling involves Nusselt number and therefore can only be applied *aposteori*. With this scaling the rms horizontal velocity fluctuation, $\langle u^2\rangle, = 1.3$, rms vertical velocity fluctuation, $\langle w^2\rangle, = 1.26$ and rms temperature fluctuation about the horizontal mean, $\langle T'^2\rangle, = 2.7$ have been normalized to a value close to unity. The present simulation yield an ensemble averaged Nusselt number of 23.5. Based on the corresponding length and time scales, the non-dimensional horizontal grid spacing $\Delta x = \Delta y = 0.03$, the non-dimensional vertical grid spacing $\Delta z = 0.01$ and the non-dimensional time step $\Delta t = 0.0004$. the corresponding non-dimensional Kolmogorov length (η) and time scales (τ) are 0.0008 and 0.025 respectively. Thus there are two grid spacings resolving the smallest relevant length scale[12] and there are 62 times steps resolving the Kolmogorov time scale. Moreover, the horizontal and vertical energy spectra for both velocity and temperature exhibit four to five orders of magnitude decay, indicating adequate spatial resolution[13].

5. Eigenfunction calculation. According to Lumley[3,4] the coherent structures of the flow can be obtained from the eigenfunctions of the two-point correlation tensor or covariance matrix. For the thermal convection problem the covariance matrix is defined as

(10) $$K_{ij}(\mathbf{x},\mathbf{x}') = \langle u_i(\mathbf{x})u_j^*(\mathbf{x}')\rangle \quad i,j = 1,2,3 \text{ and } 4$$

where $u_i = (u, v, w, T')$ is the fluctuating flow field and includes temperature fluctuations as well. The superscript $*$ indicates complex conjugate. Horizontal homogeneity implies translational invariance, which in turn simplifies the horizontal dependence of the covariance matrix as follows:

$$K_{ij}(\mathbf{x}, \mathbf{x}') = K_{ij}(x - x', y - y', z, z') \tag{11}$$

In equation (10), the angle brackets indicate an ensemble average. Here the ensemble average is approximated as an average over horizontal planes and an average over discrete sampling time instants. Here the sampling times are 200 time steps apart and the complete turbulent convection data-base consists 50,000 time steps. Ideally it is desirable to have these sampling times separated by more than an eddy turn over time, so that the corresponding sample flow fields are uncorrelated. The numerical data-base covers 22 eddy turn over times, where eddy turn over time is estimated[14] as $(\mathrm{RaNuPr})^{1/3}$. A sixteen fold symmetry present in the problem is exploited to further expand the database to obtain a better approximation for the ensemble average[15,16]. The sixteen fold symmetry includes an eight fold symmetry in the horizontal plane resulting from the square plan form (diehedral group D_4) and a vertical symmetry about the mid plane. For example a reflectional symmetry about the $y - z$ vertical plane results in a brand new flow field $(-u, v, w, T')$ which satisfies both the governing equations and the boundary conditions.

The eigenvalues $(\boldsymbol{\lambda}_n)$ and eigenfunction $\boldsymbol{\Phi}^{(n)} = (\boldsymbol{\phi}_1^{(n)}, \boldsymbol{\phi}_2^{(n)}, \boldsymbol{\phi}_3^{(n)}, \boldsymbol{\phi}_4^{(n)})$ of the covariance matrix can be computed as

$$\int K_{ij}(\mathbf{x}, \mathbf{x}') \boldsymbol{\phi}_j^{(n)}(\mathbf{x}') d\mathbf{x}' = \boldsymbol{\lambda}_n \boldsymbol{\phi}_i^{(n)}(\mathbf{x}) \tag{12}$$

The kernel K_{ij} is Hermitian, non-negative and square integrable on physical grounds. Therefore the eigenfunctions form a complete set of orthonormal basis and the flow field can be expanded as

$$u_i(\mathbf{x}, t) = \sum_n a_n(t) \boldsymbol{\phi}_i^{(n)}(\mathbf{x}) \tag{13}$$

where the summation is over all possible values of n. By definition, the eigenfunctions are frozen in time and all the temporal variation appears in the eigen coefficients $a_n(t)$, which are defined as

$$a_n(t) = \int u_i(\mathbf{x}, t) \boldsymbol{\phi}_i^{(n)}(\mathbf{x}) d\mathbf{x} \tag{14}$$

The horizontal homogeneity reduces the eigenfunctions to be simple sinusoidals in x and y. Thus

$$\boldsymbol{\Phi}^{(n)}(\mathbf{x}) = \boldsymbol{\Psi}^q(z; k_1, k_2) \exp\left[i(k_1 x + k_2 y) \frac{2\pi}{L} \right] \tag{15}$$

where $\boldsymbol{\Psi}^q(a; k_1, k_2)$ is the qth eigen mode corresponding to the horizontal wave number combination k_1, k_2. Therefore the index n for the eigenfunction and the eigenvalue represents the triplet (k_1, k_2, q), corresponding to the two horizontal wave numbers and the vertical quantum number. It also follows from the property of the kernel that the eigenvalue $\boldsymbol{\lambda}_n$ represents the average energy in mode n provided the eigenfunctions are normalized as follows:

$$\int \boldsymbol{\Phi}^{(n)}(\mathbf{x}) \cdot \boldsymbol{\Phi}^{(n)*}(\mathbf{x}) d\mathbf{x} = 1 \tag{16}$$

6. Eigen mode results. The top ten most energetic eigen modes are listed in table 1. The percentage model energy corresponding to the lower Rayleigh number simulation (Ra = 46000) is shown in parenthesis. The ordering of the eigenvalues is different in the two simulations. The ten most energetic modes account for 67.5% of the total energy at $70\mathrm{Ra}_c$, whereas at the higher Rayleigh number they account for only 34.1% of the total energy. This clearly indicates that, though eigenfunction expansion is optimal in capturing energy, as Rayleigh number increases, more and more modes are required to represent certain percentage of the total energy. Inclusion of symmetry in the ensemble average leads to degenerate eigenvalues. For example $(k_1, k_2, q) = (1, 0, 1)$ mode has the same energy as $(-1, 0, 1), (0, 1, 1)$ and $(0, -1, 1)$ modes. These four modes represent roll structure aligned along the x or y direction and each contribute one fourth of the net 24.56% energy.

n	k_1	k_2	q	Degeneracy	% Energy
1	1	0	1	4	24.56 (38.91)
2	1	1	1	4	3.23 (9.96)
3	0	1	2	4	2.00 (4.53)
4	1	2	1	8	0.95 (3.81)
5	0	2	1	4	0.71 (2.98)
6	0	2	2	4	0.68 (1.38)
7	1	2	2	8	0.57
8	1	3	1	8	0.56
9	0	1	3	4	0.46 (1.82)
10	0	2	3	4	0.39

Figure 1 shows the four components of the most energetic $(1, 0, 1)$ mode. Since this mode represents a role structure aligned along the y axis, the u and w components are non-zero, which results in a non-zero temperature distribution in the vertical direction. On the other hand, since there is no motion in and out of the $x-z$

plane the y component of the velocity is identically equal to zero. Figure 2 shows the components of the second most energetic $(1,1,1)$ mode. This mode is four fold degenerate and represents roll structure aligned along the diagonals. The u and v velocities combine to produce zero in and out motion along the axis of the roll. The eigenfunctions corresponding to the lower Rayleigh number is also included in Figure 2 and is marked by the dash line. Though the eigenfunctions compare well qualitatively, the low and high Rayleigh number results differ in two aspects. First, the high Rayleigh number simulation exhibits a much thinner thermal sub-layer indicating higher Nusselt number. Second, in the high Rayleigh number case, the modal energy in the thermal component is significantly smaller than that in the low Rayleigh number case.

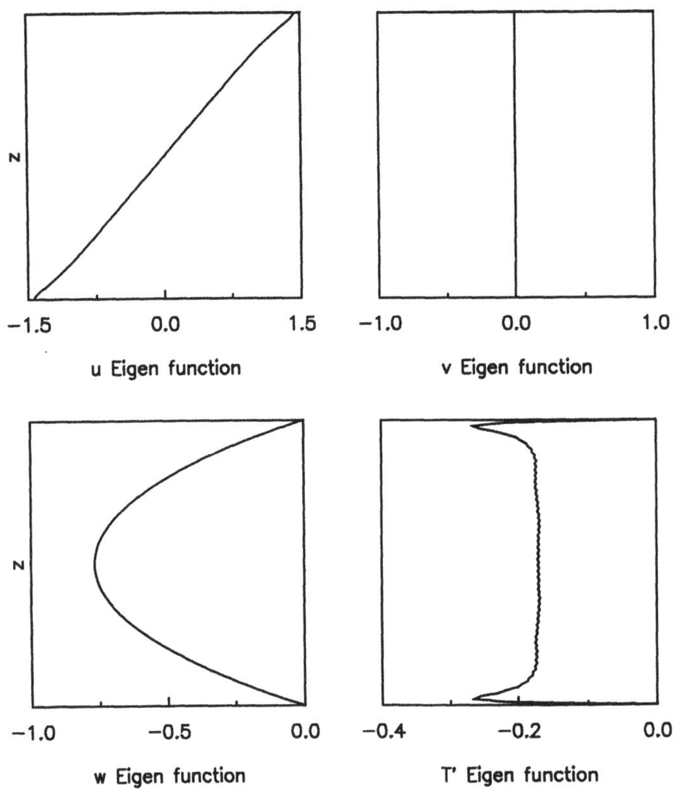

Figure 1: u, v, w and T' components of the most energetic, $(1,0,1)$ eigen mode.

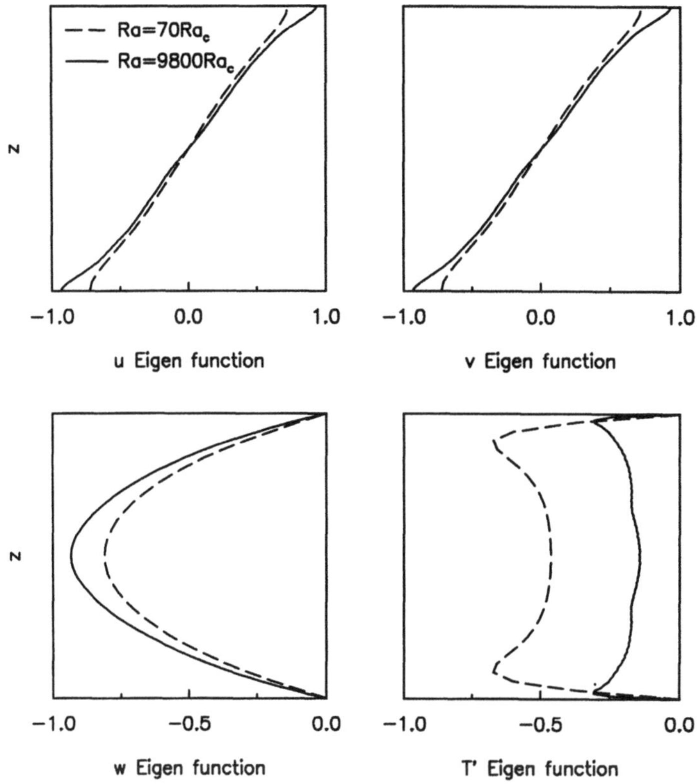

Figure 2: u, v, w and T' components of the second most energetic, $(1,0,1)$ eigen mode.

Figure 3 shows the u component of the $(0,1,2)$ eigen mode. This mode (a sloshing mode) represents pure shearing motion in the x direction and the corresponding eigen components for ν, w and T' are identically equal to zero. Figure 4 depicts the roll structure in the $x - z$ plane for the $(1,0,1)$ mode and the corresponding isotherms are marked by dashed line. Figure 5 shows the corresponding roll structure for the $(1,1,1)$ mode along with the isotherms. Unlike figure 4, here the abscissa is along the diagonal of the horizontal $x - y$ plane.

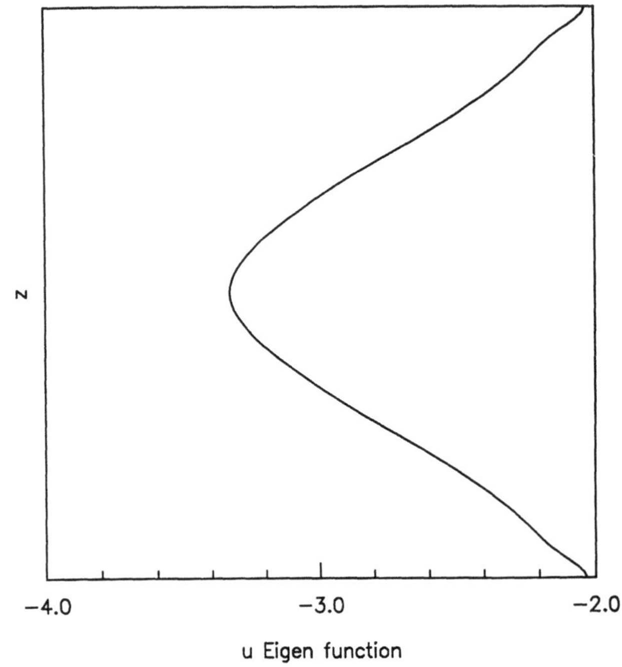

Figure 3: u component of the third most energetic $(0,1,2)$ pure shear eigen mode.

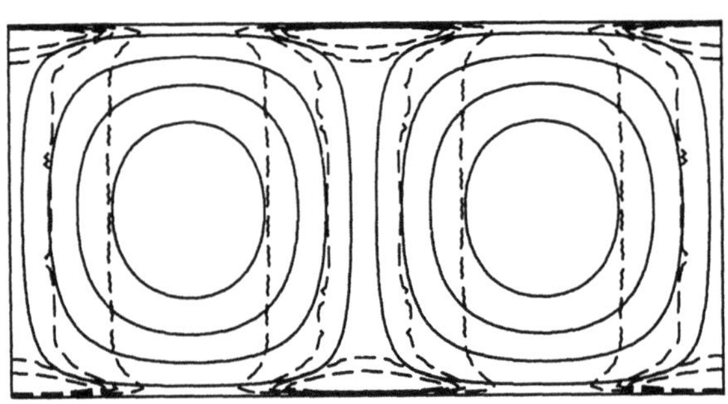

Figure 4: Stream function and isotherms for $(1,0,1)$ mode

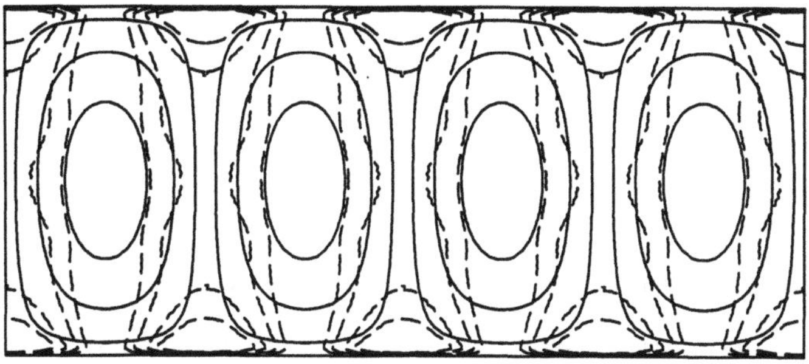

Figure 5: stream function and isotherms for $(1,1,1)$ mode

7. Data reduction and flow reconstruction. At any time instant the flow field can be represented in many ways. The flow field can be represented in real space as the grid point velocity and temperature, or it can be represented in spectral space as the corresponding Fourier coefficients. The third option is to represent the flow field in the eigenfunction space by projecting the flow field on to the eigenvectors (see Eq. 14). Truly there are as many Fourier coefficients and eigen coefficients as there are grid point variables. A complete summation over all these coefficients will faithfully reconstruct the flow field. For example Eq. 13 shows the reconstruction of the flow field from the eigen coefficients.

Data analysis, such as a three dimensional visualization of the dynamics of the plume structure, require a knowledge of the flow at every Kolmogorov time scale or so. Saving the entire flow field would then require hundreds of giga bytes of computer storage for the present simulation. Efficient data compression is imperative and can be achieved by retaining only the M most energetic eigen coefficients and approximately reconstructing the flow field by truncating the summation in equation 13. The level of approximation improves as M approaches the total number of grid point variables. But for a given M, by definition the eigen function expansion is the optimal representation in terms of the energy content.

Colour plate 1 (see Plates section) shows the surface of constant total temperature equal to 0.25, computed from the complete flow field at six instants $200\delta t$ apart. Here the total temperature is normalized to be 0.5 at the bottom boundary and -0.5 at the top boundary. Though the plume structure is clearly evident in these frames, the temporal evolution is not clear since these frames are far apart in time. Colour plate 2 (see Plates section) shows similar isothermal surface at six different time instants, each $4\Delta t$ apart. These surfaces are generated from a reconstructed flow field based on only 1231 most energetic eigen modes (not including degeneracy). Temporal evolution of the plumes can be tracked following these frames. Unlike the frames of plate 1, the plume structure is smooth and void of small-scale features. Since eigenfunctions optimize energy content, they tend to emphasize the energy containing large scale structure. Therefore successful data compression by eigenfunction (based on two point correlation) decomposition can be achieved if properties of large-scale features need to be addressed. In the present context, both surface and volumetric rendering techniques are used, to extract large-scale coherent structures from the reconstructed flow field.

REFERENCES

[1] F. HESLOT, B. CASTING, AND A. LIBCHABER, *Phys. Rev. A*, 36 (1987), p. 5870.
[2] B. CASTING, G. GUNARATINE, F. HESLOG, L. KADANOFF, A. LIBCHABER, S. THOMAS, X.Z. WU, S. ZALESKI, AND G. ZANNETTI, *J. Fluid Mech.*, 204 (1989), p. 1.
[3] J.C. LUMLEY, *Stochastic Tools in Turbulence*, Academic Press, New York (1970).
[4] J.C. LUMLEY, *Coherent Structures in Turbulence*, in Transition and Turbulence (ed. R.E. Meyer), Academic Press, New York (1981).
[5] R.F. BLACKWELDER, AND R.E. KAPLAN, *J. Fluid Mech.*, 76 (1976), p. 89.
[6] R.A. ANTONIA, *Ann. Rev. Fluid Mech.*, 13 (1981), p. 131.
[7] R.J. ADRIAN, *Proceedings of the Fourth Biennial Symposium on Turbulence in Liquids*, University of Missouri-Rolla Sep. (1975).
[8] R.J. ADRIAN, AND P. MOIN, *J. Fluid Mech.*, 190 (1988), p. 531.
[9] H. TARMAN, *Ph.D. Thesis*, Brown University, Providence (1989).
[10] L. SIROVICH, M.R. MAXEY, AND H. TARMAN, *Proceedings of the Sixth Symposium on Turbulent Shear Flows*, (ed. B.E. Launder) Springer-Verlag, Berlin (1988), p. 68.
[11] P.G. DRAIZIN, AND W.H. REID, *Hydrodynamic Stability*, Cambridge University Press (1985).
[12] G. GRÖTZBACH, *J. Comput. Phys.*, 49 (1983), p. 241.
[13] S. BALACHANDAR, M.R. MAXEY, AND L. SIROVICH, *J. Sci. Comput.*, 4 (1989), p. 219.
[14] J.W. DEARDOFF, AND G.E. WILLIS, *J. Fluid Mech.*, 28 (1967), p. 675.
[15] L. SIROVICH, *Quar. Appl. Math.*, 45 (1987), p. 573.
[16] L. SIROVICH, *Quar. Appl. Math.*, 45 (1987), p. 583.

SPHERICAL SYMMETRY-BREAKING BIFURCATIONS AND THERMAL CONVECTION IN THE EARTH'S MANTLE

PASCAL CHOSSAT AND CHERYL A. STEWART

Abstract. Perhaps the strongest influence on the pattern of thermal convection in the earth's core and mantle is the spherical symmetry of the boundary conditions. This symmetry quantizes the solutions of the convection equations, and has a profound effect on the range of solutions available. The solutions appear as a discrete rather than a continuous set of modes, and changes in the pattern as a function of Rayleigh number and aspect-ratio occur as well defined transitions in the relative amplitudes of the modes, known as bifurcations. These descrete modes are reflected in the geophysical data which indicate the pattern of thermal convection in the earth's mantle– the global geoid, dynamic topography, seismic tomography, hotspot volcanic distribution, plate motions and continental distributions. These data should therefore indicate which branches of solution have been chosen by the earth's mantle. In order that this choice of solution branch reveal anything about the physics of the earth's interior, it is necessary to understand which solutions are available for the earth's mantle to choose from, and how various physical conditions affect these.

We start with a review of the linear stability results for thermal convection in a spherical shell, and show how this sets the stage for the analysis of the nonlinear, finite-amplitude problem via the center manifold reduction. We then review the mathematical results which have been obtained over the past ten years which reveal the effect of spherical symmetry on the steady-state single-mode bifurcation structure of the nonlinear thermal convection problem at spherical harmonic orders l=1-3. We further indicate the mode interactions between the axisymmetric solutions of spherical harmonic orders l=1 and l=2. The geophysical data which indicate the low-order spherical harmonic modes of thermal convection currently active in the earth's mantle are then reviewed. The spectra of these geophysical data have peaks in spherical harmonic orders l=1-6 with a strong spatial resonance between the nonaxisymmetric solutions of l=2,6 for the deep thermal structure of the earth's mantle, and peaks in orders $l = 4, 5$ for the plate-motions at the earth's surface. These data suggest the analytical study of some mode-interaction problems based on an observationally and theoretically justified separation between the "slow" $l = 2, 6$ subspace and the "fast" $l = 3 - 5$ subspace.

Introduction. Thermal convection in the earth's mantle has long been recognized as the physical mechanism responsible for continental drift. It supplies the mechanical energy dissipated by earthquakes, and the thermal energy dissipated by volcanoes. An excellent review of the historical discoveries in earth science which led to this conclusion can be found in Oxburgh and Turcotte [1978], and some of the more recent developments in this area are covered by Olson *et al.* [1990].

One of the many outstanding problems in mantle geodynamics is the time-evolution of the pattern of thermal convection in the earth's mantle. The patterns of thermal convection in the earth's mantle may be classified and understood in terms of the bifurcation structure of the mathematical problem that the physical problem reduces to. Furthermore, many aspects of this bifurcation structure can be deduced directly from the symmetry properties of a spherical shell. The purpose of this paper is to bring together, for the first time, the mathematical theory of thermal convection in spherical shells and the geophysical data which provide clues as to the current pattern of thermal convection in the earth's mantle.

The point of view adopted here is that examining the bifurcation structure of a thermal convection problem yields useful and valid results; furthermore, the bifurcation structure of the problem is very strongly influenced by the symmetries of the governing equations and boundary conditions. The methods of determining the bifurcation structure of the nonlinear problem in the presence of the symmetries of the governing equations and boundary conditions have allowed the quantitative prediction of transitions in the patterns and time-dependent dynamics Taylor-Couette problem [Chossat and Iooss, *in preparation*], and for thermal convection in two- and three-dimensional Cartesian coordinates. There is therefore strong reason to believe that a similar line of inquiry will increase our understanding of thermal convection in the earth's mantle and core.

We first review bifurcation structure of the nonlinear 2- and 3-dimensional Cartesian problem. It has been found, for instance, that stating the Nusselt number, or the onset of period-doubling in a 2-dimensional Cartesian numerical experiement is *not* a rigorous test of the accuracy of a model, since these quantities change drastically and suddenly with the aspect-ratio of the cells [e.g., Toomre *et al.* 1982]. Not only does the Rayleigh number of the onset of period-doubling change with the change in aspect-ratio, but also, whether or not period-doubling occurs changes with the change in aspect-ratio. Very close to the critical wavenumber and integer multiples of the critical wavenumber, for instance, periodic orbits bifurcate to symmetric sets of invariant two-tori, rather than undergo period-doubling [Stewart, 1990]. At aspect-ratios away from this case, period-doubling is observed [Lennie *et al.*, 1988, Vincent and Yuen, 1988]. When two wavenumbers have equal linear stability and are integer multiples of each other, mixed-mode solutions are not possible for the flow; two distinct stable fixed-points coexist with a well-defined boundary between their basins of attraction [Segel, 1962; Guckenheimer and Knobloch, 1983]. At infinite Prandtl number ($\nu/\kappa \to \infty$ where ν is kinematic viscosity and κ is thermal diffusivity) this basin boundary breaks down at higher Rayleigh number, admitting chaotic solutions [Stewart and Turcotte, 1989]. The remarkable thing about this is that at infinite Prandtl number,the conditions of the earth's mantle, the Lorenz equations–a single-mode truncation of the same problem [Saltzman, 1962; Lorenz, 1962; Sparrow, 1982] –admit only stable fixed-points, i.e. stable steady states, up to infinite Rayleigh number [McGuinness and Fowler, 1983]. This is an indication that, at infinite Prandtl number, too low a truncation order leads to spurious stability; whereas the opposite has been found for certain finite and zero Prandtl number cases [Marcus, 1981].

When two wavenumbers have equal linear stability and are integer multiples of each other and travelling wave solutions are allowed, there exist structurally stable heteroclinic orbits among the fixed-points [Armbruster *et al.* 1987]. A heteroclinic connection between two fixed points can cause a numerical experiment to hover around one fixed point with one Nusselt number for quite some time, and then suddenly shift to another fixed-point characterized by a different convective pattern and a different Nusselt number. This behavior has been observed in very high resolution numerical experiments at very high Rayleigh-number and aspect-ratio [Gunkel and Hansen, 1989]. Furthermore, when this phenomenon occurs in the numerical experiments, it involves a shift from a convective pattern which is dominant in a single wave number, $k + 1$, to the solution dominant in the wave number k. The analysis of Armbruster *et al.* was a center-manifold analysis of the nonlinear interaction between *only* the k and $k+1$ solutions; no other modes were included. This convergence between the high resolution numerical experiments and the center manifold analysis of relatively few modes lends strength to the claim that the essential dynamics of a convective system can be captured by a low-order representation.

The problem of thermal convection in a three-dimensional planar region has also been subject to significant attention [e.g. Whitehead and Busse, 1971; Cross and Newell, 1984; Golubitsky *et al.*, 1984; McKenzie 1988]. In the 3-dimensional Cartesian problem, the shape of the lateral boundary conditions can be varied arbitrarily or removed entirely. In a spherical shell, as in the 2-dimensional Cartesian case, only one parameter characterizes the geometry of the boundary conditions: the the ratio of inner radius to outer radius of the shell. All other aspects of the pattern of convection in the spherical shell are strongly influenced by the peculiar effects of the symmetries of the sphere itself.

Just as in the 2-dimensional Cartesian problem, the codimension one and codimension two bifurcations must be understood in order to relate the theory of thermal convection to the observations which indicate the patterns of thermal convection in planetary interiors. These bifurcations are the sudden changes in convective pattern, and the sudden changes in number of branches of solution–the number and type of possible convective patterns which can be realized by the system–as both the Rayleigh number and aspect-ratio of a spherical shell are varied independently and together. Understanding the bifurcation structure of a system of equations that represent, say, thermal convection in the earth's mantle, can effectively complement the results of high-resolution numerical experiments [Glatzmaier, 1986; Bercovici *et al.*, 1989; Glatzmaier, 1990]. Numerical experiments converge to the most stable attracting sets. The bifurcation structure indicates the existence and stability of *all* branches of solution, stable and unstable, as well as all transitions to time-dependent behavior. The structure of the transitions to time-dependent behavior then provides a structure for understanding the evolution of trajectories on higher-dimensional, e.g., chaotic, attractors. It can then be seen what physical conditions are necessary for choosing one branch of solution–one pattern or sequence of patterns–over another.

How does the spherical symmetry constrain the bifurcations of stable and unstable modes of convection as one varies the Rayleigh number and ratio of inner to outer radius? How do the stable and unstable manifolds of unstable solutions interconnect? How do related modes interact with time? How is this bifurcation structure modified when the effects of an infinitesimal Taylor number or exponentially temperature-dependent viscosity are considered? The methods for obtaining the answers to these questions using the symmetry subgroups of the sphere are presented in this paper. We show that these methods have significant potential for increasing our understanding of thermal convection in planetary interiors, starting with convection in the earth's core and mantle.

We first review the linear stability problem, and projections of the nonlinear problem based on the center manifold theorem. We then discuss the use of the symmetries of the original system to deduce the bifurcation structure, and review the basic results of the bifurcations of the thermal convection problem in the presence of spherical symmetry which have been obtained in the past ten years. We then review the observational evidence for the boundary conditions present and the modes of thermal convection which are active in the earth's core and mantle. We then state some hypotheses about how the patterns of thermal convection in the earth's core and mantle fit into the bifurcation structure of of thermal convection in a spherical shell. We suggest the means by which these hypotheses may be tested.

Linear and Nonlinear Stability. In Chandresekhar [1961] is reported the stability limits of the linear stability problem in a spherical shell. The multiple solutions to the governing equations are first described in terms of expansions in the spherical harmonics,

$$Y_l^m(\theta,\phi) = e^{im\phi} Pa_l^m(\cos\theta) \tag{1}$$

where ϕ is the longitude in radians and θ is the colatitude and $Pa_l^m(\cos\theta)$ is the associated Legendre polynomial of order l and degree m. The radial variation in the flow is expressed by the Elbert function,

$$C(\eta,r) = J_{-(l+1/2)}(\alpha_j\eta)J_{l+1/2}(\alpha_j r) - J_{l+1/2}(\alpha_j\eta)J_{-(l+1/2)}(\alpha_j r) \tag{2}$$

where $J_{\pm(l+1/2)}$ is Bessel's function of the first kind, and η is the ratio between the inner and outer radii. The α_j are a series of numbers related to the aspect-ratio η. The α_j are chosen so that the Elbert function will satisfy the boundary conditions [Chandrasekhar and Elbert, 1953]. The number of zeros of the Elbert function in the spherical shell depends on the choice of α_j; therefore, the velocities and temperatures can be expanded in these functions with the terms α_j.

These functions, the spherical harmonics and the Elbert functions, are the eigenfunctions of the governing equations when the boundary conditions are constant temperature and no-slip. The velocity fields are then expressed as linear combinations of these functions. The linear stability depends on the spherical harmonic order, l, but not the spherical harmonic degree, m.

Figure 1 depicts the linear and nonlinear stability of modes order $l = 1 - 10$, where both the inner and outer boundaries have free-slip and constant-temperature boundary conditions, and the ratio of inner to outer radius is that of the earth's mantle, $r_i/r_o = 0.54$. The vertical axis is Rayleigh number, and the horizontal axis is spherical harmonic order. The dark solid line is the line of linear stability; at Rayleigh numbers below this line, convection is not possible; above this line, convection is possible. The third dimension, axis **A**, indicates the amplitude of nonlinear convection. The dotted lines indicate the unstable branches of solution in the nonlinear theory. These indicate the subcritical pitchfork bifurcations which occur for odd spherical harmonic order, and the transcritical bifurcations which occur for even spherical harmonic order, for the axisymmetric patterns of convection. The instability of the axisymmetric solutions is a result of Chossat and Lauterbach [1989].

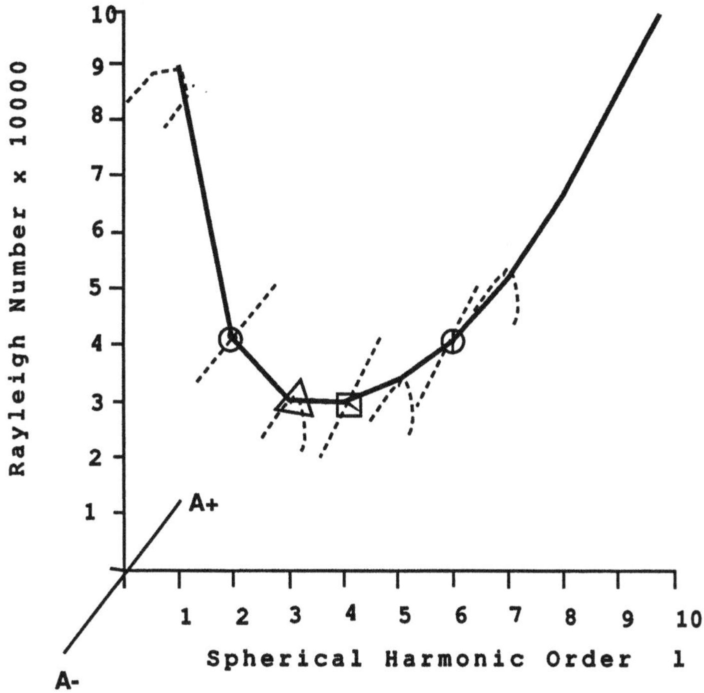

Figure 1 Linear and nonlinear stability for thermal convection in a spherical shell, aspect-ratio $r_i/r_o j = 0.55$. Heavy solid line is linear stability curve interpolated from Chandrasekhar [1962] for free-slip, constant temperature inner and outer boundary conditions. **A+** and **A−** are the positive and negative values for the coefficient of the $Y_l^{m=0}(\theta, \phi)$ pattern bifurcating from the quiescent basic state. Light solid lines are possible stable branches and light dotted lines are possible unstable branches. Note that even-l solutions have unstable solutions bifurcating transcritically from the first instability, and that odd-l solutions bifurcate as either supercritical or subcritical pitchforks. The triangle indicates a tetrahedral/tetrahedral solution is possible for $l = 3$, and the square indicates that a cubic/octahedral solution is possible for $l = 4$. The triangle and square also indicate a mode-interaction at $l = 3, 4$. Open circles indicate the $l = 2, 6$ mode interaction; a dodecahedral/icosahedral pattern of upwelling/downwelling is possible for $l = 6$.

The most important thing to notice about the linear stability of the spherical shell of aspect ratio $r_i/r_o = 0.54$. are the degeneracies at $l = 3, 4$ and $l = 2, 6$. The modes $l = 3$ and $l = 4$ (triangle and square in Figure 1) have the same critical Rayleigh number as each other, and moreover, are found at the minimum critical Rayleigh number for the earth's mantle. This means that as the Rayleigh number is raised through the critical value, both $l = 3$ and $l = 4$ solutions bifurcate simultaneously. Mode interactions are therefore expected with the fastest dynamics confined to the $l = 3,4$ subspace, as all of the other modes have negative eigenvalues at this point. An $l = 4$ mode is found for the horizontal divergence of the plates [Forte and Peltier, 1988], but $l = 3$ is absent. One general result of the nonlinear stability theory reviewed below is that when mode interactions occur between odd order and even order patterns, pure mode branches are expected for the even order but not the odd order mode. This, and the fact that the dimensionality of the $l = 4$ space is far higher than the $l = 3$ space contribute to the absence of the $l = 3$ mode.

The other degeneracy apparent from just the linear stability problem is between $l = 2,6$, (open circles in Figure 1). Because both of these even-order modes become unstable at the same Rayleigh number and aspect-ratio, mode-interactions are expected at these orders as well. These are both even order, and moreover, are observed in the correlations among the global geoid, dynamic topography, seismic tomography and hotspot distribution [Richards *et al.*, 1988; Cazenave, *et al.* 1989]. We make further comments on this fact after examining some aspects of the nonlinear problem.

Projections of the Nonlinear Problem

In a simple eigenmode expansion, a set of nonlinear ODEs are derived by substituting the eigenfunctions of the linearized equations back into the perturbation equations. For the Boussinesq-Oberbeck equations in a spherical shell, the resulting set of equations has the form,

$$\frac{d\mathbf{A}}{dt} = F(\mathbf{A}, \mu)$$

or, more explicitly,

$$\begin{aligned}
\dot{A}_{l,m,\alpha} =& \mathbf{C}^{\mathbf{hfw}}_{l,m,\alpha} + \mu(R,\eta)^{\mathbf{Lin}}_{(l,m,\alpha)} A_{l,m,\alpha} \\
& + \mathbf{C}^{quadratic}_{(l',m',\alpha'),(l'',m'',\alpha'')} A_{(l',m',\alpha')} A_{(l'',m'',\alpha'')} \\
& (-L \le l' \le L)(-L \le l'' \le L) \\
& (-l' \le m' \le l')(-l'' \le m'' \le l'') \\
& (\alpha_1 \le \alpha' \le \alpha_N)(\alpha_1 \le \alpha'' \le \alpha_N) \\
& + \mathbf{C}^{cubic}_{(l',m',\alpha'),(l'',m'',\alpha'')(l''',m''',\alpha''')} A_{(l',m',\alpha')} A_{(l'',m'',\alpha'')} A_{(l''',m''',\alpha''')} \\
& (-L \le l' \le L)(-L \le l'' \le L)(-L \le l''' \le L) \\
& (-l' \le m' \le l')(-l'' \le m'' \le l'')(-l''' \le m''' \le l''') \\
& (\alpha_1 \le \alpha' \le \alpha_N)(\alpha_1 \le \alpha'' \le \alpha_N)(\alpha_1 \le \alpha''' \le \alpha_N)
\end{aligned}$$

$+higher\ order\ terms$

where $A_{l,m,\alpha}$ is the amplitude of the spherical convective mode associated with the spherical harmonic pattern of degree l and order m, and Elbert function of degree

l and order α. (Expansions in Chebychev polynomials for the radial variation have been favored for spectral numerical schemes [Glatzmaier, 1986]; however, the choice of the Elbert function is perhaps more natural to the problem with free-slip constant temperature boundary conditions because it solves the radial variation of the linear part of the governing equations exactly. The advantage of using the Chebychev polynomials is that a wider variety of boundary conditions can be treated.)

The nonhomogeneous terms are $\mathbf{C}^{\mathbf{h}fw}_{l,m,\alpha}$, and are nonzero only if the governing equations include an internal heating term or external driving function. (Nonhomogeneous terms are useful in studies of the earth's thermal history [Davies, 1980] as well as mixing problems in which a geochemical heterogeneity has a specific source-pattern.) The linear coefficients are $\mu(R,\eta)^{\mathbf{Lin}}_{(l,m,\alpha)}$, and depend directly on the Rayleigh number R and aspect-ratio η. The quadratic coupling coefficients are $\mathbf{C}^{quadratic}_{(l',m'),(l'',m'')}$ and are only important in mode-interaction problems. When the bifurcation of a single mode is considered, the quadratic terms are necessarily zero.

In a simple eigenmode expansion, the eigenfunctions of the linear equations are simply substituted back into the governing equations to obtain the nonlinear coefficients $\mathbf{C}^{quadratic}_{(l',m'),(l'',m'')}$. Because the governing equations only involve nonlinear terms of the type $\mathbf{u}\cdot\nabla u$ and $\mathbf{u}\cdot\nabla T$, a straight eigenmode expansion produces only quadratic nonlinearities in the resulting set of ordinary differential equations. In this case, there are no theorems which prove the validity of this approximation when it is truncated to a limited set of modes.

When only the modes involved in the truncation are active, i.e. the eigenvalues of all other modes are negative, the parameters are near the linear stability limits, and within this eigenmode approximation, a center manifold of the may be found [Guckenheimer and Knobloch, 1983]. What is meant by "center manifold" is that within the truncated set of ODEs defined by the eigenmode expansion, a rigorous reduction of the dimensionality of the finite set of ODEs can be made by expanding the coefficients of the eigenmodes in Taylor series around the known fixed-points of the ODEs. The modes which are not directly involved with the bifurcation at hand are thus written in terms of the amplitudes of the relevant modes. In this manner, rather than arbitrarily truncating the infinite set of ODEs resulting from the eigenmode expansion, the other modes *are* taken into account, inasmuch as they affect the time-derivatives of the modes which are known to be directly involved in the particular bifurcation under closer scrutiny. This is what produces the terms of cubic order, whose coefficients are $\mathbf{C}^{cubic}_{(l',m',\alpha'),(l'',m'',\alpha'')(l''',m''',\alpha''')}$ in the above equation. For single modes bifurcating, it is these cubic order terms which determine the type of bifurcation, whether it is, for instance, transcritical or pitchfork.

The derivation of these amplitude equations on the center manifold of the governing equations is discussed in Coullet and Spiegel [1983]. This method has been applied to the problem of thermal convection in a spherical shell by Chossat and Giraud [1983] and by Friedrich and Haken [1986]. These methods have also been extensively applied to the Taylor-Couette problem, and have been tremendously successful in predicting experimentally-observed transitions in this extensively studied problem [Chossat and Iooss, 1990]. It has been argued, correctly, that the center manifold reduction is only valid in a neighborhood of the basic state. The first basic state to be analyzed is usually the marginally stable state, with the critical parameters – Rayleigh number and aspect-ratio, in the case of thermal convection – near the linear stability limits.

However, the basic state for the perturbation analysis need not be the quiescent, conductive state. It is just as valid to perturb about known steady-states and even time-periodic solutions, to find out about the secondary bifurcations in the system as the parameters are further varied. The known steady-states and time-periodic solutions need to be found first, and occur as primary bifurations from the marginally stable states. Why is it, instance, that one numerical simulation will choose one sequence of patterns and the same simulation, with slightly different initial conditions, will choose another–modulo a symmetry (e.g. reflection through the origin, rotation through k/π...). Why is it that the same simulation will choose a new sequence of patterns with time when a parameter (e.g. Rayleigh number, aspect-ratio) is varied? A new mode of the pattern of convection is added with a change in Rayleigh number or aspect-ratio, These modes are added one or two symmetries or modes at a time. The purpose of examining the symmetry subgroups of the sphere, and their successive bifurcations is to establish the properties of *all possible* sequences under the simplest conditions. As parameters are added, the mode selection is infludenced primarily in the *number* of such modes that can go unstable at one time.

To look at the details of each such symmetry-breaking bifurcation, the number of independent equations which contribute to the bifurcation are limited by the number of parameters which are being varied to produce the bifurcation [see for instance Guckenheimer and Holmes, 1983; Golubitsky *et al.*, 1988]. These are the amplitude equations, which represent the stable manifolds around the basic state. Within a convecting spherical shell, spherical harmonic orders l are added one at a time as Rayleigh number is increased. This increases the number of degrees of freedom of the problem by $2l+1$. The existence and stability of the new solution branches within these new degrees of freedom can be established on the basis of their symmetry properties alone–the topic of the subsequent section. The reduction is performed at the point in parameter space at which there a finite number of modes are at marginal stability.

Busse [1974] and Busse and Riahi [1982] found some solutions for spherical harmonic order 3, 4 and 6, which form tetrahedral/tetrahedral (T/T), cubic/octahedral (C/O) and dodecahedral/icosahedral (D/I) patterns of convection, i.e. the platonic solids and their duals. The platonic solids are depicted in Figure 2. The convective patterns which correspond to these geometric figures have upwelling plumes at the corners and downwelling at the center of each face – or, for the pattern corresponding to the dual figure, upwelling at the center of each face, and downwelling at the corners. These are the patterns which maximize the heat-flow; however, it is clear from experiments on flat layers that the configuration chosen by the system is rarely the configuration which will maximize the heat-flow. While these are fixed-points of the system, it is unkown whether they are stable fixed-points. The numerical simulations of Glatzmaier [1990] indicate that solutions with these symmetry groups do appear, but are part of time-dependent flows. Because the time-dependent flows follow attractors which arise from the stable and unstable steady-states found in the bifurcation analysis, what is needed is a more complete understanding of the bifurcation structure which gives rise to these branches of solution and time-dependent attractors.

Symmetry-breaking bifurcations. Here we consider the ODE

$$\frac{d\mathbf{A}}{dt} = F(\mathbf{A}, \mu), \qquad \mathbf{A}(t) \in V \tag{4}$$

which results from the initial system of equations, via the center manifold reduction. The vector variable $\mathbf{A}$ represents the set of amplitudes of the spherical harmonic functions $Y_l^m(\theta\phi)$ in the eigenmode expansion of the velocity of fluid in the spherical shell, $\mathbf{U}(r, \theta, \phi)$. These $\mathbf{A}$ as they vary through time, represent the evolution of a perturbation of the basic state. The thermal field of the basic state analyzed here is a radially symmetric conduction profile, and the flow field is quiescent. Thus it contains *all* of the symmetries of O(3), the group of rotations on the sphere.

We emphasize again that this equation reproduces the dynamical behaviour of the initial system in a neighborhood of the basic state. This necessarily restricts the current analysis to a Rayleigh number which is close to the critical value. However, only when the behavior of the system is known in a neighborhood of the basic state, can secondary perturbations from the known nontrivial states of the system be understood. Here we investigate the first principles of pattern formation in spherical shells; hence we first consider perturbations from a quiescent, conductive state. We therefore assume $F(0, \mu) = 0$.

The linear space V is spanned by the eigenmodes which are responsible for the unstability of the basic state, at criticality. It follows that V is spanned by spherical harmonics $Y_l^m(\theta, \varphi), -l \leq m \leq l$, where l is either a single fixed integer or can take several values l_1, l_2, ... The "single value" case is expected to occur when only one parameter is varied, typically the Rayleigh number, as in Figure 1. In this case the space V is said *irreducible* of weight l.

If two parameters are allowed to vary simultaneously–e.g. the Rayleigh number and the aspect ratio– then two different values for l can coexist. Mode interaction takes place when the aspect-ratio and the Rayleigh number are such that modes with two different spherical harmonic degree, l_1 and l_2 both have zero eigenvalues. This happens near aspect-ratio 0.3, i.e. the aspect-ratio of the fluid outer core. This mode-interaction is generically expected in the core, while higher-order mode-interactions (e.g. $l = 3, 4, l = 2, 6$) are expected for the aspect-ratio of the mantle ($r_i/r_o = 0.55$). In order to reach such a point in parameter-space, one generically has to vary two parameters– Rayleigh number and aspect ratio–at the same time. Mode-interaction is therefore said to be a *codimension two* bifurcation phenomenon. In this case, we write the multi-parameter $\mu = (\mu_1, ..., \mu_k) \in R^k$, where $k = 2$ and (μ_1, μ_2) are directly related to the physical parameters Rayleigh number and aspect-ratio.

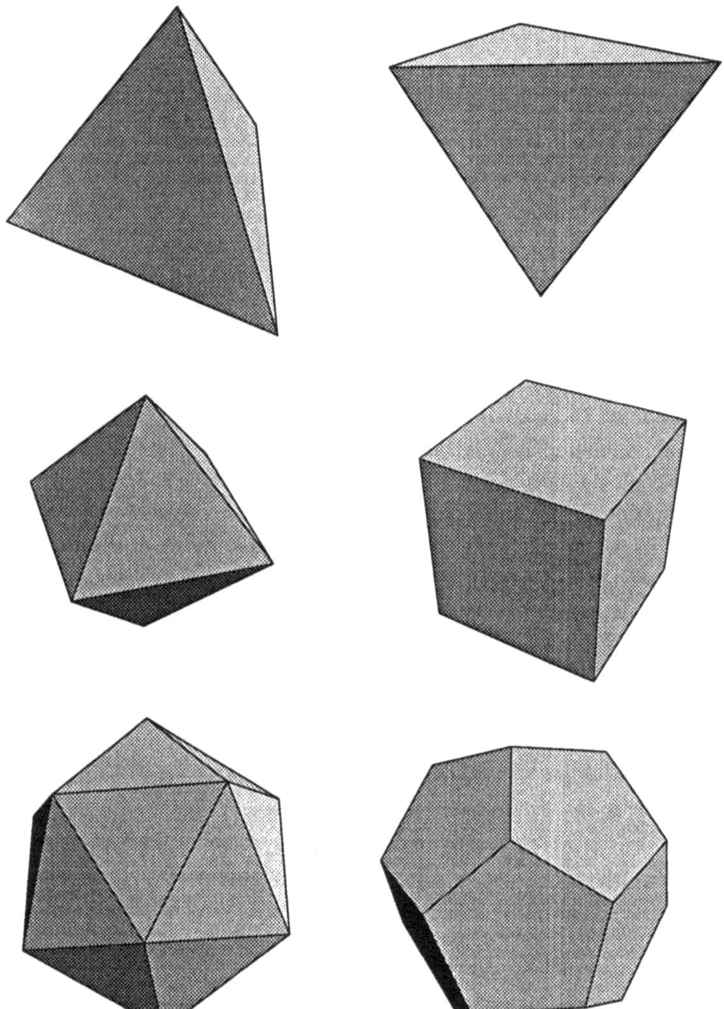

Figure 2 Patterns of convection in a spherical shell correspond to the platonic solids. Here are shown the three platonic solids and their duals. The tetrahedron is self-dual; the octahedron and cube are duals; the icosahedron and dodecahedron are duals. The regular convective patterns in a spherical shell have upwelling at the corners and downwelling at the faces of each platonic solid.

The next important fact for the bifurcation analysis is that the group invariance of the initial equations propagates to equation (1). Equivalently we can say that F is *O(3)-equivariant* : if $T_g, g \in O(3)$, is the representation of O(3) in V, then

$$F(T_g\mathbf{A},\mu) = T_gF(\mathbf{A},\mu), \qquad (\mathbf{A},\mu) \in V \times R^k.$$

One first consequence of the equivariance property is that the map F is constrained to have a peculiar structure. For example suppose that V is irreducible. Then any linear map in V which commutes with T_g is just a scalar multiple of the identity.

In fact there exists a family $v_1, ..., v_p$ of invariant polynomials and a family $F_1, ..., F_q$ of equivariant polynomial maps such that *any* equivariant smooth map in V has the form

$$\hat{F} = \alpha_1(v_1, ..., v_p)F_1 + ... + \alpha_q(v_1, ..., v_p)F_q,$$

where the α_j are smooth functions of v_1,...,v_p. In principle this equivariant structure of F can be determined by a pure algebraic algorithm.

However the calculations are very long and the number of terms can be huge even for quite low-dimensional representations ($l \geq 3$).

Hence in practice only the terms which are needed for complete determination of the bifurcation diagram in the Taylor expansion of F are calculated. The second consequence of equivariance of F, although quite elementary, is of crucial importance in the analysis of symmetry-breaking bifurcations.

First we give a few definitions: given $x \in V$, the *isotropy subgroup* of x is the group of all elements in O(3) which let **A** fixed:

$$\Sigma_{\mathbf{A}} = \{g \in O(3)/T_g\mathbf{A} = \mathbf{A}\}.$$

The $O(3)$-*orbit* (or simply the orbit) of **A** is the set of points which are obtained by letting the group $O(3)$ act on **A**:

$$O(\mathbf{A}) = \{T_g\mathbf{A}, \mathbf{A} \in O(3)\}.$$

The algebraic chain-rule shows that the isotropy of any point **B** in $O(\mathbf{A})$ is a conjugate group to $\Sigma_{\mathbf{A}} : \Sigma_{\mathbf{B}} = g\Sigma_{\mathbf{A}}g^{-1}$. We say that two points in the same orbit, and more generally two points with conjugate isotropy subgroups, have the same *isotropy type* . Now, let Σ be an isotropy subgroup of $O(3)$ (or more generally of the symmetry group of the problem), and define the fixed-point subspace associated to Σ, to be the space of all points in V which are fixed by Σ. Note that this includes points with "higher" isotropy than Σ:

$$V^{\Sigma} = \{\mathbf{A} \in V/T_g\mathbf{A} = \mathbf{A}, \forall \quad g \in \Sigma\}.$$

Then an algebraic calculation shows that for any equivariant map F: $V \to V$, we also have

$$V^{\Sigma} \to V^{\Sigma}.$$

It follows that, given an isotropy subgroup Σ , we can look for solutions of (1) with isotropy (i.e. symmetry) Σ , by restricting the equation to the subspace V^{Σ}.

Notice that orbit types are partially ordered by set inclusion, therefore fixed-point subspaces are also partially ordered by set inclusion with the "reversed" order to that of the corresponding groups.

Our approach to the problem is therefore to perform the following steps:

1) compute the "equivariant structure" of $F(\mathbf{A}, \mu)$

2) determine the orbit types of the group representation in V and the corresponding fixed-point subspaces

3) solve equation (1) restricted to these subspaces.

If for an isotropy subgroup Σ, the dimension of $V^{\Sigma} = 1$, then (1) restricted to V^{Σ} is a scalar equation, hence usual results about existence of bifurcated branches at simple eigenvalue hold in this subspace. This is known as the equivariant branching lemma. These techniques are explicated in greater detail in Golubitsky *et al.* [1988].

Before we go to the next section, it is useful to fix notations for the isotropy subgroups occuring in the various representations of O(3). A complete list of these subgroups was obtained by L. Michel [1980] and by Ihrig and Golubitsky [1984]. This list together with a description of the corresponding fixed-point subspaces can be found in Chossat, Lauterbach and Melbourne [1990]. It must also be noted that unfortunately, the notation in the two last references is different from the Schoenflies notation for subgroups of $O(3)$. Therefore, the Schoenflies notation is given in square bracktets where it differs from the notation used in these references and here.

1) **Subgroups of $SO(3)$**

- Continuous subgroups: $SO(2)$ [C_∞], $O(2)$ [D_∞].

- Planar subgroups: Z_n [C_n: n-fold rotations], D_n [direct symmetries of the regular n-prism].

- Exceptional groups: T, O, I [the direct symmetries of the tetrahedron, cube and icosahedron respectively].

2) **Subgroups of O(3) containing the symmetry s: $\mathbf{A} \to -\mathbf{A}$**

Any group of the form $\Sigma \cup s\Sigma$ where Σ belongs to type 1. Defining the two-element group $Z_2^c = \{Id, s\}$, these groups can be written as $\Sigma \times Z_2^c$.

3) **other subgroups of $O(3)$**

- Continuous subgroup: $O(2)^-$ complete symmetry group of the cylinder [$D_{\infty h}$].

- Z_{2n}^-: if n is even, the group generated by the rotation of angle $\frac{\pi}{n}$ around an axis followed by the reflection κ through the plane perpendicular to this axis [S_{2n}]. If n is odd, the group generated by the rotation $\frac{2\pi}{n}$ and κ [C_{nh}].

- D_n^z: the complete symmetry group of an n-pyramid [$C_{n\nu}$].

- D_{2n}^d: if n is odd, the complete symmetry group of the n-prism [D_{nh}]. If n is even, the complete symmetry group of the twisted n-prism [D_{nd}].

- O^-: complete symmetry group of the tetrahedron. This notation is due to the fact that this group is isomorphic to O. [Schoenflies notation T_d].

A simple example: $l = 1$

Here we suppose that V is the 3-dimensional representation of $O(3)$, i.e. O(3) acts by its natural action in $\mathbf{R}^3$: $T_g \simeq g \in O(3)$. Every axis ζ in $\mathbf{R}^3$ passing through the origin is clearly a fixed point subspace. The corresponding isotropy subgroup is composed of rotations around ζ and reflexions through any plane containing ζ. This group is isomorphic (but not conjugate !) to $O(2)$ and it is denoted $O(2)^-$. There are no other isotropy subgroups, apart from $O(3)$ itself (which fixes the origin). This implies that there is only one non-trivial (i.e. non-zero) orbit type, therefore all the dynamics and bifurcation diagram is given by that of the equation restricted to the axis ζ. Notice that the symmetry through the origin $\mathbf{A} \to -\mathbf{A}$ is an element of $O(3)$. This means that the restriction of $F(\mathbf{A}, \mu)$ to ζ is the equivariant map $f(\mathbf{A}, \mu)$) and is *odd* in $\mathbf{A}$.

It results from this analysis that the problem is completely reduced to a pitchfork bifurcation of equilibria, and the principle of exchange of stability holds in this case. For the complete problem in V, stability means "stability of the group orbits of equilibria". The bifurcated equilibria have the symmetry group $O(2)^-$ and the orbits are spheres in V. This result was obtained by Babskii & Sklovskaia [1971] in the context of the onset of convection in a self-gravitating fluid sphere.

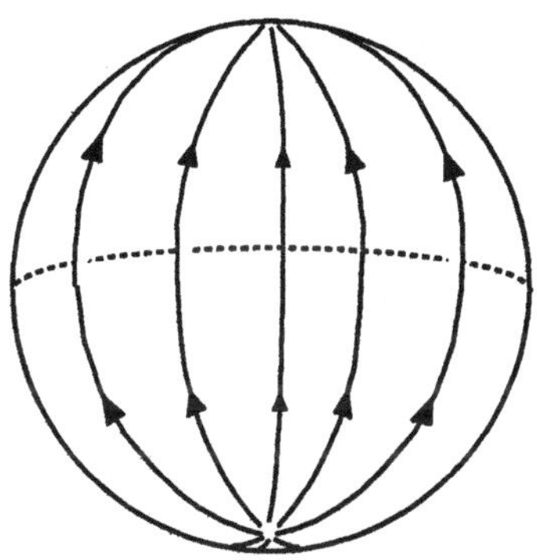

Figure 3 Pattern of flow on outer surface of $l = 1, m = 0$ convective pattern; the two other values of $m = \pm 1$ produce the same pattern of convection with a permutation of the axes in physical space defined in spherical coordinates (r, θ, ϕ) as $(r, 0, \phi)$,$(r, \theta, 0)$ and,$(r, \theta, \pi/2)$, and in Cartesian coordinates (x, y, z) as $(x, 0, 0)$,$(0, y, 0)$ and $(0, 0, z)$.

Figure 3 depicts the flow on the surface of the sphere for the solution at $l = 1$. The upwelling and downwelling material is at the poles. Like for any spherical solution, these poles can be rotated to pass through any two antipodal points on the sphere. In the $l = 1$ solution, and in fact, for any solution with *odd* l, inversion through the origin also produces a physically equivalent flow. Conversely, a solution with *even* l when inverted through the origin produces a new flow which is still a solution but which is not physically equivalent to the first one; we will return to this point in the discussion of the $l = 2$ case.

Steady-state bifurcations with one parameter. We noticed in the previous section that when one parameter only is allowed to vary, the bifurcation equation (1) takes place in a space V isomorphic to the space spanned by the spherical harmonics $Y_l^m(\theta,\varphi), -l \leq m \leq l$, l being a fixed positive integer. In this section we describe successively the two next "simple" cases after $l = 1$, i.e. $l = 2$ and $l = 3$. The latter is more complicated since it occurs in a higher dimensional space, but it is also very different from the former because even and odd values of l lead to qualitatively different dynamical behaviour. This will also be illustrated in the third subsection, where we show some results which are valid for large families of values of l. It must also be noticed that so far, the problems with $l > 3$ have not yet been completely investigated, although significant progress has been made for the cases $l = 4$ and $l = 5$ in Chossat *et al.*.

Bifurcation at $l = 2$

In this case, V is a 5-dimensional space. It can be shown however [Golubitsky, *et al.*, 1988] that:

1) The non-trivial isotropy subgroups for this representation are $O(2) \times Z_2^c$, and $D_2 \times Z_2^c$. These subgroups are associated with fixed-point subspaces L and P respectively. L is associated with the axisymmetric convective patterns which are invariant under the transformations $O(2) \times Z_2^c$ and is a one-dimensional fixed-point subspace. P is associated with the nonaxisymmetric convective patterns which are invariant under the transformations $D_2 \times Z_2^c$ and is a two-dimensional fixed-point subspace.

2) when O(3) acts on the plane P, the union of transformed $\bigcup_g T_g P$ fills all V up. Hence all the dynamics and bifurcation is contained in the equation restricted to P;

3) the subspace P is invariant under the action of a subgroup of O(3) which is equivalent to the group of symmetry of the triangle in the plane P. This implies notably that the axis L possesses two copies L' and L" in P, and these three axis are the symmetry axis of an equilateral triangle. In terms of O(3) action, one passes from L to L' (and, respectively, L") by the action of a rotation of 90 degrees around an axis ξ' (resp. ξ") perpendicular to the axis of rotational (i.e. O(2)) symmetry for the isotropy subgroup associated with L. The axes ξ' and ξ" make an angle of 90 degrees between them.

These facts allow the complete solution of the bifurcation problem in this case: in general the only bifurcated equilibria in P are those appearing on the axes L, L' and L". They correspond therefore to *axisymmetric* solutions since they are invariant under the action of $O(2)$, hence of any rotation around a given axis. More precisely, the equation on one axis of symmetry has the form

$$\frac{d\mathbf{A}}{dt} = \sigma(\mu)\mathbf{A} + b\mathbf{A}^2 + c\mathbf{A}^3 + h.o.t.$$

where $\sigma(\mu)$ is the eigenvalue crossing 0 at $\mu = 0$. We can assume here that $\frac{d\sigma}{d\mu}(0) > 0$ because this is the case for the problem of onset of convection which was described in section 1.

If the coefficient $b \neq 0$, then the bifurcation is *transcritical*. Moreover it can be shown that the branches of equilibria are *unstable* on both sides of the bifurcation point.

If $b = 0$, then the bifurcation is one-sided, and if in addition $c < 0$, then it is supercritical. In this case, one branch is stable and the other is unstable. In Figure 4 is shown the axisymmetric pattern of convection at $l = 2, m = 0$. In this case, there is a plume-like upwelling at each pole and a sheet-like downwelling along the equator. Notice that if the sign on this mode is reversed, the resulting flow has a sheet-like upwelling at the equator and a plume-like downwelling at each pole. These two cases are significantly different from each other, but are both contained in the one-dimensional fixed-point subspace L.

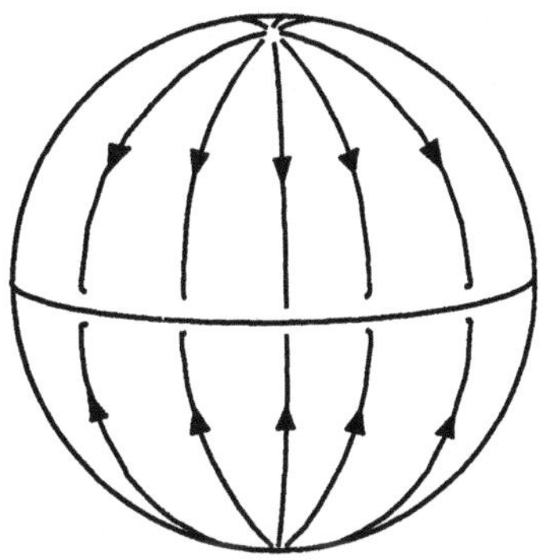

Figure 4 Pattern of flow on outer surface for $l = 2, m = 0$ convective pattern. Note that this flow has two plume-like upwellings at the poles, and a single sheet-like downwelling along the equator. Reversing the sign on this pattern produces sheetlike upwellings and plume-like downwellings.

If b is close to 0 but not equal to 0, then the bifurcation is slightly transcritical and unstable on both sides, with a turning point which allows one branch to "bend back" and recover stability. A multi-parameter analysis (b being the second parameter) would allow to make a complete analysis of the degenerate situation $b \approx 0$ [Golubitsky and Schaeffer, 1982].

In figure 5 we show a typical phase portrait of equation (1) in the plane P when $b \neq 0$.

Bifurcation at $l = 3$

This case was first investigated by Busse [1974]. He found the three branches of equilibria associated with the maximal isotropy subgroups $O(2)^-$, O^- and D_6^d. In fact these are the only possible branches in this bifurcation problem [Chossat *et al.* 1990]. In the last reference it was also proved that the system is *gradient-like*in this case, i.e. that all the asymptotic dynamics is either escaping to infinity or converging to one of these equilibria. Assuming supercritical bifurcation, either equilibria with isotropy D_6^d or O^- can be stable, depending on the sign of a coefficient of cubic

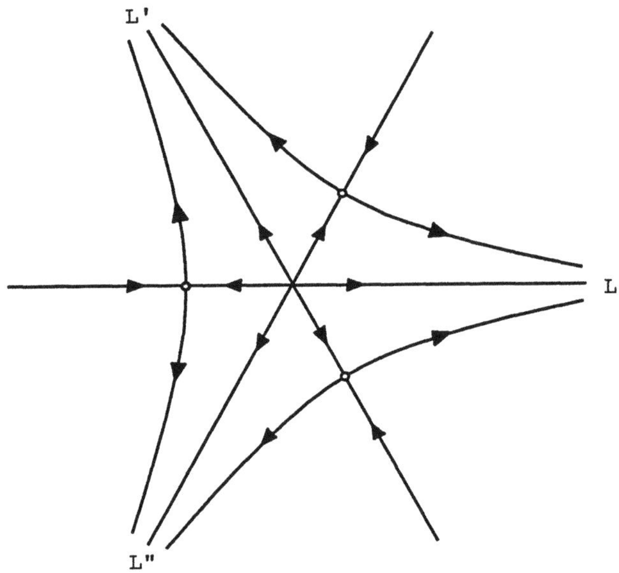

Figure 5 Phase portrait for $l = 2$ in subspace P. This shows the fixed-point at the origin and the three nontrivial fixed-points (open circles) for the axisymmetric solutions along the axes L, L', L'' Notice that these axes are stable manifolds of the fixed points and the unstable manifolds are in the direction of the non-axisymmetric solutions.

order terms. Notice that the axisymmetric solutions (isotropy $O(2)^-$) are never stable. The isotropy O^- is in fact the full tetrahedral symmetry.

A selection of general results

For each value of l, the representation of $O(3)$ in V possesses a characteristic (and finite) set of orbit types. However certain orbit types appear at every value of odd l, while some others appear at every value of even l. Here we treat two examples of this sort.

1) **Axisymmetric equilibria**

An *axisymmetric* point in V is one which is invariant under any rotation around a given axis. In other words, its isotropy group contains $SO(2)$. We have already seen that for $l = 2$, only such equilibria do bifurcate in general. When l is odd this happens with the orbit type $O(2)^-$, while for even l it occurs with orbit type $SO(2) \cup sSO(2)$. Moreover in any case, the associated fixed-point subspaces are one-dimensional, and therefore, thanks to the "equivariant branching lemma" stated in the previous section, branches of axisymmetric equilibria always exist. When l is odd, the bifurcation is one-sided (supercritical in the cases of interest), while when l is even, the bifurcation is transcritical (subcritical and supercritical branches exist). In addition when $l > 1$, these solutions are *always*unstable, unless if in the transcritical case, the subcritical branch bends back (in which case it may recover stability after the turning point, as in the example $l = 2$). All these facts are proved in Chossat *et al.* [1990].

2) **Equilibria with l-fold symmetry**

Another interesting class of orbit types is that of l-fold symmetries (when $l > 2$). When l is odd, the isotropy group is D_{2l}^d, while when l is even, it is $D_l \cup sD_l$, except for $l = 4$, where D_l is only part of O (octahedral) isotropy. In the first case the associated fixed-point subspaces are one-dimensional, while in the second case it is two-dimensional. The existence of bifurcated equilibria with this kind of symmetry results from the "equivariant branching lemma" when l is odd, and it has been also proven in Chossat [1983] when l is even (first known example of bifurcation with submaximal isotropy). When l is even these solutions are transcritical but always unstable. However when l is odd, it has been shown by Barany and Melbourne [1990] that such solutions can be stable, if supercritical, under a "generic" condition (sign of a coefficient).

3) **Instability of equilibria when l is even**

In the examples which have been reviewed so far, bifurcated branches in the case of even l have always been found unstable. There is a general result about this: it was shown in Ihrig and Golubitsky [1983] that for even l, solutions occuring in one-dimensional fixed-point subspaces are unstable near the bifurcation point. The reason is that in this case, the presence of quadratic terms in the bifurcation equations forces one eigenvalue at least for the linearized operator to have the "wrong" sign (there is no such result for odd l because then the bifurcation equation does not contain quadratic terms). This result was extended to solutions with isotropy associated to fixed-point subspaces of arbitrary imension by Chossat *et al.* [1990]. This means that in order to find bounded local dynamics in such problems with l even, one must consider the "degenerate" situation where the quadratic terms in the bifurcation equation have a coefficient close enough to 0, in order to allow higher order terms to dominate and give all the eigenvalues the right sign. This is typically what happens in the case $l = 2$.

Interaction of spherical modes in steady-state bifurcation. It has been noticed in the previous sections that a coexistence of two different values of l, say l_1 and l_2, is possible if several parameters are allowed to vary independantly. For example in the problem of the onset of convection in a spherically symmetric model, the study of the neutral stability curve for the basic state of pure conduction shows that for certain values of the aspect ratio (radius ratio) of the spherical shell, the criticallity is reached at the same time for the modes l_1 and $l_2 = l_1 + 1$. In this problem, the interaction of these two modes is therefore likely to occur when both the Rayleigh number and the aspect ratio are considered as free parameters. It is important at this point to remark that the main effect of mode interaction in bifurcation with spherical symmetry is to introduce ***non-trivial dynamics***. In the previous section, only equilibria have appeared as bifurcated solutions. This is not just due to the fact that bifurcation occurs at a 0 eigenvalue (steady-state bifurcation), because the high dimension of the space V, which is just the eigenspace for this 0 eigenvalue, would in general allow more complicated time-dependant solutions to bifurcate. However the equations truncated at cubic order have a gradient structure, which prevents or makes more difficult such possibility Recently a time-periodic solution with long period has been found in the case $l = 5$ by Chossat [1990].

The mode interaction $l_1 = 1$ *and* $l_2 = 2$

When the space V is spanned by spherical harmonics with ***different*** l's , nontrivial time-dependant dynamics, especially intermittency, can be observed. The following results are taken from Armbruster and Chossat [1990]. Friedrich and Haken [1986] also studied this mode-interaction problem and discovered numerically some of the phenomena which are described here.

Now V has dimension 8, and is decomposed along the representations of weight $l = 1$ and $l = 2$ as follows: $V = V_1 \oplus V_2$, where dim $V_1 = 3$ and dim $V_2 = 5$. Figure 6 shows the lattice of orbit types, which shows the set inclusion of the isotropy subgroups. By following a path in the diagram, the isotropy increases when going upwards. On the right of the lattice are indicated the dimensions of the related fixed-point subspaces.

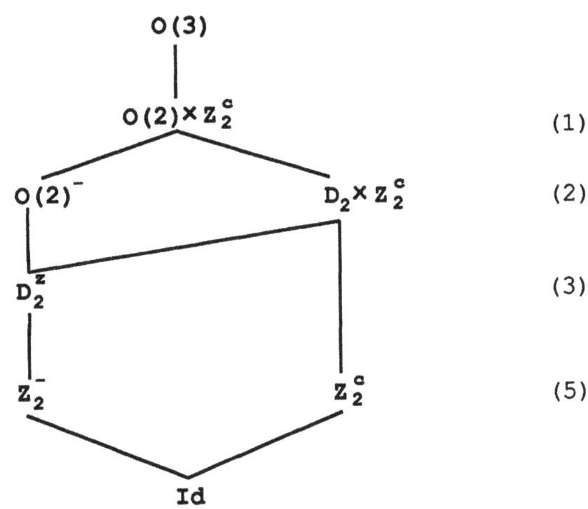

Figure 6 Lattice of isotropies for $l = 1, 2$ mode interaction for axisymmetric solutions only.

An examination of this figure leads us to formulate the following remarks: 1) the pure $l = 2$ isotropy subgroups remain unaffected by the presence of $l = 1$ modes. This implies that the axisymmetric equilibria which are found in the case $l = 2$ do still exist as solutions of the present problem. They will play a key role in the subsequent analysis. 2) W denotes the 3-dimensional fixed-point subspace associated with the isotropy D_2^z. This space is the sum of two planes P and P_1, which are respectively the fixed-point subspaces for the isotropy groups $D_2 \times Z_2^c$

(pure $l = 2$) and $O(2)^-$. The intersection of P_1 and P is the fixed-point subspace associated with the maximal isotropy subgroup $O(2)$. This axis of intersection between P_1 and P is denoted by L, and contains the pure $l = 2$ axisymmetric equilibria.

An important hypothesis assumed in the following is that the coefficient of the quadratic term of the equations restricted to the fixed-point subspace Z_2^c (pure $l = 2$ part) is *small*. Hence the branch of pure $l = 2$ equilibria has a turning point in the vicinity of the bifurcation point.

The results

Every solution listed below can be stable under certain conditions on the coefficients in the bifurcation equations.

1) The two branches of axisymmetric equilibria with pure $l = 2$ mode correspond to two different group orbits, i.e. to physically different solutions. They are named α-cells and β-cells by Friedrich and Haken [1986].

2) Mixed-modes equilibria can also bifurcate in the plane P_1, from either the origin or one of the pure $l = 2$ equilibria. Figure 7 shows a typical phase portrait in P_1. These solutions are still axisymmetric but mix $l = 1$ and $l = 2$ components. These equilibria can bifurcate to a limit cycle (time periodic solutions) in P_1.

3) Non-axisymmetric mixed-mode equilibria can also bifurcate from the previous equilibria in the 3-space W. When the parameters are suitably varied, they can undergo a Hopf bifurcation in W, and a "strange attractor" can be created by heteroclinization of the limit cycle on different equilibria in W. While this is just a numerical observation and we do not pursue the analysis in this direction.

4) The plane P is the same subspace associated with $l = 2$ as in the previous section. We recall that it contains three invariant axes: L and its two copies L' and L" obtained in V by the action of two rotations R^+ and R^- by 90 degrees around axes perpendicular to the axis of symmetry of the isotropy group $O(2) \times Z_2^c$ of L. In this plane, non-axisymmetric equilibria can branch off the α-cells and β-cells.

We now show how intermittent-like behaviour can naturally appear in this problem. The phase portraits in P shown in Figure 7 and P_1 shown in Figure 8 can *coexist*.

In words, a "heteroclinic" connection (trajectory) can exist: 1) in P from the β-cells on the axes L' and L" to the α-cell on L (Figure 7); 2) in P_1 from the α-cell on L to the β-cell on L (Figure 8); 3) in P from the β-cell on L to the α-cells on L' and L" (Figure 7). This realizes a *heteroclinic cycle.* Indeed the α-cells on L' and L" are copies by the two rotations R^+ and R^- of the α-cell on L, therefore they must have the same dynamics. In particular they are connected to the *beta*-cells on the axes L' and L" by trajectories lying into the two copies of the plane P obtained through the transformations R^+ and R^-. One remarkable feature of this object is that it is *structurally stable*, i.e. it persists existing after small perturbations in the equations, as long as these perturbations respect the spherical symmetry. If certain conditions are fullfilled, the heteroclinic cycle can also be asymptotically stable (attracting). If such conditions were realized in the physical problem, the observed behaviour would

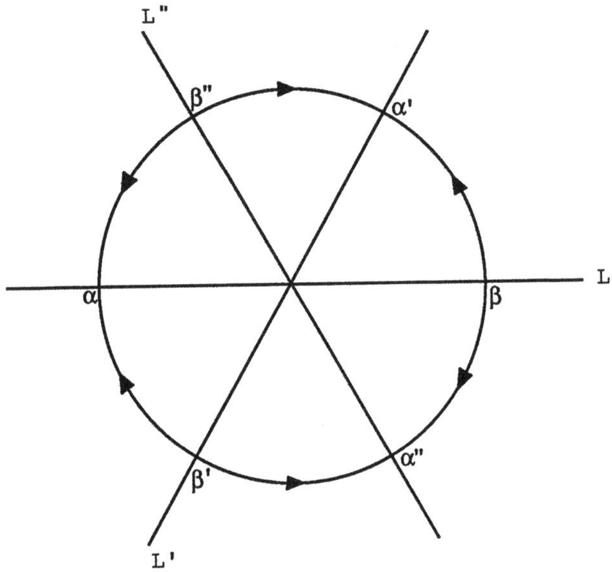

Figure 7 Phase portrait for $l = 1, 2$ mode interaction with spherical harmonic degree in subspace P when quadratic coefficient **b** is small but not zero.

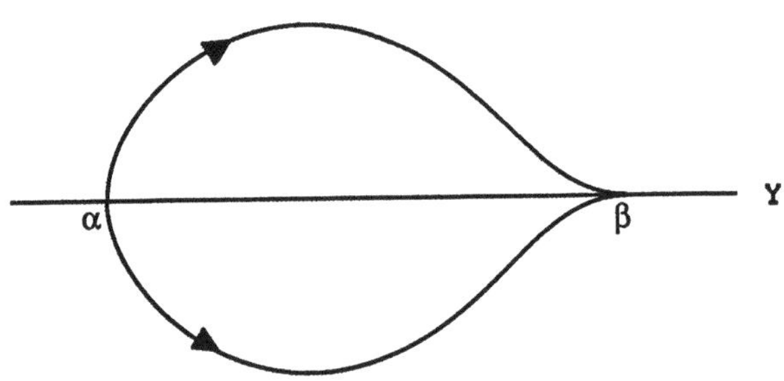

Figure 8 Phase portrait for $l = 1, 2$ mode interaction in subspace P_1.

be the following: the flow would spend long periods of time very close to the states of pure $l = 2$ equilibria, suddenly switching from one to the other via mixed-modes patterns, and following the order defined by the heteroclinic cycle. This would be due to the unavoidable low-level fluctuations which exist in real systems. A time series record of this dynamics would show random, intermittent-like pole reversals.

Other heteroclinic cycles can also appear in this problem: connections between α-cells and mixed-mode equilibria, or even between α-cells and limit cycles in P_1. In the latter case, the dynamics is likely to be chaotic, as suggested by both theory and numerical simulations. A fourth type of heteroclinic cycle has been found, which connects the non-axisymmetric equilibria appearing in P (type 4 above) among themselves. A first conclusion of this overview is that mode interaction with spherical symmetry can induce very complicated (and interesting) dynamical behaviours. The numerical work of of Friedrich and Haken [1986] suggests that this approach could be relevant for applications in geo- or astro-physics.

Thermal Convection in the Earth's Mantle

In this section, we first review the boundary conditions, Rayleigh number, and other relevant parameters for thermal convection in the earth's mantle. We then review the observational evidence for the pattern of these convective cells. Because this pattern has peaks in low-order spherical harmonic modes, we show that the pattern of convection in the earth's mantle is easily interpreted within the framework of the results generated by the foregoing bifurcation theory, without having to look at an unreasonably large number of secondary bifurcations to higher-order modes.

The whole mantle has aspect-ratio $r_i/r_o = 0.545$, the lower mantle has an aspect-ratio of $r_i/r_o = 0.612$ and the upper mantle has an aspect-ratio of $r_i/r_o = 0.891$. The whole mantle is bounded by the core-mantle boundary below and the earth's surface above, the lower mantle is bounded by the core-mantle boundary below and the 670km discontinuity above, and the upper mantle is bounded by the 670 km discontinuity below, and the earth's surface above.

The core mantle boundary can be taken to be free-slip and constant temperature. This is because the core has a much lower viscosity than the mantle, and a much faster time-scale of thermal convection. The earth's surface can also be taken to be a free-slip constant temperature boundary condition for the same reason. The 670 km discontinuity can be taken to be free-slip because this internal boundary is characterized by at least a phase transformation and perhaps some moderate chemical stratification of iron. Phase transformations in situ are rarely sharp boundaries, particularly when the material has many chemical components and several phases, only one of which is undergoing a transition. The spatial and mechanical heterogeneities of a phase transformation is generally associated with viscosity drops of orders of magnitude known as *transformational superplasticity* [Rubie, 1983], justifying the use of a free-slip boundary condition at the 670 km discontinuity.

A review of the recent progress on the large-scale circulation of the earth's mantle by Olson *et al.* [1990] indicates that the degree to which thermal convection traverses the 670 km seismic discontinuity in the earth's mantle is still basically unknown. Therefore, it is reasonable to model the earth's mantle both as a thick spherical shell with aspect-ratio 0.54, as well as two thermally and dynamically coupled spherical shells, the lower and upper mantle, respectively. The point of view adopted here is that of whole mantle convection, because the spectrum of convection exhibited by the data seems to be characteristic of some mode-interactions expected for the aspect-ratio of the whole mantle. This does not rule out a significant component of layered mantle convection. While the suite of modes discussed here, evident in the data, are favored by the geometry of the whole mantle, there is also data which suggests that there is also a secondary scale of convection confined to the upper mantle, arranged in patches of rolls subparallel to the motions of the plates. Patches of rolls evolving in a thin spherical shell subject to nonhomogeneous boundary conditions is a very interesting problem, and not further discussed here.

The Rayleigh number of the earth's mantle, estimated to be between 10^6 and 10^8 is far higher than the critical value predicted by linear stability for a constant viscosity fluid. Furthermore, numerical simulations for a constant viscosity fluid in a spherical shell at these Rayleigh numbers indicate that most of the convective energy is in modes higher than l=10 [Bercovici, *et al.*, 1989; Glatzmaier, 1990]. These two facts are obviated by the very strong temperature-dependence of viscosity in the earth's mantle, which can raise the critical Rayleigh number predicted by linear stability theory by orders of magnitude. Looking at the primary and secondary bifurcations from the marginally critical state is not an unreasonable approach. The geoid, seismic tomographic, dynamic topographic and hotspot volcano distributions and divergence and curl of surface motions on the earth itself indicate that low-order ($l < 7$) modes are active. Since it is the low-order ($l < 7$) modes that bifurcate from the marginally stable state, examining the properties of these bifurcations produces valuable insights on convection in the earth's mantle.

The spatial structure of thermal convection in the earth's mantle is revealed by the hotspot vocano distribution, absolute plate motions, the geoid, the dynamic topography, the surface divergence and radial vorticity fields. The hotspot volcano distribution and absolute plate motions [Crough and Jurdy, 1980] indicate that the Y_2 and Y_1 modes are active in the mantle as well as the core.

Correlations among hotspot distributions, dynamic topography and the earth's geoid [Cazenave, *et al.* 1989] suggest a strong spatial resonance between the Y_2^2 and Y_6^m convective modes. These indicate the global lateral density variations which drive thermal convection in the earth's mantle. The flow resulting from this thermal drive is observed on the earth's surface as plate tectonics. The surface divergence and radial vorticity quantify the distributions of lithospheric plates, and their movements normal and transverse to their boundaries [Forte and Peltier]. The surface divergence field has peaks at spherical harmonic degrees $l = 4$ and $l = 5$, and the radial vorticity field has a peak at spherical harmonic degree $l = 5$. Furtherermore, the continental masses themselves have a broadly $l = 3$ distribution, on two faces of a tetrahedron. More details about these data are next presented, organized in terms of increasing spherical harmonic order.

$l = 1, 2$: Hotspot Volcano Distributions and Absolute Plate Velocities

The raw distribution of hotspot volcanoes on the earth has a peak in the Y_0^1 [Crough and Jurdy, 1980; Stefanik and Jurdy, 1990]. Most hotspot volcanoes occur on the Pacific half of the globe. This is also the half the globe which is dominated by oceanic lithosphere, the half which is nearly completely surrounded by subduction zones and the half which has the fastest plate velocities. Criss and Hoffmeister [1990] has recently suggested that these facts are qualitative evidence of a "unicell" of thermal convection in the earth's mantle. It should be noted here that it is impossible to obtain meaningful correlations of the hotspot volcano distribution with the geoid or dynamic topography in this case, because the Y^1 mode of the global geoid and global topography are simply offsets in the earth's center of mass. Since the topography and the geoid can only be measured relative to the center of mass, any geoid or topographic anomaly in the Y^1 mode due to a single large plume of thermal convection is undetectable with these data.

It is interesting to note, however, that even in the high-resolution of the numerical experiments in a 3-dimensional spherical shell [Bercovici, *et al.*; Glatzmaier, 1990], the convective pattern is characterized by two large antipodal upwellings, one larger than the other. This low-order $l = 1, 2$ signal appears despite the fact that the simulations are carried out at Rayleigh numbers ($R = 10^5$) far above marginal stability. This further suggests that there is something special about the geometry of a spherical shell and the primary and secondary bifurcations of the thermal convection when constrained by this geometry that strongly influences the supercritical finite-amplitude convective pattern. The numerical work of Friedrich and Haken [1986] on the $l = 1, 2$ interaction, and the analysis of Chossat and Armbruster [1990] indicate that $l = 1, 2$ mode interaction produces aperiodic oscillations in both.

An unresolved issue is whether the presence of modes $l = 1, 2$ in the spectrum of hotspot distribution and absolute plate motion are a natural result of the dynamics of thermal convection in a spherical shell, or whether it is because of some spatial forcing at the core-mantle boundary due to the motions of the core. Because the motions in the core have a much shorter time-scale than the mantle, and because the core has a much lower viscosity than the mantle, the core-mantle boundary is usually taken as providing a constant-temperature free-slip boundary condition for convection in the earth's mantle. However, it is possible that the activity of the Y_2 and Y_1 modes in the the core provides uneven heating at the core mantle boundary, thus exciting the Y_2 and Y_1 modes of convection by destabilizing the hot thermal boundary layer at the CMB. The action of the thermal and mechanical boundary layers above and below the core-mantle boundary gives rise to the deformation of the CMB which then influences the further evolution of the thermal and mechanical boundary layers. These dynamic effects of two-layer convection have been explored numerically in cartesian geometry and in axisymmetric spherical geometry. The fact that this internal boundary layer itself is spherically symmetric in three dimensions, however, indicates that the time-evolution of the dynamics above and below the CMB and the shape of the CMB itself is strongly constrained by the structure of the symmetry subgroups of O(3).

$l = 2, 6$:*Correlations with the global geoid*

Upwelling from the core mantle boundary (CMB) is the most easily resolved in terms of the spherical harmonic patterns of convection, because the upwelling is in the form of two major plumes, one at the Afar triangle region, and one in the South Pacific [Cazenave, *et al.*, 1989; Richards *et al.*, 1988]. Associated with these large low-density anomalies are a discernable lows in the seismic velocity [Woodhouse and Dziewonski, 1984; Dziewonski, 1984] combined with geoid and dynamic topographic highs [Richards and Hager, 1984; Hager *et al.* 1985]. The fact that the geoid is more elevated at these sites, rather than depressed, is an indication that the earth's surface is uplifted by strong dynamic convective upwelling caused by the density anomaly. This is shown by correlation with the *dynamic* topography. The dynamic topography is that which has the crustal topography removed, i.e. that portion of the topography which is due to known lateral variations of density in the crust within 30 km of the surface and the isostatic compensation thereof [Cazenave *et al.*, 1989].

The correlations among global geoid, seismic tomography, dynamic topography and hotspot distribution performed by of Cazenave *et al.* [1989], as well as those of Hager *et al.* [1985] therefore indicate a that the pattern of convection in the earth's mantle is dominated by a $Y_2^2(\theta, \phi)$ pattern. A later geodynamic model of Forte and Peltier [1988] that explicitly takes into account the plate motions at the surface indicate a peak in the correlation spectrum at Y_2^2 between the geoid and dynamic topography. The models of Richards *et al.* [1988] indicate that the Y_2^2 density anomalies primarily concentrated in the lower mantle. Furthermore, lateral variations in seismic velocities indicate a Y_2^2 density structure pattern in the lower mantle, although this depends fairly sensitively on the velocity model chosen for the upper mantle [Richards *et al.*, 1988].

Convection cells in the lower mantle arises primarily from the thermal instability at the core-mantle boundary (CMB). It is no coincidence that the seismically-determined topography at the core-mantle boundary also has a peak in the spectrum at Y_2^2, although different seismological studies find a different sign of the anomaly. However, the seismic anomaly so determined may merely be a variation in the wavespeed due to a chemical boundary layer at the core-mantle boundary. While there have been models of the dynamic deformation of the lower boundary of convection in the lower mantle [Hansen, 1989], the coupling between the convective system in the core and the convective system in the lower mantle is still largely unknown.

Another peak in the correlations among global geoid, dynamic topography hotspot distribution and seismic tomography is in a dodecahedral/ icosahedral Y^6 mode, i.e. $Y_m^6, \{-3 \leq m \leq 3\}$ [Cazenave, *et al.*, 1989]. This is the exceptional symmetry subgroup exhibited by an ordinary soccer-ball which has twelve pentagonal faces (the faces of a dodecahedron) and twenty hexagonal faces (the faces of an icosahedron). Normally, an icosahedron has twenty triagonal faces; however, on a soccer-ball, the corners of these triangles are cut off by the pentagonal faces of the dodecahedron. Twenty triangles with their corners cut off forms twenty hexagons. The fact that this works out so nicely is due to the fact that the twelve corners of the icosahedron correspond exactly to the twelve faces of the dodecahedron. This is what is meant by the statement that the dodecahedron is the dual of the icosahedron, and vice versa. Similarly, the tetrahedron is its own dual, and the dual of a cube is the octahedron.

The pattern exhibited by the global geoid correlations has a hot-spot at the center of each of the twelve pentagonal faces of the dodecahedron, and a cold-spot at the center of each of the twenty faces of the icosahedron. The remarkable thing about this pattern is that it is the high-order pattern of thermal convection in a sphere that maximizes the heat-flow for spherical harmonic pattern $l = 6$ [Busse, 1974]. In this pattern, also, two of the upwellings of the dodecahedral pattern correspond to the upwellings of the Y_2^2 pattern, and to the upwelling indicated by the hotspot-distribution. The correspondence between the upwellings of the $l = 2$ and $l = 6$ patterns suggests a strong spatial resonance condition.

There are two factors which contribute to this resonance. One is a degenerate bifurcation at the aspect-ratio of the whole mantle, whereby both the $l = 2$ and $l = 6$ modes have zero eigenvalues at the same Rayleigh number at marginal stability. While the Rayleigh number of the earth's mantle is far from the point of marginal stability, the observational fact remains that these two modes are present and active. The center manifold that is present by virtue of the equal linear stability of the $l = 2$ and $l = 6$ modes, however, is unstable in the $l = 3, 4, 5$ directions. This suggests that the modal analysis of the mantle convection problem may be split into two subspaces, a "slow" manifold defined by the $l = 2, 6$ modes, and a "fast" manifold defined by the $l = 3, 4, 5$ modes.

The other relevant fact about the $l = 2, 6$ geoid/tomography/topography/hotspot convective pattern is that the upwellings of the $l = 2, m = 2$ mode correspond to antipodal upwellings of the dodecahedral $l = 6$ pattern. Another way of stating this is that both the dodecahedron $l = 6$ and the $l = 2, m = 2$ spherical harmonic patterns share the element $-s$. The branch of solution exhibited by the earth contains both of these modes, in a combination which reflects their common symmetry. It would be interesting to know whether his common symmetry element establishes a formal resonance condition in the equations governing the mode interaction. Furthermore, this resonance may be so attractive geometrically that it may not be strongly affected by mass flux through the 670 km seismic discontinuity, or the degree of splitting of these two resonant modes, $l = 2$ dominant in the lower mantle and $l = 6$ dominant in the upper.

$l = 3, 4$: *Lithospheric heterogeneity*

The $l = 3$ mode of thermal convection in the earth's mantle is notable primarily by its absence. $l = 3$ is absent from the plate-motion spectrum, and it is absent from the global geoid, dynamic topography, seismic tomography and hotspot distribution spectra. The remarkable thing about the absence of the $l = 3$ mode is that, judging from the linear stability alone, this, along with the $l = 4$ mode, should be one of the most active modes of convection in the earth's mantle. One possible explanation for the absence of the $l = 3$ mode is the dearth of actual solutions possible in this mode. As noted above, the only stable bifurcation from marginal stability at $l = 3$ is the tetrahedral mode of convection; the axisymmetric branches are unstable, and no other branches even exist. What could serve to suppress the $l = 3$ tetrahedral pattern?

Low-density continental crust, being nearly impossible to subduct, is distributed in a roughly tetrahedral pattern on the earth's surface [Carey, 1976]. Asia is currently in the center of one face of this tetrahedron, Europe and Africa reside on another, the Americas on a third, and Antarctica on a fourth. The corners of this tetrahedron correspond to the oceans: the Arctic, the Atlantic, the Pacific and the Indian. Continental masses distributed in a tetrahedral pattern would effectively "turn off" the $l = 3$ tetrahedral pattern of convection because of the refusal of the low-density continents to subduct, and the relative difficulty with which strong continental lithosphere is rifted. This mechanism can only suppress the $l = 3$ tetrahedral pattern of mantle convection for as long as the continents maintain this distribution.

Woodhouse and Dziewonski [1984] note that if the continents are not taken into account in their tomographic inversions of the upper mantle, they obtain an apparent $l = 3$ density heterogeneity in the mantle; whereas if the inversion is adjusted for the presence of the continents, there is no apparent $l = 3$ density heterogeneity in the upper mantle. This is a possible indication that it is, in fact, the presence of the continents that suppresses the $l = 3$ mode of convection in the mantle. However, it could be that the continents tend to be located in an $l = 3$ tetrahedral arrangement *because* the $l = 3$ mode of convection is absent from the spectrum of mantle convection.

There are geometric explanations for the absence of the $l = 3$ mode from thermal convection in the earth's mantle. One is that the spatial resonance peculiar to the $l = 3, 4$ mode interaction force a spherical shell to irrevocably choose between one mode or the other. In fact, in the analysis presented here, it can be shown that whenever an odd and an even mode bifurcate at the same time, the resulting branches of solution cannot have a "pure" odd mode: the branches of solution are necessarily mixed-mode odd and even, or pure even mode. Additionally, there are some peculiarities and shared symmetries of the $l = 3$ tetrahedral mode and $l = 4$ cubic mode. Given this analysis, it is also reasonable to expect that the $l = 3, 4$ fast-mode interaction involves some complicated time-dependence, whereby the convective system is sometimes dominant in the mixed $l = 3, 4$ tetrahedral/octahedral mode, and sometimes dominant in the $l = 4$ pattern.

$l = 4, 5$: *Divergence and Curl Spectrum at the Surface*

The spectrum of horizontal divergence and curl (radial vorticity) of mantle convection at the surface is calculated from the known motions of the tectonic plates [Forte and Peltier, 1986; 1989]. These are defined as

$$\nabla_H \cdot \mathbf{U} = \frac{1}{\sin\theta}\left(\frac{\partial}{\partial\theta}(U_\theta \sin\theta) + \frac{\partial}{\partial\phi}U_\phi\right)$$

and

$$\hat{r} \cdot (\nabla \times \mathbf{U}) = \frac{1}{\sin\theta}(\frac{\partial}{\partial\theta}(U_\phi \sin\theta) - \frac{\partial}{\partial\phi}U_\theta)$$

respectively, where $\mathbf{U} = (U_\theta, U\phi)$ is the plate motion with respect to colatitude and longitude. The horizontal divergence at the surface indicates the pattern of upwelling and downwelling of mantle convection apparent at the surface, while the radial vorticity indicates the degree of local twisting between plates. The horizontal divergence has a very pronounced peak at spherical harmonic degrees $l = 4, 5$ while the radial vorticity has a pronounced peak at $l = 5$. There is a less pronounced peak at $l = 2$ for both fields. The radial vorticity is generated entirely by the transfer of convective energy from the poloidal to the toroidal component of the flow. These are therefore affected greatly by the temperature dependence of viscosity as well as the distribution of the continents.

The actual motions of plates on the surface represent the fast timescale of thermal convection in the earth's mantle, because they are associated with the overturn of the convective cells. Furthermore, these modes $l = 4, 5$ have positive eigenvalues at the point where modes $l = 2, 6$ have zero eigenvalues. Even though the Rayleigh number of the earth's mantle is such that all of these modes have positive eigenvalues in the linear analysis, the $l = 3, 4, 5$ modes always have more positive eigenvalues than the $l = 2, 6$ modes. The place to examine the first splitting between the fast and slow subspaces is the point at which the slow subspace has zero eigenvalues and the fast subspace has positive eigenvalues. This corresponds to the points denoted by open circles in Figure 1.

These facts are consistent with separating the problem into "fast", i.e. positive eigenvalue modes $l = 3, 4, 5$ associated with convective overturn and the "slow" $l = 2, 6$ modes associated with the thermal anomalies in the mantle. This is analogous to examining the plate motions in the "hotspot reference frame," as this assumes that the plate motions and associated convective overturn in the mantle takes place on a faster timescale than the motions of the hotspots themselves.

Summary of modes $l = 1 - 6$ in the earth's mantle

To summarize, we have modes $l = 4, 5$ in the spectrum of plate motions, and a notable absence of the $l = 3$ tetrahedral mode from the convective spectrum combined with an $l = 3$ tetrahedral distribution of low-density continental crust. We have the dominance of the $l = 2, 6$ modes in correlations between the geoid, the seismic tomography, dynamic topography and hotspot distributions, which indicates that the deeper thermal structure of th earth's mantle contains a resonance between the $l = 6$ dodecahedral "soccer-ball" mode of convection, and the $l = 2, m = 2$ "clover-leaf" mode of convection. There is a marked difference between the spectra of the actual plate motions–which evolve on a fast time-scale, as evidenced by the history of the continents–and the spectrum of the deeper thermal structure of the mantle–which contains modes which necessarily evolve on a slower time-scale, as indicated by the eigenvalues of the linear stability problem. This difference indicates that splitting the problem between the "fast modes" $l = 3, 4, 5$ associated with continental distributions and plate-motion spectrum, and the "slow modes"

$l = 2, 6$ associated with a more slowly evolving thermal structure of the interior, may be a reasonable way to reduce the problem of thermal convection in the earth's mantle to more analytically reasonable subspaces. Within each of these subspaces, the symmetries, bifurcations and invariant manifolds of the $l = 3, 4$ and $l = 2, 6$ mode interactions remain largely unexplored. The justification for this separation of subspaces is both mathematical and geophysical: the $l = 3, 4, 5$ modes have positive eigenvalues at the point of $l = 2, 6$ mode interaction. Furthermore, that several modes in the $l = 3, 4, 5$ subspace are active on the shorter ***convective overturn timescale*** is evident in the $l = 4, 5$ horizontal divergence spectrum–i.e. the tectonic motions of the lithospheric plates–whereas the $l = 2, 6$ pattern is evident in the more slowly-evolving deep thermal structure.

Discussion. What is the pattern and time-evolution of the thermal convection which gives rise to the observed plate tectonic motions, as well as the global geoid, dynamic topography, seismic tomography and hotspot distributions? How are these affected by the physical and geometric parameters Rayleigh number and aspect ratio $A = r_i/r_o$? How are these affected by the temperature-dependence of viscosity? What other geophysical phenomena might result from an aperiodic nonlinear exchange between the higher-order modes of mantle convection?

Caldera eruptions, supercontinent formation and breakup, and the mixing/isolation of geochemical reservoirs all are appropriate phenomena for comparison to nonlinear dynamic models of thermal convection, because they each have some observable time-dependent behavior, and an identifiable pattern potentially driven by deep nonlinear exchange between the modes of thermal convection. Ash layers indicate that large caldera eruptions, such as Tambora in the South Pacific and Yellowstone in America recur aperiodically. Is the recurrence cycle of caldera eruptions related to recurring dynamic peaks in the amplitude of some primary mode of thermal convection in the earth's mantle? The results from the two-dimensional Cartesian studies would suggest so, as thermal convection within these boundary conditions undergoes successive Hopf bifurcations which give rise to stable periodic orbits. This has not been established for thermal convection in spherical shells, however. It may be that the periodic geodynamic cycles result from heteroclinic cycles like those found for the $l = 1, 2$ mode interaction, rather than periodic orbits resulting from Hopf bifurcation.

The paleogeographic record indicates that that supercontinents have formed and broken up several times in the plate-tectonic history of the earth. Is the cycle of formation and breakup of supercontinents related to nonlinear dynamic oscillations among the modes of thermal convection? What rate and pattern of mixing of chemical heterogeneities do the modes produce during their evolution? What are the physical processes that lead to the mantle's choice of solution branch? Will a planetary body with slightly different initial conditions undergo a qualitatively completely different thermal history? Was the earth on a different solution branch or attractor during its early history, say, 3 or 4 billion years ago, due to the higher temperatures and different levels of differentiation and accretion? The geological record indicates that during the Archean, there were more tectonic plates, and that the plates themselves were smaller, hotter, more mobile and less rigid. Furthermore, the distinction between oceanic and continental lithosphere was not nearly as clear, making the outer boundary condition of thermal convection in the earth's mantle more spherically symmetric.

These questions serve to illustrate the need for models of time-dependent three-dimensional convection in a sphere which provide insight into the physical and geometrical reasons that convection in spherical shells aquires the patterns and time-evolution of patterns that it does. The first step to understanding the pattern and evolution of thermal convection in the earth's mantle is is to work out the bifurcations of successively higher-order patterns from the most symmetric basic state. On this topic, there are two very interesting open questions which are tractable within the theoretical framework of symmetry-breaking bifurcations presented here. The first is the dynamics of the $l = 3,4$ mode interaction, as compared with (and possibly influenced by) the dynamics of the $l = 1,2$ mode interaction in the core. It is very likely that the absence of the $l = 3$ mode is a result of the very fundamental physics of symmetry breaking bifurcations in the sphere. It may well be that the absence of the $l = 3$ pattern does not require any special mechanism involving the nonhomogeneity of the boundary conditions due to, for instance, the presence of the continents or the temperature dependence of viscosity.

Perhaps even more interesting than the $l = 3,4$ mode interaction is the $l = 2,6$ mode interaction. First, there is much observational evidence supporting the existence of both modes of thermal convection in the earth's deep interior. Second, the correspondence of upwelling of the $l = 2, m = 2$ pattern with the antipodal upwellings of the dodecahedral $l = 6$ pattern suggests strong spatial resonance which will in all likelihood be revealed by the lattice of the isotropy subgroups shared by these two patterns. Third, this is an even-even mode interaction which means that, unlike the even-odd mode interactions, it involves two transcritical bifurcations, rather than one transcritical and one pitchfork bifurcation. This makes the $l = 2,6$ interaction fundamentally different from the $l = 3,4$ or the $l = 1,2$. Fourth, this problem requires separating the fast $l = 3,4,5$ from the slow $l = 2,6$ subspaces.

Acknowledgements

C.S. thanks Don Turcotte and Claude Allègre for lively discussions. This work was supported by a NATO-NSF Postdoctoral Fellowship (CS) and C.N.R.S. (P.C.)

REFERENCES

Armbruster, D., and P. Chossat, *Heteroclinic Orbits in a Spherically Invariant System* Physica D, (to appear) .

Armbruster, D.,J. Guckenheimer and P. Holmes, *Heteroclinic cycles and modulated travelling waves in systems with O(2) symmetry*, Physica 29D, (1987), pp. 257–282.

Armbruster, D. and R. Rand, *Perturbation Methods, Bifurcation Theory and Computer Algebra*, (1987) Springer-Verlag.

Babskii V.G and I.L. Sklovskaia, *On convection in a self-gravitating fluid sphere with internal heating*, PMM, 35, 6, (1971) pp. 951–965.

BERCOVICI, D., G. SCHUBERT AND G. GLATZMAIER, *Three-Dimensional Spherical Models of Convection in the Earth's mantle*, Science 244, (1989) pp. 950–955.

BUSSE, F. H., *Pattern of Convection in Spherical Shell*, J. of Fluid Mech. 72, (1975) pp. 67–85.

BUSSE, F. H. AND N. RIAHI, *Pattern of Convection in Spherical Shells II*, J. of Fluid Mech. 123, (1982) pp. 283–301.

BUSSE, F.H. AND J. A. WHITEHEAD, *Instabilities of Convection Rolls in a high Prandtl number fluid*, J. of Fluid Mech. 47, (1971) pp. 305–320.

BUSSE, F.H. AND J. A. WHITEHEAD, *Oscillatory and collective instabilities in large Prandtl number convection*, J. Fluid Mech. 66, (1974), pp. 67–79.

CAREY, S. W. *The Expanding Earth* Elsevier, (1976).

CAZENAVE, A., A. SORIAU AND K. DOMINH, *Global Coupling of Earth surface topography with hotspots, geoid and mantle heterogeneities*, Nature, 340, (1989), pp. 54–57.

CHANDRASEKHAR, S., *Hydrodynamic and Hydromagnetic Instability*, Clarendon Press., (1961).

CHANDRASEKHAR, S. AND DONNA ELBERT, *The roots of* $J_{-l-1/2}(\alpha_j\eta)J_{l+1/2}(\alpha_j r) - J_{l+1/2}(\alpha_j\eta)$ $J_{-l-1/2}(\alpha_j r)$ Proc. Camb. Phil. Soc. 49,, (1953), pp. 446–448.

CHOSSAT, P., *Bifurcation and Stability of Convective Flows in a Rotating or Non-Rotating Sphere*, SIAM J. of Appl. Math., 37, (1979), pp. 624-647.

CHOSSAT, P, *Le Problème de Bénard dans une Couche Spherique*, Thèse d'Etat, Université de Nice, (1982).

CHOSSAT, P., *Solutions avec symétrie dihedrale dans les problèmes de bifurcation invariantes par symétrie spherique*, CRAS series I, (1983), pp. 639-442.

CHOSSAT, P., *Bifurcations en présence de symétrie dans les problèmes classiques de l'hydrodynamique*, J. Méc. Théor. Appl. Numéro Special, (1984), 157-192.

CHOSSAT, P., *Branching of rotating waves in a codimension 1 problem with spherical symmetry*, Preprint Universite de Nice, (1990).

CHOSSAT, P. AND J. P. GIRAUD, *Calcul des coefficients de bifurcation dans un problème de convection en symétrie spherique*, J. Méc. Théor. Appl., 2, (1983), pp. 799-828.

CHOSSAT, P. AND G. IOOSS, *The Couette-Taylor Problem*, Springer, 1990.

CHOSSAT, P. AND R. LAUTERBACH, *The instability of Axisymmetric solutions in problems with spherical symmetry*, SIAM J. Math. Anal., 20, (1989), pp. 31–38.

CHOSSAT, P., I. MELBOURNE AND R. LAUTERBACH, *Steady-state bifurcation with O(3) symmetry*, Arch Rat. Mech. Analys., (to appear).

CHRISTENSEN, U, *Time-dependent Convection in Elongated Cells*, Geophys. Res. Lett. 14, (1987), pp. 220–223.

COULLET, P. H. AND E. A. SPIEGEL, *Amplitude Equations for Systems with Competing Instabilities*, SIAM J. Appl. Math., 43, (1983), pp. 776–821.

CRISS, R. E. AND A. HOFFMEISTER, *Unicellular convection of the mantle beneath the continents: implications for heat flow, geotherms, stratification and tectonics*, EOS, 71, (1990), p. 632.

CROSS, M.C. AND A. C. NEWELL, *Convection Patterns in Large Aspect Ratio Systems*, Physica 10D, (1984), pp. 299–328.

CROUGH, T. AND D. JURDY, Earth and Planet. Sci. Lett. 48, (1980), pp. 15–22.

CURRY, J.H., J. R. HERRING, J. LONCARIC, AND S. A. ORSZAG, *Order and Disorder in two- and three-dimensional Bénard Convection*, J. Fluid Mech. 147, (1984), pp. 1–38.

DAVIES, G.F., *Thermal Histories of Convective Earth Models and Constraints on Radiogenic Heat Production in the Earth*, J. Geophys. Res., 85, (1980), pp. 2517–2530.

DZIEWONSKI, A.M, *Mapping the Lower Mantle: Determination of Lateral Heterogeneity in P Velocity up to Degree and Order 6*, J. of Geophys. Res. 89, (1984), pp. 5929–5952.

FORTE, A. M. AND W. R. PELTIER, *Plate Tectonics and Aspherical Earth Structure: The importance of Poloidal-Toroidal Coupling*, J. Geophys. Res. 92, (1987) pp. 3645-3679.

FORTE, A. M. AND W. R. PELTIER, *Lateral Heterogeneity and the Geoid: The importance of the Surface Kinematic Constraints*, Mathematical Geophysics: A Survey of Recent Developments in Seismology and Geodynamics, ed. N. J. Vlaar, G. Nolet, M.J.R. Wortel and S. A. P. L. Cloetingh, 1988, pp. 291–323.

FOWLER, A.C. AND M. J. MCGUINNESS, *A Description of the Lorenz attractor at high Prandtl number*, Physica 5D, (1982), pp. 149–182.

FRIEDRICH, R. AND H. HAKEN, *Static, Wavelike and chaotic thermal convection in Spherical Geometries*, Phys. Rev. A, 34, (1986), pp. 2100–2120.

GLATZMAIER, G.A., *Numerical Simulations of Mantle Convection– Time-dependent, 3-dimensional, Compressible, Spherical Shell*, Geophysical and Astrophysical Fluid Dynamics, 42, (1986), p. 223.

GLATZMAIER, G.A., J. Fluid Mech., *in press*, (1990).

GOLUBITSKY, M. AND D. G. SCHAEFFER, *Bifurcations with O(3) Symmetry Including Applications to the Bénard problem*, Comm. Pure and Appl. Math., 35, (1982), pp. 81–111.

GOLUBITSKY, M., I. STEWART AND D. G. SCHAEFFER, *Singularities and Groups in Bifurcation Theory volume II*, Springer-Verlag, (1988).

GOLUBITSKY, M., J.W. SWIFT AND E. KNOBLOCH, *Symmetries and Pattern Selection in Rayleigh-Bénard Convection*, Physica 10D, (1984), pp. 249–276.

GUCKENHEIMER, J. AND P. J. HOLMES, *Nonlinear Oscillations, Dynamical Systems and Bifurcations of Vector Fields*, Springer-Verlag, 1983.

GUCKENHEIMER, J. AND E. KNOBLOCH, *Convective transitions induced by a varying aspect ratio*, Phys. Rev. A, 27, (1983) pp. 408-417.

GUNCKEL, D. AND U. HANSEN, *Scaling Relationships in Time-Dependent Convection* European Converence on Mantle Convection, La Londe des Maures, France, (1989).

HAGER, B.H., R. W. CLAYTON, M. A. RICHARDS AND A. M. DZIEWONSKI, *Lower Mantle Heterogeneity, dynamic topography and the geoid*, Nature, 313, (1985), pp. 541–545.

HANSEN, U., *Dynamical Implications for the CMB from recent experiments on the Pressure-dependence of Thermal Expansivity*, European Conference on Mantle Convection, La Londe des Maures, France (1989).

IHRIG, E. AND M. GOLUBITSKY, *Pattern Selection with O(3) Symmetry*, Physica D13, (1983), pp. 1–33.

JONES, C. A. AND M. R. E. PROCTOR, *Strong Spatial Resonance and Travelling waves in Benard Convection* , Phys. Lett. A, 5, (1987), pp. 224-228.

LENNIE, T.B., D. P. MCKENZIE, D. R. MOORE AND N. O. WEISS, *The Breakdown of Steady Convection*, Jour. Fluid Mech. 188, (1988), pp. 47–85.

LORENZ, E.N, *Deterministic Non-periodic Flow*, J. Atmos. Sci. 20, (1963), pp. 130–141.

MCKENZIE, D., *The Symmetry of convective transitions in space and time*, Jour. Fluid Mech., 191, (1988), pp. 287–339.

MALKUS, W. V. R. AND G. VERONIS, *Finite Amplitude Cellular Convection*, J. Fluid Mech., 4, (1958), p. 225.

MARCUS, P.S., Effects of Truncation in Modal Representations of Thermal Convection, J. of Fluid Mech., 103, (1981), pp. 241–255.

OLSON, P., P. G. SILVER AND R. W. CARLSON, *The large-scale structure of convection in the earth's mantle*, Nature, 344, (1990), pp. 209–215.

OXBURGH, E.R. AND D. L. TURCOTTE, *Mechanisms of Continental Drift* Reports on Progress in Physics, 41, (1978), pp. 1249–1312.

RICHARDS, M.A. AND B. H. HAGER , *Geoid Anomalies in a Dynamic Earth*, J. Geophys. Res., 89, (1984), pp. 5987–6002.

RICHARDS, M. A., B. H. HAGER AND N. H. SLEEP, *Dynamically Supported Geoid Highs over Hotspots: Observation and Theory*, J. Geophys. Res., 93, (1988), pp. 7690–7608.

RUBIE, D. C., *Reaction-enhanced ductility: the role of solid-solid univariant reactions in deformations of the crust and mantle*, Tectonophysics 96, (1983), pp. 331–352.

SALTZMAN, B., *Finite Amplitude free convection as an initial value problem*, J. Atmos. Sci., 19, (1962), pp. 329–341.

SATTINGER, D. H. AND O. L. WEAVER, *Lie Groups and Algebras with Applications in Physics, Geometry and Mechanics*, Springer-Verlag, (1986).

SEGEL, L.A., *The Non-linear Interaction of Two Disturbances in the Thermal Convection Problem*, J. Fluid Mech.,14, (1962), pp. 97-114.

SPARROW, C., *The Lorenz Equations: Bifurcations, Chaos, and Strange Attractors*, Springer-Verlag, (1982).

STEFANIK, M. AND D. JURDY, *Patterns of Hotspot Distribution* EOS 70, (1990), p. 1351.

STEWART, C. A. AND D. L. TURCOTTE, *Chaotic Thermal Convection in the Earths mantle I: Some trajectories and bifurcations*, J. Geophys. Res., 94, (1989), pp. 13,707-13,717.

STEWART, C. A. AND D. L. TURCOTTE, [submitted to J.G.R] *Chaotic Thermal Convection in the Earth's mantle II: Higher-order truncations and Lyapunov exponents*, (1989).

TOOMRE, J., D. O. GOUGH AND E. A. SPIEGEL, *Time-dependent solutions of multimode convection equations*, J. Fluid Mech., 125, (1982), pp. 99–122.

VINCENT, A. AND D. A. YUEN, *Thermal attractor in chaotic convection with high-Prandtl-number fluids*, Phys. Rev. A, 38, (1988), pp. 328–334.

WOODHOUSE, J.H. AND A. M. DZIEWONSKI, *Mapping the Upper Mantle: Three-dimensional modelling of Earth Struicture by inversion of seismic waveforms*, J. Geophys. Res., 89, (1984), pp. 5953–5986.

ZEBIB, A., A. K. GOYAL AND G. SCHUBERT, *Convective Motions in a Spherical Shell*, J. Fluid Mech., 152, (1985), pp. 39–48.

CONVECTION AND CHAOS

A.C. FOWLER*

Abstract. We discuss two types of chaotic behaviour exhibited by high Prandtl number convection, that is 'phase chaos' and plumes. In mantle convection, these differing aspects of the motion find their expression in the migration of subduction zones and hot spots, respectively. The analysis of plumes in a single convection cell can be attempted in the framework of Howard's 'bubble' model of convection, using an asymptotic analysis based on a similar method applied to the Lorenz equations. In the partial differential equation case, this leads us in principle to an approximate Poincaré map for the flow. However, we find that Howard's assumption of differing time scales for the processes of growth and flow instability is in error, and (for a single developing plume) the thermal regime is likely to be periodic. For the case of a large aspect ratio, where multiple plume development can take place, the corresponding Poincaré map should lead to a chaotic distribution of plumes in space and time.

For cellular convection, we give a synopsis of the recent scaling theory of the Chicago group. There is a mean flow in the cell, fuelled by thermal plumes which erupt from the boundary layer as they are advected across the cell. The theory is strictly applicable, however, *only to Prandtl numbers of* $O(1)$.

Cellular ('phase') chaos at large Rayleigh number can be modelled using a set of ordinary differential equations for variables which describe the size and location of slowly varying convection cells. The differential equations are parametrised using quasi-stationary boundary layer theory. The same method can in principle be extended to three dimensions, and represents a paradigm for the study of time-dependent motions of the earth's lithospheric plates.

1. Turbulence in convection

The general picture that we have of the transition to turbulent convection derives from experiments of Krishnamurthi and Busse (Krishnamurthi 1970, Busse and Whitehead 1974, 1976), building on earlier work by Malkus (1954). For fluids with Prandtl number $\sigma \gtrsim O(1)$, this picture is as follows: two-dimensional rolls become unstable to three-dimensional (superimposed) rolls as the Rayleigh number Ra is increased. At higher Ra, time-dependence sets in, the flow becoming periodic and eventually chaotic.

This general picture was augmented by experimental physicists (Ahlers, Libchaber, Gollub and co-workers), beginning in the late seventies, who added such topics of contemporary interest as period-doubling and frequency locking (Libchaber and Maurer 1982), thus relating time-dependent convection to the phenomena exhibited by chaotic solutions of ordinary differential equations. These experiments were done in small aspect ratio boxes, and it is because of this that the fluid behaves like a finite-dimensional system, at least for low enough Rayleigh number. The spectrum of available modes is discrete, and at low Rayleigh number, few are excited. At high Rayleigh number, the length scale of "coherent features" (e.g. plumes) of the flow shrinks, and the box size becomes of decreasing significance.

*Mathematical Institute, Oxford University, 24-29 St. Giles', Oxford OX13LB, ENGLAND.

(i) Phase chaos

In large aspect ratio boxes, however, finite dimensional notions are more or less irrelevant. In this article we wish to focus on two particular aspects of convection in large boxes (or alternatively, at high Rayleigh number) where the dynamics is fundamentally infinite-dimensional, and in effect occupies a continuous spectrum of wavenumbers. The most obvious example is 'phase chaos', as studied, for instance, by Gollub et al (1982). Here the idea is that at Rayleigh numbers Ra not far above the critical value Ra_c, two-dimensional convection rolls may exhibit a slow large-scale irregular behaviour due to an incoherence of phase between widely separated rolls. That is, the flow is locally regular and two dimensional, but its amplitude and phase, described by a complex amplitude function A, can vary over long space and time scales. This idea is exactly that for which nonlinear stability theory is designed, and expansions of the flow equations lead to equations for A which may be typically represented by the complex Ginzburg-Landau equation

$$A_t = \alpha A + \beta |A|^2 A + \gamma \nabla^2 A \tag{1.1}$$

(which is not actually appropriate to convection (Newell and Whitehead 1969), though it does properly apply to instabilities in some other systems). Depending on the values of α, β, γ, chaotic solutions of (1.1) can appear through the possible mechanisms of spontaneous defect generation (Coullet et al. 1989) or 'homoclinic bursts' (Newell et al 1988), neither of which phenomenon seems easily related to the behaviour of finite-dimensional systems.

This idea, that chaotic solutions can be manifested by large scale evolution of a local, quasi-regular structure, can be extended to high values of the Rayleigh number, at least at high Prandtl number. Busse and Whitehead (1974) showed that there is an instability of bimodal three-dimensional steady convection (which may be thought of as two superimposed rolls) to a 'spoke-pattern', where a quasi-cellular pattern bounded by sheets of descending fluid has a hub of upwelling hot fluid connected to the periphery by sheets of rising hot fluid. This pattern was studied further by Whitehead and Parsons (1978). The spoke pattern evolves irregularly in space and time on a slow time scale, and while the evolution appears random, the local structure is regular. However, low Rayleigh number expansions can not sensibly be applied, and some other analytic technique is necessary.

(ii) Plumes

A different way in which chaotic behavior is manifested in convecting fluids occurs when the Rayleigh number is increased in a box of finite dimension. Thermal boundary layers develop at the base and at the top, and as the cell dimension effectively grows, the thermal boundary layers develop instabilities which erupt as plumes from the boundary layer. At low Rayleigh number, these instabilities may occur as ripples which ride round the cell on the thermal boundary layers, but at higher Rayleigh number, they can break off. This eventually leads to a horizontal spatial incoherence, where plumes break away from the top and bottom at irregular locations. Visualisation of such plumes has been made by Sparrow et al. (1970), for example. A second degree of spatial incoherence can occur when upwelling thermal plumes do not connect to the opposite boundary. In this case they may mix into a mean circulation in the core of the cell, as has been observed by Krishnamurthi and Howard (1981). At high Prandtl number, it may be that such plumes are more isolated events, both in space and time, though this is not certain. Thus if one seeks "coherent structures" in turbulent convection, candidates may be the 'mushroom-cap' heads of thermals, as described by Griffiths (1986), for example.

2. Mantle convection

A good deal of interest exists in the geophysical community in understanding convection, and some observations on the nature of mantle convection will be made here. Convection in the earth's mantle has been reviewed by Hager and Gurnis (1987) and Silver et al. (1988). The literature is enormous, but little concrete understanding of the problem has been gained since the early review by Turcotte and Oxburgh (1972). Constant viscosity convection provides a useful paradigm, but must be (though it is usually not) viewed with a critical perspective, since there are fundamental geophysical variations which must alter the nature of the flow enormously. Chief amongst these are chemical inhomogeneity, phase changes (perhaps) and variable viscosity.

(i) Subduction and plate migration

Despite this, certain features of the flow can be taken as analogues of laboratory experimental features. In particular, subducting lithospheric plates are the descending cold boundary layers of convection 'cells'. The flow is asymmetric because of the enormous temperature dependence of viscosity: thus the cold, subducting plates are thick and stiff, while corresponding ascending plumes are likely to be less positively buoyant. The slabs descend as more or less coherent sheets. Thus, while the details are different, the slabs organise the geometry of the flow in a cellular way. It is known that the cells (bounded by plate margins) are time dependent on the earth, and one ultimate goal of plate tectonics must be to account for and predict this dependence. It is essentially obvious (because of the lateral asymmetry) that as subduction begins, and slabs sink into the mantle, that the subduction zone will migrate backwards towards the ridge. This may of course be compensated by ridge movement, but the point is that secular movement is to be expected, and one ultimate way of modelling this may be by parametrising the cellular structure in the way that is done here. Certainly, the evolution of plate boundaries bears a surficial resemblance to phase-incoherent cellular convection in laboratory fluids.

(ii) Hot spots

There are numerous regions of anomalous heat flow on the earth, often associated with volcanism; the two obvious examples being Hawaii and Iceland. The plume hypothesis (Morgan 1971) associates these with upwelling thermal plumes in the mantle. Laboratory versions of such plumes in high Prandtl number fluids have been studied by Griffiths (1986) and Olson and Singer (1985); they retain a strong spherical cap-like structure with a thin umbilical tail beneath.

If the whole mantle convects in a more or less coherent fashion (something which is by no means clear), then such plumes could represent instabilities in a basal thermal boundary layer, by analogy to laboratory convection. Mantle plume hot spots seem to be largely independent of plate boundaries, and it may be possible to view the two phenomena as being largely independent. By understanding how thermal plumes are erupted from basal boundary layers, we may therefore be led towards an understanding of the behaviour of hot spots in the mantle.

3. Plumes

The basic feature about plumes is that they form localized regions in which the temperature is anomalously high (upwelling) or low (downwelling). As such, their basic description involves a boundary layer analysis of convection at high Rayleigh number. Several authors have undertaken such an analysis of steady high Rayleigh number convection; most notable in a geophysical context are the papers by Turcotte and Oxburgh (1967) and Roberts (1979), the latter of which is the best treatment of this subject. We begin by recalling this basic theory. For simplicity in the exposition we will suppose that the Prandtl number is infinite, and will also consider only the case of free slip boundaries.

(i) Boundary layer theory for steady convection

Dimensionless equations governing Boussinesq convection in two dimensions may be written

$$\begin{aligned} \nabla\cdot\mathbf{u} &= 0, \\ \frac{1}{\sigma}\frac{d\mathbf{u}}{dt} &= -\nabla p + \nabla^2\mathbf{u} + Ra\,T\mathbf{k}, \\ \frac{dT}{dt} &= \nabla^2 T, \end{aligned} \tag{3.1}$$

where T is temperature, $\mathbf{u}$ is the velocity, p is pressure, $\mathbf{k}$ is a unit vector pointing vertically upward, and d/dt represents the material derivative $\partial_t + \mathbf{u}\cdot\nabla$. In two dimensions, $\mathbf{u} = (u, v)$, and we define a stream function ψ by $u = -\psi_y$, $v = \psi_x$ (notice the sign convention used here), and the vorticity $\omega = v_x - u_y = \nabla^2\psi$. We take the Prandtl number σ to be infinite, and will suppose that the Rayleigh number Ra is large. Eliminating p, we have

$$\begin{aligned} \omega &= \nabla^2\psi, \\ \nabla^2\omega + \frac{1}{\delta}T_x &= 0, \\ T_t + \psi_x T_y - \psi_y T_x &= \delta^2\nabla^2 T, \end{aligned} \tag{3.2}$$

where we have rescaled ψ and ω with $Ra^{2/3}$, and defined

$$\delta = Ra^{-1/3}; \tag{3.3}$$

this ensures that (for free slip boundaries), both ψ and ω are $O(1)$ in the flow.

Suitable boundary conditions to apply are that

$$\psi = \omega = 0 \text{ on } x = 0, a;\ y = 0, 1, \tag{3.4}$$

and

$$\begin{aligned} T_x &= 0 \text{ on } x = 0, a; \\ T = 1/2 \text{ on } y &= 0, T = -1/2 \text{ on } y = 1. \end{aligned} \tag{3.5}$$

The basic boundary layer theory studied (most comprehensively) by Roberts (1979) is now summarised briefly. The flow consists of a central *core*, two thermal *plumes* at the sidewalls, and two *thermal boundary layers* at the top and bottom. The plumes and thermal boundary layers are connected at the corners, where one can show that the local thermal structure is simply advected round.

A. Core

Here $T = 0$ to all (algebraic) orders of δ, and therefore

$$\nabla^4 \psi = 0, \tag{3.6}$$

with appropriate conditions being $\psi = 0$ on all boundaries, $\omega = 0$ on top and bottom, but $\omega = O(1)$ (to be found) at the sides; the non-zero vorticity (which drives the flow) is generated in the thermal plumes at the side walls.

B. Plume

We can suppose the core flow is clockwise, so that $\psi > 0$ and $\omega < 0$, and consider the plume near $x = 0$. Suitable boundary layer scales are

$$x \sim \delta, \quad T \sim 1, \quad \omega \sim 1, \quad \psi \sim \delta, \tag{3.7}$$

whence we find (at leading order) that the rescaled variables satisfy

$$\begin{aligned} \Psi &\sim v_p(y) X, \\ \omega &\sim -\int_0^X T\, dX, \end{aligned} \tag{3.8}$$

whence the core vorticity at the wall is

$$\omega_\infty \sim -\int_0^\infty T\, dX. \tag{3.9}$$

The temperature satisfies

$$\Psi_X T_y - \Psi_y T_X \sim T_{XX}, \tag{3.10}$$

and in Von Mises variables y, Ψ,

$$T_y \sim v_p(y) T_{\Psi\Psi}, \tag{3.11}$$

where v_p is the plume velocity determined by matching to the core flow. Conservation yields $\int_0^\infty T\, d\Psi = \int_0^\infty v_p T\, dX = C$ is a constant, and thus, using (3.9), the sidewall boundary condition for the core flow at $x = 0$ is

$$\psi_x \psi_{xx} = \omega_\infty v_p = -C. \tag{3.12}$$

The core flow can now be solved provided C is determined; this requires solutions of the thermal boundary layer equation.

C. Thermal boundary layer (top)

Putting

$$y = 1 - \delta Y, \quad \psi = \delta\Psi, \quad \omega = \delta\Omega, \tag{3.13}$$

we find

$$\Psi \sim u_s(x)Y, \tag{3.14}$$

where u_s is the surface velocity determined from the core flow, and then in Von Mises variables x, Ψ,

$$T_x \sim u_s T_{\Psi\Psi}. \tag{3.15}$$

Thus

$$\frac{\partial T}{\partial s} \sim u\frac{\partial^2 T}{\partial \Psi^2} \tag{3.16}$$

all the way round the boundary, where u is the core-determined speed, and s is the distance coordinate. Boundary conditions for T are that $T \to 0$ as $\Psi \to \infty$, and alternatively $T = 1/2$, $T_\Psi = 0$, $T = -1/2$, $T_\Psi = 0$ on the four sides of the flow at $\Psi = 0$; T is required to be periodic in s.

D. Solution Strategy

Solve the core flow with $C = 1$, denote the solution with an overhat: $\hat{\psi}$. Thus $\psi = C^{1/2}\hat{\psi}$, $u = C^{1/2}\hat{u}$. The solution of (3.16) is thus

$$T = \widehat{T}(s, C^{-1/4}\Psi), \tag{3.17}$$

where $\widehat{T}$ is the solution when $C = 1$. Thus $\hat{\psi}$ and $\widehat{T}$ are successively solved, and then C is determined from the value of $\int_0^\infty T\,d\Psi$ in the plume, whence

$$C = \left\{\int_0^\infty \widehat{T}\,d\Psi\right\}^{4/3}, \tag{3.18}$$

which completes the solution.

(ii) The Howard 'bubble' model of turbulence

The description of convection by boundary layer theory at high Rayleigh number is idealistic, but not without value. As Ra is increased, convection in real fluids at quite large Prandtl numbers undergoes a transition to steady three-dimensional flow, then to time-dependent, and eventually 'chaotic' or 'turbulent' flow (meaning here simply disordered in space and time). Nevertheless, even when the motion is irregular, the basic structure of a well-mixed, relatively isothermal interior with thermal boundary layers remains appropriate. As explained earlier, the time-dependence arises through the twin features of irregular meandering of cell boundaries, and irregular eruption of plumes from the thermal boundary layers.

An influential paper which seeks to give an approximate quantitative description of this latter phenomenon is that of Howard (1966). This description (sometimes called the 'bubble' model) involves the idea that, if the Prandtl number is large, then convective instability can occur on a time scale which is short compared to that of thermal diffusion. Thus, he visualised turbulent convection at high Rayleigh number *and* high Prandtl number as consisting of a relaxational oscillation, in the slow phase of which, conductive thermal boundary layers grow into a relatively stagnant, isothermal layer of fluid. At some point, the thermal boundary layers become convectively unstable. They then detach rapidly to form a plume, which ascends rapidly through the fluid, and recreates a well-mixed isothermal state.

There is plenty of quantitative experimental evidence which is consistent with this remarkable description (Katsaros et al. 1977, Tamai and Asaida 1984, Tritton et al. 1980), and its conception has become extremely popular in the geophysical literature (e.g. Kenyon and Turcotte 1983). The quantitative theory rests on the notion of a locally-defined 'critical Rayleigh number'. The idea here is that, for a growing boundary layer of dimensionless thickness ϵ, say, the effective local Rayleigh number describing instability is given roughly by $Ra/2\epsilon^3$ (the factor of two coming from the temperature difference between the base and the interior). If one supposes that instability occurs when this is equal to some appropriate critical value Ra_c, then one predicts $\epsilon \sim (Ra/2Ra_c)^{-1/3}$, and thus the Nusselt number (averaged in time) will be $Nu \sim (Ra/Ra_c)^{1/3}$ which is not inconsistent with some experimental results.

This idea of a locally defined Rayleigh number has been widely misappropriated in the geophysical literature, and has foisted such misguided concepts as 'small-scale convection' (Parsons and McKenzie 1978). Howard's idea has never been placed in a consistent mathematical framework. Here we will try to show how this can be done, and in so doing, give a more precise description of the boundary layer instability.

(iii) The Lorenz equations and the Lorenz-Howard connection

The Lorenz equations

$$\begin{aligned} \dot{X} &= -\sigma X + \sigma Y, \\ \dot{Y} &= (r - Z)X - Y, \\ \dot{Z} &= XY - bZ. \end{aligned} \tag{3.19}$$

form both a paradigm for chaos in ordinary differential equations, and have an association with convection, insofar as they represent the amplitudes X, Y, Z of a three-mode Fourier truncation of two-dimensional Boussinesq convection; particularly, X represents the velocity (or vorticity) field, Y represents the horizontal temperature variation, and Z represents the horizontally averaged temperature. Lorenz (1963) found that at $r = 28$, $\sigma = 10$, $b = 8/3$, solutions behaved in an irregular chaotic manner, and the trajectories consisted of alternating excursions of X and Y (and jumps in Z), followed by longer, quiescent phases where X, Y are small and Z decays exponentially. He also showed that the dynamics could be predicted from a one-dimensional contraction of the Poincaré map. The map is one-dimensional because of the large dissipation, which is due to σ being large. Complicated dynamics occurs when r is large, and in later work (Sparrow 1982), it was shown that chaotic motion arose through the occurrence of homoclinic bifurcations associated with homoclinic orbits of the flow.

The approximate dynamics realized by the equations when r and σ are both large suggests that a partial analysis of the equations might be possible. This is, in fact, the case, and Fowler and McGuinness (1982, 1983) showed that when $r \sim \sigma >> 1$, the solutions were characterised by the striking oscillations described above. One can then *derive* an approximate Poincaré map for the system, whose predictions of chaotic behaviour agree both quantitatively and qualitatively with numerical computations.

There is an interesting connection between these relaxational solutions of the Lorenz equations, and the Howard 'bubble' model. Notice first that in the Lorenz equations, σ is the Prandtl number, and r is the 'reduced' Rayleigh number Ra/Ra_c. If one recalls that X represents the velocity (or vorticity) field, while Z represents the horizontally averaged temperature, then the rapid excursions of X and Y correspond to a rapid overturning of the cell, while the slow decay of Z and the small values of X and Y during the quiescent phase correspond to the slow thermal diffusion from the boundaries in the absence of convection. In other words, the Lorenz equations are doing their best to represent Howard's conceptual model at high Ra and σ.

This suggests the following strategy: we seek to give a mathematical description of the bubble model by emulating the analysis which successfully copes with the Lorenz equations. The main complication is that the solutions we seek are those of a relaxation boundary layer flow, and involve singularly perturbed regions in both space and time. We will give a description of one part of this analysis in the following section. First we recall the salient features of the Lorenz analysis (for details, see Fowler and McGuinness (1982)).

In the slow phase, X, Y, are *exponentially small*, but crucially non-zero. Z decays slowly, and X, Y are governed by a *linear* equation with slowly varying coefficients (due to Z), which can be trivially solved using the WKB method. At first, X, Y also decay (rapidly), but eventually they grow. Since their growth rate is rapid, they become of order one suddenly, and there is then a rapid excursion in which thermal diffusion (decay of Z) is irrelevant, but decay of X and Y occurs due to the viscosity (i.e. due to the dissipation term in the X equation). This nonlinear part of the solution only needs to be solved in terms of its initial and final states, which thus gives a map from one side to the other. Solution of the linear slow phase then permits the determination of a *Poincaré map* for the flow. The analysis is, in fact, very similar to that involved in homoclinic bifurcation analysis (Sparrow 1982). Our ultimate aim for the equations of Boussinesq convection is thus to *derive* an approximate Poincaré map. It may be quite possible to *prove* results on strange *attractors* in this way.

4. Homoclinic connections for boundary layer instabilities

(i) Boundary layer instability

We begin by recalling the basic non-dimensional equations:

$$\begin{aligned} \frac{1}{\sigma}\dot{\omega} &= RaT_x + \nabla^2\omega, \\ \omega &= \nabla^2\psi, \\ \dot{T} &= \nabla^2 T, \end{aligned} \tag{4.1}$$

where $\dot{T} = dT/dt$, $\dot{\omega}$ are the advective derivatives. Let us suppose that diffusive thermal boundary layers grow to a distance of order ϵ, and that within this developing boundary layer exponentially small vorticity and buoyancy disturbances eventually grow exponentially on a time scale $t \sim \nu$. In addition, we suppose the natural length of such disturbances is $O(\delta) \lesssim 1$ (we might suppose that $\delta \sim \epsilon$ for very high Ra, but $\delta \sim 1$ at lower Ra).

Rescale the variables as

$$t \sim \nu, \quad y \sim \epsilon, \quad x \sim \delta, \quad \psi \sim \epsilon\delta/\nu, \quad \omega \sim \delta/\epsilon\nu, \tag{4.2}$$

where the choice of ψ scaling is in order for the Jacobian advective term to remain comparable to the time derivative. We define

$$Ra = \frac{\delta^2}{\epsilon^3\nu}, \quad \beta = \epsilon/\delta; \tag{4.3}$$

the equations become (in the rescaled variables)

$$\begin{aligned} \frac{1}{\sigma b}\dot{\omega} &= T_x + \nabla_\beta^2\omega, \\ \omega &= \nabla_\beta^2\psi, \\ \dot{T} &= b\nabla_\beta^2 T, \end{aligned} \tag{4.4}$$

where

$$b = \nu/\epsilon^2, \quad \nabla_\beta^2 = \frac{\partial^2}{\partial y^2} + \beta^2\frac{\partial^2}{\partial x^2}. \tag{4.5}$$

Before choosing b, let us set out a strategy based on Howard's model, and analogous to the treatment of the Lorenz equations. If $\psi << 1$, then the horizontally average temperature $\overline{T}$ satisfies (approximately)

$$\overline{T}_t = b\overline{T}_{yy}, \tag{4.6}$$

with $\overline{T} = 1/2$ on $y = 0$, $\overline{T} \to 0$ as $y \to \infty$. In fact we take (4.6) as the definition of $\overline{T}$, put

$$T = \overline{T} + \widetilde{T}, \tag{4.7}$$

so that

$$\widetilde{T}_t + \psi_x \overline{T}_y + (\psi_x \widetilde{T}_y - \psi_y \widetilde{T}_x) = b\nabla_\beta^2 \widetilde{T}, \tag{4.8}$$

and the vorticity equation in full is

$$\frac{1}{\sigma b}[\omega_t + (\psi_x \omega_y - \psi_y \omega_x)] = \widetilde{T}_x + \nabla_\beta^2 \omega. \tag{4.9}$$

Thus our ansatz will be that ψ, ω, $\widetilde{T}$ are all exponentially small, so that the following linear approximation holds:

$$\begin{aligned} \frac{1}{\sigma b}\omega_t &= \widetilde{T}_x + \nabla_\beta^2 \omega, \\ \omega &= \nabla_\beta^2 \psi, \\ \widetilde{T}_t + \psi_x \overline{T}_y &= b\nabla_\beta^2 \widetilde{T}. \end{aligned} \tag{4.10}$$

If we now pursue a strict analogy to the Lorenz equations (where the equations for X, Y, Z correspond to those for ω, $\widetilde{T}$, $\overline{T}$ respectively), we would choose $b = 1/\sigma << 1$, so that $\overline{T}$ would be a function of the slow time t/σ, while ω and $\widetilde{T}$ would evolve on the fast time. Let us postpone this assumption for the time being. For simplicity, suppose $\delta >> \epsilon$, so that $\beta << 1$. We have

$$\begin{aligned} \frac{1}{\sigma b}\omega_t &= \widetilde{T}_x + \omega_{yy}, \\ \omega &= \psi_{yy}, \\ \widetilde{T}_t + \psi_x \overline{T}_y &= b\widetilde{T}_{yy}. \end{aligned} \tag{4.11}$$

Elimination of $\widetilde{T}$ and ω yields the equation

$$\left(\frac{\partial}{\partial t} - b\frac{\partial^2}{\partial y^2}\right)\left(\frac{1}{\sigma b}\frac{\partial}{\partial t} - \frac{\partial^2}{\partial y^2}\right)\psi_{yy} + \psi_{xx}\overline{T}_y = 0, \tag{4.12}$$

or (neglecting a term of $0(1/\sigma)$)

$$\frac{1}{\sigma b}\psi_{yytt} - \psi_t^{iv} + b\psi^{vi} + \psi_{xx}\overline{T}_y = 0, \tag{4.13}$$

where superscript Roman numerals indicate partial derivatives with respect to y. Relevant boundary conditions for conditions of free slip at the base are

$$\begin{aligned} \psi = \psi^{ii} = \psi^{iv} &= 0 \quad \text{at } y = 0, \\ \widetilde{T},\, \psi_y &\to 0 \quad \text{as } y \to \infty. \end{aligned} \tag{4.14}$$

Now we find that if $b = 1/\sigma$ (and we neglect the third term in (4.13)) then the solution grows unstably for all t. That is, there is no period of decay, and the strategy adopted for the Lorenz system fails. Instead, suppose we choose

$$b = 1, \tag{4.15}$$

so that

$$\nu = \epsilon^2, \quad \delta^2/\epsilon^3 = Ra. \tag{4.16}$$

Then the boundary layer equation governing the growth of ψ is , to leading order,

$$\psi_t^{iv} = \psi^{vi} + \psi_{xx}\overline{T}_y. \tag{4.17}$$

Solutions are exponential in x but not in t (since $\overline{T}_y$ is time-dependent), thus

$$\psi = \phi e^{ikx}, \tag{4.18}$$

and

$$\phi_t^{iv} = \phi^{vi} + k^2(-\overline{T}_y)\phi. \tag{4.19}$$

An initially developing conductive thermal profile has

$$\overline{T} = \frac{1}{2}\operatorname{erfc}(y/2t^{1/2}), \quad -\overline{T}_y = \frac{1}{2\sqrt{\pi t}}e^{-y^2/4t}. \tag{4.20}$$

Some idea of the nature of solutions of (4.19) can be obtained by using the quasi-static approximation, in which $\overline{T}$ varies slowly with t (Robinson 1976). Then

$$\phi \sim \exp\left[\int^t \mu(t)\,dt\right] f(y) \tag{4.21}$$

gives the relevant WKB approximation, and

$$\mu f^{iv} = f^{vi} + k^2(-\overline{T}_y)f. \tag{4.22}$$

The eigenvalue μ can be determined by a variational principle, which is obtained from

$$\mu = \frac{\rho k^2 \int_0^\infty (-\overline{T}_y)|f|^2\,dy - \int_0^y |f'''|^2\,dy}{\int_0^\infty |f''|^2\,dy}. \tag{4.23}$$

Since $-\overline{T}_y \to 0$ as $t \to 0$ for $y > 0$, it is clear from (4.23) that $\mu < 0$ for small values of t, but as t increases, μ will become positive. There will thus be an initial decay of ψ towards zero, followed by a resurgence, and the eruption of the next plume.

In fact, a more complete analysis shows that both terms in b in (4.13) need to be included; the solutions for ψ have multiple scale behaviour in y, and a thorough analysis is necessary.

(ii) Plume eruption dynamics

Suppose an erupting thermal plume is (relative to the boundary layer scales in (4.14), with $b = 1$) of thickness δ_p. Then $x \sim \delta_p$, $\psi \sim 1$, whence the plume velocity $v_p \sim 1/\delta_p >> 1$; if the plume rises in stagnant fluid, then (if $\delta_p << \beta$) $\omega \sim \beta^2/\delta_p^2$, and also the vorticity-buoyancy balance implies that $\beta^2\omega/\delta_p^2 \sim 1/\delta_p$; thus

$$\delta_p \sim \beta^{4/3}, \tag{4.24}$$

and the plume dynamics are described by the approximate equations (where we take $y \sim \beta^{-2/3}$, $t \sim \beta^{2/3}$, in order to obtain the vertical scale over which heat conduction acts):

$$\begin{aligned} 0 &= T_x + \omega_{xx} \\ \omega &= \psi_{xx}, \\ \dot{T} &= T_{xx}. \end{aligned} \tag{4.25}$$

A further integration yields

$$\begin{aligned} \psi_{xxx} &= -T, \\ T_t + \psi_x T_y - \psi_y T_x &= T_{xx} \end{aligned} \tag{4.26}$$

and one must seek numerical solutions; the important point from the point of view of the thermal boundary layer is that the plume ejection is *rapid* if $\beta << 1$. This leads us to a schematic description of the dynamics of the boundary layer, which is consistent with Howard's bubble model. This is the following: we solve the linear equation (4.17) for the evolution of the stream function field in the boundary layer, starting from a (small) initial value $\psi_0(x)$. There are two distinct cases. Firstly, suppose the domain is $(0, 2\pi)$, and assume that ψ develops a maximum within the cell (i.e. use (4.18) with $k = 1$); when $\psi = 0(1)$, nonlinear terms become important and lead (we suppose) to the rapid ejection of a plume. Howard's idea would then be that after the plume ejection, the average temperature profile reverts to $\overline{T} = 0$ in $y > 0$, and ψ evolves as before, till ejection of the next plume. This leads naturally to periodicity in plume emission with a periodicity of $O(1)$, which in dimensional terms is of order $(d^2/\kappa)\epsilon^2 = (d^2/\kappa)(Ra/\delta^2)^{2/5}$. Chaos is essentially irrelevant here. The second, more relevant case is where $\delta << 1$ (as we suppose), so that the boundary layer length scale is large; then multiple plumes can form. We divide the x axis into 'catchment areas', divided by stagnation points (where $\psi_y = 0$). Then initial evolution of a stream function $\psi_0(x)$ will lead to ψ first becoming $O(1)$ in one of the catchment areas. The rapid plume eruption then takes place, and we could suppose that in this catchment area, ψ rapidly drops to zero, and gives us a new initial $\tilde{\psi}_0(x)$. In this way, we can in principle derive a Poincaré map taking $\psi_0 \to \tilde{\psi}_0$. Such a map is conceptually very similar to the way one would construct a Poincaré map for , e.g., the complex Ginzburg-Landau equation, cf. Newell et al. (1988). However, because the time scale of evolution of the instability is the same as the interval between plume ejection, then really $\psi = O(1)$ all the time, and the formalisation of the idea is not asymptotically valid. It is likely that homoclinic trajectories connecting $\psi = 0$ to itself exist, but in practice they would not be obtained. Nevertheless, the discussion here raises the interesting idea that chaos in extended systems may arise through the phase indeterminacy of the homoclinic structures (here, plumes), and this can be compared to the homoclinic bifurcation analysis for partial differential equations on an infinite domain (Fowler 1990).

(iii) The KZZ model

The above effort to analyse Howard's scaling analysis may in fact be rendered obsolete by the scaling analysis done by the Chicago group (Kadanoff, Zaleski and Zanetti being those responsible) (Castaing et al. 1989). Here we paraphrase this model, and comment on its applicability. In particular, it requires the Prandtl number σ to be $O(1)$, and is hence ***inapplicable*** to geophysical flows. We begin with the dimensionless equations (4.1) in stream function/vorticity variables:

$$\begin{aligned} \frac{1}{\sigma}\dot{\omega} &= RaT_x + \nabla^2\omega, \\ \omega &= \nabla^2\psi, \\ \dot{T} &= \nabla^2 T; \end{aligned} \tag{4.27}$$

equivalently,

$$\begin{aligned} \frac{1}{\sigma}\dot{\mathbf{u}} &= -\boldsymbol{\nabla}p + RaT\mathbf{k} + \nabla^2\mathbf{u}, \\ \boldsymbol{\nabla}\cdot\mathbf{u} &= 0, \\ \dot{T} &= \nabla^2 T. \end{aligned} \tag{4.28}$$

The KZZ argument is the following. There is a basal thermal boundary layer of thickness $\epsilon \sim 1/\text{Nu}$, where Nu is the Nusselt number, a 'mixing region' of height δ, in which plumes erupted from the boundary layer mix with the core flow. Suppose that eddies retaining a temperature excess Δ_c (over the average, zero) move with a typical velocity u_c in the core (note that the quantities δ_c, u_c used here are ***dimensionless***, scaled as in (4.28)). Suppose that plumes ascend from the basal boundary layer with velocity u_p and have temperature excess δ_p. Then we have

$$u_c \sim (\sigma Ra\Delta_c)^{1/2}; \tag{4.29}$$

this represents a balance of acceleration and buoyancy in the core of the cell; it assumes that the Reynolds number of the flow is large, specifically that

$$u_c \gtrsim \sigma. \tag{4.30}$$

If we integrate $(4.28)_3$ over a closed volume V, and assume that T is statistically stationary, then the divergence theorem implies

$$\int_S (Tu_n - \frac{\partial T}{\partial n})\, ds = \text{Nu}, \tag{4.31}$$

where S is any horizontal line in the cell. If hot parcels rise and cold parcels descend, this gives

$$\text{Nu} \sim \Delta_c u_c \sim \Delta_p u_p. \tag{4.32}$$

A third balance arises by balancing vorticity and buoyancy in the upwelling plumes. *Assuming* these have representative thickness ϵ, *but have temperature excess* $O(1)(!)$, then

$$u_p \sim \epsilon^2 Ra. \tag{4.33}$$

Note the seeming contradiction in this statement (plumes had temperature excess Δ_p, which as we shall see is *not* $O(1)$). Closure of the model is effected by assuming either $u_c \sim u_p$ (the first KZZ prescription) or $\Delta_p \sim \Delta_c$ (the second); evidently these are equivalent.

The KZZ theory is not, as it stands, an entirely happy one; its best feature is that it gives the exponents

$$\Delta_c \sim Ra^{-1/7}, \quad u_c \sim Ra^{3/7}, \quad \text{Nu} \sim Ra^{2/7}, \tag{4.34}$$

observed in the experiment. We shall therefore assume it is correct, and try to give a prescription for the flow geometry which can be consistent with the assumptions. A further drawback of the theory is that the 'mixing length' seems irrelevant to the determination of the scales (4.34).

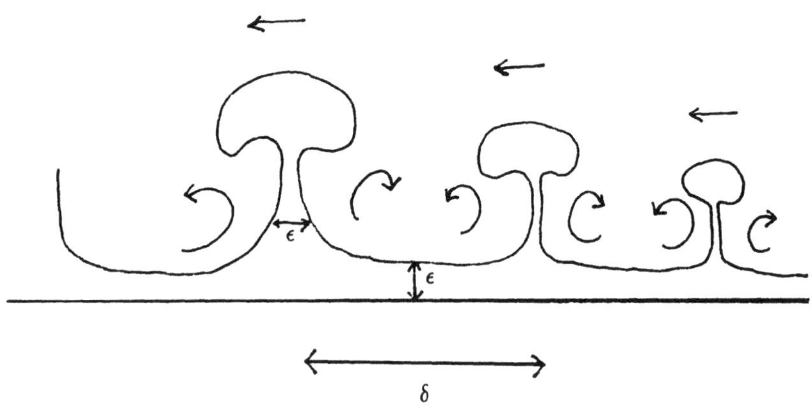

FIGURE 4.1

We now offer a cartoon of the flow which may be consistent with the KZZ model (see Figure 4.1). There is a core circulation with velocity $\sim u_c$. As this sweeps past the base, a thermal boundary layer of thickness ϵ is formed. If the Prandtl number σ is $O(1)$, then there is also a shear layer of thickness $0(\epsilon)$. As the boundary layer is advected along the base, instabilities grow and erupt upwards as plumes. We suppose these plumes are separated by a horizontal distance $\sim \delta$ (this will be the mixing length). Observations of thermal plumes (e.g. Sparrow et al. 1970) suggest that they typically evolve into mushroom-shaped objects by entraining surrounding fluid. Thus the thin tails of the plumes will maintain a temperature excess $T \sim 1$ (just as in boundary layer theory), while we suppose the cap temperature is Δ. There is no real distinction between Δ_p and Δ_c: the mixing length here is a horizontal measure of plume spacing. Also, because of the viscous shear layer at the base, the core velocity u_c is comparable to the plume ascent velocity.

If we write the equations in a moving coordinate system with $\xi = x + u_c t$ (thus the core flow is clockwise) and put $\mathbf{u} = -u_c \mathbf{i} + \mathbf{u}'$, where $\mathbf{i}$ is the horizontal unit vector, we regain

(4.35) $$\frac{dT}{dt'} = \nabla^2 T,$$

here d/dt' is the corresponding material derivative. In effect, the plumes are fixed in this moving frame, and it is then plausible to *choose* the mixing length δ as the length over which the thermal boundary layer grows to $O(\epsilon)$. This gives

(4.36) $$u_c/\delta \sim 1/\epsilon^2, \quad \text{i.e. } \delta \sim \epsilon^2 u_c.$$

Next, it is because the boundary layer is a viscous shear layer that (compare boundary layer theory with free slip conditions) the plume velocity is comparable, and hence the plume tail thickness is also $\sim \epsilon$. If σ is large, then the boundary layer velocity will be small, and then the plume tail thickness will be thinner. Hence the KZZ scaling requires $\sigma \lesssim O(1)$. The upwelling plume tail generates a vorticity which drives the eddy flow between plumes of $O(u_c)$. The plumes extend to the core, and the notion of a depth of the mixing layer may be redundant.

Thus we have

(4.37) $$u_c \sim \epsilon^2 Ra,$$

and

(4.38) $$\text{Nu} \sim \Delta u_c$$

as before. This gives the KZZ scalings, and in addition that

(4.39) $$\delta \sim \epsilon^2 u_c \sim Ra^{-1/7},$$

which is also determined by KZZ.

The last scale successfully determined by KZZ is the periodicity of thermal oscillations in the cell. The data indicates that these consist of more severe fluctuations in temperature which are superimposed on top of a higher frequency oscillation which would correspond to the passage of developing plumes. At a spacing δ, with a velocity u_c, these would have a characteristic frequency

$$\omega_{p\ell} \sim u_c/\delta \sim Ra^{4/7}. \tag{4.40}$$

KZZ associate the longer periodicity with a Brunt-Väisälä frequency, which they associate with oscillations of stably stratified cooler eddies below warmer ones. The frequency is defined as

$$\omega_{BV} \sim (\sigma Ra\Delta/\delta)^{1/2} \sim Ra^{1/2}, \tag{4.41}$$

since $\delta \sim \Delta$, and corresponds to a time scale

$$t_{BV} \sim \delta^{1/2}/u_c. \tag{4.42}$$

The data is consistent with this estimate, but it is not straightforward to incorporate it in the model. For one thing, the idea of a gravity wave oscillation in a horizontally stratified medium is not obviously consistent with the picture presented here. For another, the data might suggest that the oscillations (of $O(1)$) occur in the boundary layer; nor do they necessarily look like a slow modulation of the faster oscillation – they seem to be associated with particularly sharp spikes. Our natural explanation would then be that they represent oscillations of the boundary layer itself within the quasi-steady evolving picture we have described. But a time scale t_{BV} in (4.22) is difficult to isolate in (4.28).

On the other hand, there may be problems in associating t_{BV} with a Brunt-Väisälä frequency. This frequency describes oscillations in which density variations are important in the continuity equation. If we include compressibility in (4.28), we have

$$\begin{aligned} \frac{1}{\sigma}\dot{\mathbf{u}} &= -\boldsymbol{\nabla} p + RaT\mathbf{k} + \nabla^2\mathbf{u} \\ \boldsymbol{\nabla}\cdot\mathbf{u} &= -B\dot{T} \\ \dot{T} &= \nabla^2 T, \end{aligned} \tag{4.43}$$

where

$$B = \alpha\Delta T \tag{4.44}$$

might be called the Boussinesq number. The Boussinesq approximation involves the limit $B \to 0$. If, for the moment we consider T to be a diffusionless passive scalar, then $\dot{T} = 0$, and small oscillations of (4.43) involve the basic equations

$$\begin{aligned} v_t &\sim \sigma RaT, \\ T_t &\sim -vT_y, \end{aligned} \tag{4.45}$$

whence

$$v_{tt} \sim -\sigma Ra T_y v, \tag{4.46}$$

which gives the Brunt-Väisälä frequency (4.42) (with $T_y \sim \Delta/\delta$). It is not immediately clear that the same idea will apply here, since T is not passive. Actually the diffusion time for T over a length δ is $\delta^2 \sim Ra^{-2/7} >> Ra^{-1/2} \sim t_{BV}$. Therefore, it is in fact consistent to consider oscillations on a time $O(t_{BV})$, and we suppose the mechanism is just (4.45), with oscillations taking place in the eddies above the boundary layer.

Thus the Brunt-Väisälä oscillation is consistent with the model. The eddies between plumes are described by the approximate equations

$$\begin{aligned} \frac{1}{\sigma}\dot{\mathbf{u}} &= -\nabla p + Ra T\mathbf{k}, \\ \nabla \cdot \mathbf{u} &= 0, \\ \dot{T} &= 0, \end{aligned} \tag{4.47}$$

and as illustrated, the recirculating eddies have a temperature variation Δ. This is as opposed to the isothermal temperature in steady state boundary layer theory, and may be thought of as being due to re-incorporation of the ascending plume cap, which will be eroded by shear from the downwelling eddy. Thus the core velocity will oscillate, and this will lead to a modulation of the plume emission temperature, as we suppose is observed.

This completes our discussion of the remarkable KZZ model. In summary of this section, the Howard model is not really viable as a model of convection in a circulating flow. The KZZ model represents a significant improvement, but is restricted to values of $\sigma = O(1)$. The obvious next step is to extend the KZZ model to values of $\sigma >> 1$. We leave this undertaking for a future publication.

5. Cellular chaos at high Rayleigh number

In this section we indicate a way of modelling time-dependent convection, when this takes the form of quasi-stationary convection cells whose boundaries can migrate. The type of convection we are thinking of is represented in the laboratory by spoke-pattern convection (Whitehead and Parsons 1978), and is akin to that which occurs in the earth's mantle (although the precise physics is then rather different). The idea is to parametrise the convective motion within any cell in terms of the cell dimension and Rayleigh number, and then to determine via boundary layer theory the relative velocities of the thermal plumes which divide different cells. The philosophy is thus one of combined asymptotic and numerical methods, since we hope to be able to model efficiently high Rayleigh number convection in large containers by first analysing the boundary layers. The computational problem then becomes much simpler to perform.

The strategy we adopt in this section is the following. The boundary layer theory for steady, infinite Prandtl number convection with free-slip boundaries has been summarised in section 3. We now show how this theory is modified for the case of slowly-varying cell boundaries (the fact that the boundary plumes move slowly is a consequence of the analysis). We begin by describing the theory for the simplest case of two convective cells in $0 < x < a$, $0 < y < 1$, with a simple moving plume at $x = \xi(y,t)$ which divides the cells. Later we will discuss possible generalisations to multiple cells, and to real, three-dimensional convection.

(i) Boundary layer theory for slowly varying cells

We follow the recipe given in section 3. The flow configuration is as shown in Figure 5.1, and the equations are given by (3.2).

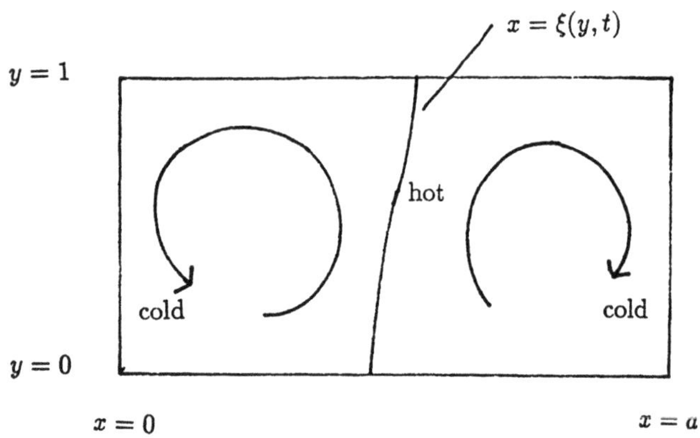

FIGURE 5.1

A. Cores

This is as before; $T = 0$ to all orders of $\delta, \omega \sim O(1)$ at the sides, and at the central plume. (We here anticipate the result that $\partial\xi/\partial t << 1$.) The stream function satisfies the biharmonic equation of Stokes flow $\nabla^4\psi = 0$ in both cores.

B. Plumes

For the stationary plumes at $x = 0$ and $x = a$, the analysis is as before, and we have the boundary conditions for the core flow:

$$\text{(5.1)} \qquad \begin{aligned} \psi_x\omega &= -C_L \text{ at } x = 0, \\ \psi_x\omega &= -C_R \text{ at } x = a; \end{aligned}$$

for the central plume, we put

$$\text{(5.2)} \qquad x = \xi + \delta X, \quad \tau = \delta t, \quad \psi = \delta\Psi, \quad \omega \sim 1, \quad T \sim 1,$$

and anticipate that $\xi = \xi(y,\tau)$ (as is necessary to obtain a sensible balance below); then we have, at leading order,

$$\Psi \sim X v_p(y), \tag{5.3}$$

$$(1+\xi_y^2)\omega_X + T \sim 0, \tag{5.4}$$

whence

$$(1+\xi_y^2)[\omega]_{\xi-}^{\xi+} \sim -\int_{-\infty}^{\infty} T\,dX \tag{5.5}$$

gives a boundary jump condition for the two core flows.

The temperature equation is quasi-static on the time scale $\tau = O(1)$, and T satisfies

$$-\xi_\tau T_X + \Psi_X T_y - \Psi_y T_X \sim (1+\xi_y^2) T_{XX}. \tag{5.6}$$

In von Mises coordinates y,Ψ, we get

$$T_y - \xi_\tau T_\Psi \sim v_p(1+\xi_y^2) T_{\Psi\Psi}, \tag{5.7}$$

whence we have

$$\int_{-\infty}^{+\infty} T\,d\Psi = v_p \int_{-\infty}^{\infty} T\,dX = C(\tau) \tag{5.8}$$

(note also C_L, C_R are functions of τ), so that (5.5) is

$$(1+\xi_y^2) v_p [\omega]_{\xi-}^{\xi+} = -C. \tag{5.9}$$

C. Thermal boundary layers

As before, we find

$$T_s \sim u T_{\Psi\Psi}, \tag{5.10}$$

where s is arc length, and u is the tangential velocity.

D. Solution strategy

We have to solve $\nabla^4\psi = 0$ in each core, with $\psi = 0$ on the boundaries of each cell, $\omega = 0$ on top and bottom, $\psi_x\psi_{xx} = -C_{L,R}$ at the left and right boundaries, and

$$\begin{aligned} &\psi = 0, \quad [\psi_x]_-^+ = 0, \\ &(1+\xi_y^2)[\psi_x\psi_{xx}]_-^+ = -C \qquad \text{at } x = \xi. \end{aligned} \tag{5.11}$$

Given C_L, C_R, C *and* the free boundary ξ, these are sufficient to determine both flows. The free boundary has therefore to be determined, along with C_L, C_R, C, from the solution to the temperature equation.

This is not a very attractive proposition in general, but the whole procedure becomes much simpler if one assumes that the cells are reasonably narrow, at least formally. Roberts (1979) has given approximations for this case. Basically, $\partial/\partial x >> \partial/\partial y$, so the flow is essentially vertical. Moreover, the transit time in the plumes is much longer than in the horizontal boundary layers, and therefore the temperature T at the beginning of the boundary layer can be taken as approximately zero: hence a similarity solution can be written down. This enables us to reduce the entire problem to a single (ordinary) differential equation for the plume position ξ.

E. Practical implementation

We now restrict ourselves to the case where, at least formally, a is small, so that the cells are tall and thin. In practice, we hope such an approximation will give qualitatively accurate results even for $a \sim O(1)$. The stream function in each cell now satisfies, approximately,

$$\psi_{xxxx} \sim 0. \tag{5.12}$$

Let $-v_L$, ω_L denote the velocity and vorticity in the plume at $x = 0$, and v_R, $-\omega_R$ denote those at $x = a$. Then

$$\begin{aligned} \psi &= -v_L x + \frac{1}{2}\omega_L x^2 + A_L x^3, \qquad x < \xi, \\ \psi &= v_R(a-x) - \frac{1}{2}\omega_R(a-x)^2 + A_R(a-x)^3, \qquad x > \xi, \end{aligned} \tag{5.13}$$

and it follows that we require, from the boundary conditions, that

$$\begin{aligned} v_L\omega_L &= C_L, \\ v_R\omega_R &= C_R, \\ -v_L\xi + \frac{1}{2}\omega_L\xi^2 + A_L\xi^3 &= 0, \\ v_R(a-\xi) - \frac{1}{2}\omega_R(a-\xi)^2 + A_R(a-\xi)^3 &= 0, \\ -v_L + \omega_L\xi + 3A_L\xi^2 &= v_p, \\ -v_R + \omega_R(a-\xi) - 3A_R(a-\xi)^2 &= v_p, \\ v_p[\omega_L + 6A_L\xi + \omega_R - 6A_R(a-\xi)] &= C, \end{aligned} \tag{5.14}$$

where we have taken $\xi_y << 1$. At this stage, we have twelve unknowns: v_L, ω_L, A_L, v_R, ω_R, A_R, v_p, C_L, C_R, C, ξ and ξ_τ; but only seven equations. To determine the rest, we must solve the temperature equations in each cell.

Now when the cells are narrow, a similarity solution becomes appropriate; if u is flow speed, then the relevant solutions are

$$T = \pm\frac{1}{2}\,\mathrm{erfc}\left[\frac{\Psi}{2\left\{\int_0^s u\,ds\right\}^{1/2}}\right], \tag{5.15}$$

where we have + for the lower boundary layers, and − for the upper ones. In addition, we change the sign of Ψ in these boundary layers to be positive in the left cell. The heat added through the lower boundary layers is then

$$h_L = \int_0^\xi -u\frac{\partial T}{\partial \Psi}\,ds \tag{5.16}$$

for the left cell, and similarly for the right cell h_R, and also for the heat losses $h_L{}'$, $h_R{}'$ through the top boundary layers. Using the relation

$$\frac{d}{ds}\int_0^\infty T\,d\Psi = -u\frac{\partial T}{\partial \Psi}\Big|_{\Psi=0}, \tag{5.17}$$

we then have that

$$C = -C_L + h_L - C_R + h_R, \tag{5.18}$$

where also

$$\begin{aligned} C_L &= -\int_0^\infty T\,d\Psi, \\ C_R &= -\int_0^\infty T\,d\Psi, \end{aligned} \tag{5.19}$$

on the left and right walls, respectively. Moreover, it can be shown that for isothermal interior temperatures equal to zero, the net buoyancy flux is zero, that is,

$$C = C_L + C_R. \tag{5.20}$$

Now consider the central plume. Here $-\infty < \Psi < \infty$, and $\Psi < 0$ denotes the 'left hand' part of the plume. If

$$\begin{aligned} {C_L}' &= \int_{-\infty}^0 T\,d\Psi, \\ {C_R}' &= \int_0^\infty T\,d\Psi \qquad \text{at } s = 1 \text{ (top of plume)}, \end{aligned} \tag{5.21}$$

then

$$\begin{aligned} -C_L &= {C_L}' - {h_L}', \\ -C_R &= {C_R}' - h'_R. \end{aligned} \tag{5.22}$$

Furthermore, it follows from (5.4) and (5.13) that, if we assume $\omega = 0$ on $\Psi = 0$,

$$\begin{aligned} v_p(\omega_L + 6A_L\xi) &= C'_L, \\ v_p(-\omega_R + 6A_R(a - \xi)) &= -C'_R. \end{aligned} \tag{5.23}$$

We have now introduced a further six unknowns: h_L, h_R, ${h_L}'$, ${h_R}'$, ${C_L}'$ and ${C_R}'$; so we are missing eleven equations. Five are given by (5.18), (5.22) and (5.23). The remainder follow from (5.19), (5.20) and (5.21), and the prescriptions such as (5.16) for the h's. Since determination of ${C_L}'$ and ${C_R}'$ involves solution of the plume equation (5.7) involving $\dot{\xi} = \xi_r$, this is how the plume integration speed is incorporated.

Adopting the similarity solutions for the thermal boundary layer temperatures, we find

$$h_L = (t_L/\pi)^{1/2}, \quad h_R = (t_R/\pi)^{1/2}, \tag{5.24}$$

where $t_L = \int_0^\xi u\,ds$ (similarly t_R) is the transit time across the left lower layer. In the case where the flow in the cell is laterally symmetric (and $\psi_x\psi_{xx} = \pm C$ at each side), Roberts finds numerically that $t_L \approx .266C^{1/2}a^{3/2}$; in our case the flow will really be asymmetric, but we nevertheless suppose that a simple weighted average is appropriate, and (hence) moreover that the transit time in upper and lower layers in the same.

Then

$$\begin{aligned} h_L' = h_L &= 0{\cdot}29\left[\frac{C_L^{1/4} + (C/2)^{1/4}}{2}\right]\xi^{3/4}, \\ h_R' = h_R &= 0{\cdot}29\left[\frac{C_R^{1/4} + (C/2)^{1/4}}{2}\right](a-\xi)^{3/4}, \end{aligned} \tag{5.25}$$

which determines the h's.

Lastly, we find C_L' and C_R' by solving the plume equation

$$T_s - UT_\Psi = v_p T_{\Psi\Psi}, \tag{5.26}$$

where $\dot{\xi} = U$, with

$$\begin{aligned} T = T_0(\Psi) &= \frac{1}{2}\text{erfc}\left[\frac{-\Psi}{2t_L^{1/2}}\right], \quad \Psi < 0, \\ &= \frac{1}{2}\text{erfc}\left[\frac{\Psi}{2t_R^{1/2}}\right], \quad \Psi > 0, \end{aligned} \tag{5.27}$$

at $s = 0$, and

$$T \to 0 \text{ as } \Psi \to \infty. \tag{5.28}$$

When $a << 1$, we only need the large time solution of this, which is (at $s = 1$)

$$T \sim \frac{1}{2\sqrt{\pi v_p}}\left\{\int_{-\infty}^{\infty} T_0(\theta)\,d\theta\right\}\exp\left[-\frac{(\Psi+U)^2}{4v_p}\right] \tag{5.29}$$

(since v_p is constant). After some algebra, we find

$$C_R' = \frac{1}{2}(h_R + h_L)\text{erfc}[U/2v_p^{1/2}] - C_R. \tag{5.30}$$

The equations (5.14), (5.20), (5.22), (5.23), (5.25) and (5.30) give seventeen algebraic equations for the eighteen unknowns v_L, ω_L, A_L, v_R, ω_R, A_R, v_p, C_L, C_R, C, ξ, $U = \dot{\xi}$, h_L,h_L', h_R, h_R', C_L' and C_R'. The equation (5.18) is redundant, being equivalent to $(5.14)_7$, (5.22), (5.23) and (5.25). Elimination of the surrogate variables thus gives, in principle, a first order differential equation of the form $\dot{\xi} = f(\xi)$. To date, solutions have not been obtained. However, since the system is first order, the flow can do very little of interest: either tend to a state with one cell, or tend to one with two cells.

However, the important point is that this procedure can be expanded easily (in principle) to flows with multiple cells. It is obvious that such flows will be determined by systems of first order ordinary differential equations. Therefore, as soon as four or more cells exist, chaotic behaviour is a possibility. We will pursue this investigation in future work, but here we wish to examine the possibility of extension of these ideas to three dimensions. Firstly, let us summarise the mechanics by which the plume is required to move.

If the cells are of different size, than in general $u_L \neq u_R$, and so the initial profile T_0 for T in the plume is skewed. Thus if $U = 0$, the diffusion of T in the plume will lead to a buoyancy transfer: for example if $u_L > u_R$, the left profile of T is steeper, and so buoyancy diffuses to the right. This causes an increase in the buoyancy on the right (thus increasing u_R), but also an increased delivery of heat at the right. Both effects will cause a change in cell size, because of the dependence of u_R on cell size (or vice versa).

(ii) Extension to three dimensions

The extension of the method outlined above to multiple plumes is obvious in principle. In order to extend it to three-dimensional flows, which is our ultimate aim, we must first do three-dimensional boundary layer theory. This has not been done previously, and here we sketch the procedure.

The equations can be written in the form

$$\begin{aligned} \boldsymbol{\nabla}\cdot\mathbf{u} &= 0, \\ p_x &= \nabla^2 u, \\ p_y &= \nabla^2 v, \\ p_z &= \nabla^2 \omega + \frac{1}{\delta}T, \\ \mathbf{u}\cdot\boldsymbol{\nabla}T &= \delta^2\nabla^2 T, \end{aligned} \tag{5.31}$$

in analogy to (3.2). In the core, $T = 0$ to all orders of δ, and we have Stokes flow

$$\begin{aligned} \boldsymbol{\nabla}p &= \nabla^2\mathbf{u}, \\ \boldsymbol{\nabla}\cdot\mathbf{u} &= 0, \end{aligned} \tag{5.32}$$

with free slip boundary conditions at top and bottom, $\mathbf{u}\cdot\mathbf{n} = 0$ at the sides, but the shear stress ($\sim \partial w/\partial n$) will be non-zero due to the plume bouyancy.

It is convenient to write

$$\mathbf{u} = (\mathbf{U}, w), \quad \boldsymbol{\nabla} = \left(\boldsymbol{\nabla}_H, \frac{\partial}{\partial z}\right), \tag{5.33}$$

where $\mathbf{U}$ is the horizontal component of velocity, and $\boldsymbol{\nabla}_H$ the horizontal gradient derivative. In the plume, $\mathbf{u}$ takes its value determined by the core flow. If n is the local normal coordinate at the cell boundary, then the vorticity-buoyancy equation $(5.31)_4$ is given by

$$\frac{\partial}{\partial N}\left(\frac{\partial w}{\partial n}\right) = T, \tag{5.34}$$

where $n = -\delta N$, and thus the core vorticity at the plume is determined by

$$\frac{\partial w}{\partial n} = \int_0^\infty T\,dN, \tag{5.35}$$

if, for example, we consider a single cell with a stationary plume structure. The energy equation can be written

$$\mathbf{u} \cdot \boldsymbol{\nabla} T = T_{NN}; \tag{5.36}$$

it follows on integrating across the layer that (using (5.35))

$$\frac{\partial}{\partial \ell}\left[U_\ell \frac{\partial w}{\partial n}\right] + \frac{\partial}{\partial z}\left[w \frac{\partial w}{\partial n}\right] = 0, \tag{5.37}$$

where U_ℓ is the tangential component of $\mathbf{U}$ at the boundary. This gives the boundary equation for the core vorticity; it must be supplemented by a prescription of $\partial w/\partial n = \int_0^\infty T\,dN$ at the base of upwelling plumes, (or at the top of downwelling plumes), which is determined by solving the thermal boundary layer equation.

In the thermal boundary layer (for example, at the base) we have $z \sim \delta$, $w \sim \delta$, and thus (with rescaled z and w)

$$\begin{aligned} \mathbf{u} \cdot \boldsymbol{\nabla} T &= T_{ZZ}, \\ W &= -Z \boldsymbol{\nabla}_H \cdot \mathbf{U}. \end{aligned} \tag{5.38}$$

We introduce a pseudo-stream function

$$\Psi = Z U_s(x, y), \tag{5.39}$$

where U_s is the speed (determined from the core flow) at the boundary, and let s be a coordinate measuring distance along flow lines at the base. Then introduction of von Mises coordinates s and Ψ (and a coordinate normal to both) gives the diffusion equation

$$\frac{\partial T}{\partial s} = U_s \frac{\partial^2 T}{\partial \Psi^2}, \tag{5.40}$$

and in fact the same equation can be used to describe T in the plumes.

A general numerical strategy to solve for the flow in a single cell might now go as follows. Guess a value for $\partial w/\partial n$ in the plumes, and solve for the core flow. The flow lines on the boundary form a one-parameter family of closed loops, along which

$$\frac{\partial T}{\partial s} = u_s \frac{\partial^2 T}{\partial \Psi^2} \tag{5.41}$$

gives the temperature field (where s is arc length and u_s is speed). From the (periodic) solution of this equation on each path, we determine $\int_0^\infty T\,d\Psi$. It is required that this equal $u_s \partial w/\partial n$, and an iterative procedure can now be implemented to adjust the original value of $\partial w/\partial n$. This is a feasible (but major) computational strategy, and its adaptation for a flow consisting of many cells will normally require the use of parallel processing.

By analogy with the two-dimensional simplification, let us now suppose the cells are tall. Rescaling variables with $\mathbf{r}_H \sim \epsilon << 1$, $\mathbf{U}_H \sim \epsilon$, $p \sim 1/\epsilon^2$, we find $p \approx p(z) + \epsilon^2 \widetilde{p}$, where

$$(5.42) \qquad \begin{aligned} p' &\simeq \nabla_H^2 w, \\ \boldsymbol{\nabla}_H \widetilde{p} &= \nabla_H^2 \mathbf{U}, \\ \boldsymbol{\nabla}_H \cdot \mathbf{U} + \frac{\partial w}{\partial z} &= 0, \end{aligned}$$

with appropriate boundary conditions. The simple solution is that $p' =$ constant, $w = w(x, y)$, $\mathbf{U} = \mathbf{0}$, thus

$$(5.43) \qquad \nabla^2 w = p', \quad w\frac{\partial w}{\partial n} = C(\ell),$$

where $C(\ell)$ is a function of position on the perimeter. Given C, p' is determined so that $\iint w\, dS = 0$, where the integral is over the cell interior. As before C is determined from the thermal boundary layer equation, now with similarity boundary conditions. Thus at the base

$$(5.44) \qquad T = \frac{1}{2}\operatorname{erfc}\left[\frac{\Psi}{2\left\{\int_0^s u_s\, ds\right\}^{1/2}}\right],$$

where s is measured along flow paths in the direction of flow. As before, this leads to an injection of buoyancy equal to $(t^*/\pi)^{1/2}$ (cf. (5.23)), where t^* is the transit time across the base, which depends on the position of the flow path. Further investigation is necessary to see whether any simple parameterisation of t^* in terms of ℓ is possible, since its form will in general depend on the cell shape as well as the aspect ratio and plume vorticity.

If such a parameterisation can be provided , then a complete description of the flow field is possible, and the extension of this to time-dependent problems is straightforward, at least in principle.

ACKNOWLEDGEMENTS

I would like to thank Avner Friedman and the staff of the IMA for hospitality during my visit, and for providing such a congenial working environment. Also for typing several drafts of the manuscript, particularly Marise Widmer. Thanks also to Simon Acomb for many useful discussions.

REFERENCES

BEHRINGER, R.P., *Rayleigh-Bénard convection and turbulence in liquid helium*, Rev. Mod. Phys., 57 (1985), pp. 657–687.

BUSSE, F.H., *Non-linear properties of thermal convection*, Rep. Prog. Phys., 41 (1978), pp. 1929–1967.

BUSSE, F.H. AND J.A. WHITEHEAD, *Oscillatory and collective instabilities in large Prandtl number convection*, J. Fluid Mech., 66 (1974), pp. 67–79.

CASTAING, B., G. GUNARATNE, F. HESLOT, L. KADANOFF, A. LIBCHABER, S. THOMAE, X-Z. WU, S. ZALESKI AND G. ZANETTI, *Scaling of hard thermal turbulence in Rayleigh-Bénard convection*, J. Fluid. Mech., 204 (1989), pp. 1-30.

COULLET, P., L. GIL AND J. LEGA, *A form of turbulence associated with defects*, Physica, 37D (1989) pp. 91–103.

CURRY, J.H., J.R. HERRING, J. LONCARIC AND S.A. ORSZAG, *Order and disorder in two- and three-dimensional Bénard Convection*, J. Fluid Mech., 147 (1984), pp. 1-38.

FOWLER, A.C., *Homoclinic bifurcations for partial differential equations in unbounded domains*, Stud. Appl. Math., 83 (1990), pp. 329–353.

FOWLER, A.C. AND M.J. M^{c}GUINNESS, *A description of the Lorenz attractor at high Prandtl number*, Physica, 5D (1982), pp. 149–182.

FOWLER, A.C. AND M.J. M^{c}GUINNESS, *Hysteresis, period-doubling and intermittency at high Prandtl number in the Lorenz equations*, Stud. Appl. Math., 69 (1983), pp. 99-126.

GOLLUB, J.P., A.R. M^{c}CARRIAR AND J.F. STEINMAN, *Convective pattern evolution and secondary instability*, J. Fluid Mech., 125 (1982), pp. 259–281.

GRIFFITHS, R.W., *Thermals in extremely viscous fluids, including the effects of temperature-dependent viscosity*, J. Fluid Mech., 166 (1986), pp. 115–138.

HAGER, B.H. AND M. GURNIS, *Mantle convection and the state of the Earth's interior*, Revs. Geophys., 25 (1987), pp. 1277–1285.

HESLOT, F., B. CASTAING AND A. LIBCHABER, *Transitions to turbulence in helium gas*, Phys. Rev. A., 36 (1987), pp. 5870–5873.

HOWARD, L.N., *Convection at high Rayleigh number*, Proc. 11th Cong. Appl. Mech., ed H. Görtler, Springer (1966) pp. 1109-1115.

KATSAROS, K.B., W.T. LIU, J.A. BUSINGER AND J.E. TILLMAN, *Heat transport and thermal structure in the interfacial boundary layer measured in an open tank of water in turbulent free convection*, J. Fluid Mech., 83 (1977), pp. 311–335.

KRISHNAMURTHI, R., *On the transition to turbulent convection. Part I. The transition from two- to three-dimensional flow*, J. Fluid Mech., 42 (1970a), pp. 295–307.

KRISHNAMURTHI, R., *On the transition to turbulent convection. Part II. The transition to time-dependent flow*, J. Fluid Mech., 42 (1970b), pp. 309–320.

KRISHNAMURTHI, R. AND L. N. HOWARD, *Large scale flow generation in turbulent convection*, Proc.Nat. Acad. Sci., 78 (1981), pp. 1981–1985.

MORGAN, W.J., *Convective plumes in the lower mantle*, Nature, 230 ((1971), pp. 42–43.

NEWELL, A.C., D.A. RAND AND D. RUSSELL, *Turbulent transport and the random occurrence of coherent events*, Physica, 33D (1988), pp. 281–303.

OLSON, P. AND H. SINGER, *Creeping plumes*, J. Fluid Mech., 158 (1985), pp. 511–531.

PARSONS, B. AND D.P. M^{c}KENZIE, *Mantle convection and the thermal structure of the plates*, J. Geophys. Res., 83 (1978), pp. 4485–4496.

ROBERTS, G.O., *Fast viscous Bénard convection*, Geophys. Astrophys. Fluid Dynamics, 12 (1979), pp. 239–272.

ROBINSON, J.L, *Theoretical analysis of convective instability of a growing horizontal thermal boundary layer*, Phys. Fluids, 19 (1976), pp. 778–791.

SPARROW, C.T., *The Lorenz equations: bifurcations, chaos, and strange attractors*, Springer-Verlag, Berlin (1982).

SPARROW, E.M., R.B. HUSAR AND R.J. GOLDSTEIN, *Observations and other characteristics of thermals*, J. Fluid Mech., 41 (1970), pp. 793-800.

TAMAI, N. AND T. ASAIDA, *Sheetlike plumes near a heated bottom plate at large Rayleigh number*, J. Geophys. Res., 89 (1984), pp. 727–734.

TRITTON, D.J., D.M. RAYBURN AND M.A. FORREST, *Convection of a very viscous fluid heated from below*, In: Mechanisms of Continental Drift and Plate Tectonics, ed. P.A. Davies and S.K. Runcorn, Academic Press (1980).

TURCOTTE, D.L. AND E.R. OXBURGH, *Finite amplitude convection cells and continental drift*, J. Fluid Mech., 28 (1967), pp. 29–42.

TURCOTTE, D.L. AND E.R. OXBURGH, *Mantle convection and the new global tectonics*, Ann. Rev. Fluid Mech., 4 (1972), pp. 33–68.

WHITEHEAD, J.A. AND B. PARSONS, *Observations of convection at Rayleigh numbers up to 760,000 in a fluid with large Prandtl number*, Geophys. Astrophys. Fluid Dynamics, 9 (1978), pp. 201–217.

STRONGLY CHAOTIC NEWTONIAN AND NON-NEWTONIAN MANTLE CONVECTION

DAVID A. YUEN* AND ANDREI V. MALEVSKY*

Summary. The topic of hard turbulent thermal convection is discussed with application to the earth sciences. These new ideas, stimulated by laboratory experiments initiated by physicists, may be useful for understanding the dynamics of magma oceans developed in the early stages of terrestrial planets. Results drawn from two-dimensional numerical simulations of base-heated thermal convection for both linear and non-linear rheologies are presented to show by visualization of the temperature, viscosity and vorticity fields in the transition to the hard turbulent regime. The vorticity fields may provide a good diagnostic measure for the transition to hard turbulence. Both the vorticity and viscosity fields in the non-Newtonian hard turbulent regime reveal small-scale motions and the contours of these heterogeneities assume a fractal-like appearance. The threshold Nusselt number, around 20, for the transition to hard turbulence is lower for base-heated non-Newtonian convection than for Newtonian solutions, thus making it possible for the Earth's upper mantle to be in a hard-turbulent state today.

I. Introduction. The importance of time-dependence in mantle convection has not received much attention in the geoscience community until the last decade with the analysis of seismic anomalies by global networks (e.g. Dziewonski, 1984; Woodhouse and Dziewonski, 1984; Tanimoto, 1990a; Inoue et al., 1990; Su and Dziewonski, 1991). From the work in seismic tomography emerges a picture which shows a complicated pattern of heterogeneity in the mantle consisting of long and short wavelength features. There are significant tomographical differences between the upper and lower mantles. The spectral power of the 3-D structure falls off rapidly in the lower mantle as suggested by recent shear-wave models (Tanimoto, 1990a, Woodward and Masters, 1991). The predominance of long-wavelength features in the lower mantle has been brought out by recent seismological studies (Tanimoto, 1990a, Gudmundsson, et al, 1990, Woodward and Masters, 1991, Su and Dziewonski, 1991). These observations cannot be explained successfully by steady-state convection with simple cellular patterns (e.g. Turcotte and Oxburgh, 1967; Jarvis and Peltier, 1982). Time-dependence in mantle convection is indicated by a variety of geological phenomena. These include migration of ridges and trenches (e.g. Jurdy, 1978), back-arc spreading (Karig, 1974), sea-level fluctuations (Gurnis, 1990) and true polar wander (Goldreich and Toomre, 1968; Gordon, 1983). Besides the type of time-dependent convection occurring in the present geological era, there are many lines of geological evidences (e.g. Nisbet, 1987, Ohtani, 1985) to suggest that there was a change in the style of plate tectonics in the early Earth about 4 billion years ago. The issues surrounding the solidification of a molten silicate mantle after

*Minnesota Supercomputer Institute, Army High Performance Computing and Research Center, and Department of Geology and Geophysics, University of Minnesota, Minneapolis, MN 55415

the Earth's initial formation have focussed attention on the dynamics of turbulent convection in "magma ocean" (Tonks and Melosh, 1990; Miller et al, 1991).

The topic of hard turbulence in thermal convection has been receiving much attention in the last few years since the initial experimental work by a group of physicists at the University of Chicago (Heslot et al, 1987, Castaing et al., 1989, Sano et al, 1989). There was a transition found for thermal convection at Rayleigh numbers between 10^7 and 10^8. Numerical 3-D simulations by Balachandar et al. (1989) and Sirovich et al. (1989) have demonstrated the statistical evidence of this transition by means of probability distribution functions. This transition is characterized by the appearance of disconnected plume structures in contrast to continuous plumes with mushroom-shaped tops found at lower Ra's (e.g. Castaing et al, 1989; Solomon and Gollub, 1991). Diapiric structures also play an important role in mantle dynamics (e.g. Whitehead and Luther, 1975, Whitehead and Helfrich, 1990). From investigations of hard turbulent convection geoscientists can better understand the nature of diapiric generation under high Ra conditions. Diapiric instabilities can also be produced locally in phase-transitions (Liu et al., 1991) and in non-Newtonian convection (Christensen and Yuen, 1989; Malevsky and Yuen, 1991a). The mode of heat-transfer may also change in turbulent convection, as the large-scale flow becomes the major contributor to the global heat transport with the small-scale disconnected plumes playing a minor role (Solomon and Gollub, 1991). This type of heat transport may be prevailing in the Earth's mantle (Davies, 1988). Thus, the phenomenon of hard turbulent convection will be very relevant for geoscientists in understanding better the complexities of the dynamical behavior of planetary interiors, particularly, in the early epoches.

The purpose of this paper is to discuss some of the germane issues on hard turbulent convection as applied to the Earth's mantle and to present results taken from numerical simulations of mantle convection in the hard turbulent regime for both linear and nonlinear rheologies.

II. Hard Turbulent Convection in the Mantle. The effects of inertia in the momentum equation can be neglected in mantle convection because the Prandtl number, which measures the ratio between diffusion of momentum and heat, is extremely high, $O(10^{22})$. Thus this notion of turbulent mantle convection may be puzzling to some fluid dynamicists not familiar with mantle convection. In fact, the advection term in the energy equation (see eqn. 3) is the major nonlinearity responsible for the thermal turbulence generated in infinite Prandtl number thermal convection. The phenomenon of disconnected plumes appearing in convection calculations was first found by Jarvis (1984) for infinite Prandtl number fluids in an aspect-ratio one box, where the aspect-ratio was defined to be the width divided by the height of the box. Diapiric structures, or isolated thermal patches, were produced at Ra of $0(10^8)$ in a base-heated configuration. Thereafter, studies of strongly time-dependent mantle convection were not taken up again until the investigation by Hansen, Yuen and Kroening (1990) on the transition to hard turbulent convection in both small and large boxes. A transition was found in the flow structures in Ra between 10^7 and 10^8 for infinite Prandtl number fluid, as to be distinguished from low Prandtl number of the helium gas medium used by the Chicago group. Many of the features such as plume-plume collisions (Houseman, 1988) and disconnected plumes (Solomon and Gollub, 1990) can also be found in three-dimensional numerical (Houseman, 1988) and laboratory (Solomon and Gollub, 1990) experiments. The existence of bursts in the time-series of the surface Nusselt (Nu) numbers has been identified in the hard turbulent regime of infinite Prandtl number convection (Vincent et al, 1991). Heat and mass transport across the transition to hard turbulent convection were studied by Hansen et. al. (1991) who found that the efficiency of mixing is not only influenced by the magnitude of the convective velocities but also by the style of convection, in particular, the changing nature of the boundary-layer instabilities as a function of Ra. Malevsky and Yuen (1991a, b) have conducted studies on hard turbulent mantle convection for both linear (1991b) and nonlinear rheologies (1991a). The threshold of the vigor of convection to the hard turbulent regime is found to be lower for non-linear rheology (1991a). There remain still many aspects of hard turbulent convection in the mantle to be examined.

III. Equations and Numerical Techniques Used in Modelling. In studying the problem of hard turbulent mantle convection we will make very simplifying assumptions and neglect the effects of mantle compressibility and internal heating which will further influence time-dependent behavior (Weinstein et al., 1989; Machetel and Yuen, 1989). Physical properties, such as viscosity and thermal expansivity, are taken to be constant. The dimensionless equations governing mantle convection are then written for an incompressible Boussinesq fluid in the infinite Prandtl number limit. They are:

$$\nabla \cdot \underset{\sim}{u} = 0 \tag{1}$$

$$\frac{\partial \sigma_{ij}}{\partial x_j} - \nabla p - RaT\widehat{e}_z = 0 \tag{2}$$

$$\frac{\partial T}{\partial t} + \underset{\sim}{u} \cdot \nabla T = \nabla^2 T \tag{3}$$

Eqn. (1), (2), and (3) represent mass conservation, momentum and energy equations. T is temperature, $\underset{\sim}{u}$ is velocity, $\widehat{e}_z$ is the unit vector aligned with gravity, p is dynamical pressure, and t is time non-dimensionalized by the thermal diffusion time across the layer depth. The Rayleigh number Ra is the only dimensionless parameter in this problem without internal heating. The deviatoric stress tensor σ_{ij} is related to the velocity components in cartesian coordinates given by

$$\sigma_{ij} = \frac{\eta}{2}\left(\frac{\partial u_i}{\partial x_j} + \frac{\partial u_j}{\partial x_i}\right) \tag{4}$$

where η is the dynamic viscosity. For Newtonian or linear rheology η is taken as a constant and is included in the definition of Ra. On the other hand, laboratory experiments (Goetze and Kohlstedt, 1973) indicate that mantle rocks can be described by a non-linear or non-Newtonian power-law rheology e.g. (Ranalli, 1987) which is characterized by non-linear dependence of the viscosity on the strain-rate elements. Because of experimental constraints, geoscientists have traditionally taken a power-law dependence in numerical modelling (e.g. Parmentier et al., 1976) of the mantle. This takes the form

$$(5) \qquad \dot{\varepsilon_{ij}} = A^{-1}\sigma^{n-1}\sigma_{ij}$$

where A is the viscosity for $n = 1$, and n is the power-law index. For mantle substances n lies between 3 and 4 (Ranalli, 1987) and σ is the second invariant of the deviatoric stress tensor. From a computational point of view σ_{ij} must be recast in terms of the velocity components. We have then

$$(6) \qquad \sigma_{ij} = A^{\frac{1}{n}}\dot{\mathcal{E}}^{\frac{1-n}{n}}\dot{\mathcal{E}}_{ij}$$

where $\dot{\mathcal{E}}$ is the second invariant of the strain-rate tensor, defined to be

$$(7) \qquad \dot{\mathcal{E}} = [\mathcal{E}_{ij} : \mathcal{E}_{ij}]^{\frac{1}{2}}$$

The simplicity of this formulation for non-linear mantle rheology comes from the type of laboratory experiments conducted today. The effective Ra is defined as

$$(8) \qquad Ra = \frac{2^{\frac{n-1}{n}}\ \alpha g d^{\frac{n+2}{n}}\ \Delta T \rho}{K^{\frac{1}{n}} A^{\frac{1}{n}}}$$

where d is the depth of the, K is the thermal diffusivity, ρ is the average density of the mantle, ΔT is the temperature drop across the mantle, g is the gravitational acceleration and α is the thermal expansivity. We note that there are differences in the definition of Ra, depending on the power-law index n. Henceforth, Ra will be used for Newtonian convection, while Ra' will be employed for non-Newtonian flows. Thus one must use another criterion for comparing Newtonian and non-Newtonian convection. A good candidate would be the surface Nusselt (Nu) number, which measures the amount of convective heat transport and is an output of the nonlinear system.

The results reported below are taken from two-dimensional numerical simulations in a rectangular domain for a purely heated from below configuration. Stress-free boundary conditions appropriate for mantle convection in the Earth (Olson and Corcos, 1980) are applied at the top and bottom boundaries, which are impermeable (vanishing vertical velocity). Reflective boundary conditions on the shear stress, horizontal velocity are horizontal temperature gradient are imposed along the vertical boundaries. The temperature is set to 0 at the top and 1 at the bottom.

The stream function approach is employed to satisfy the continuity equation (eqn. 1). By taking the curl of the momentum equation we obtain a single scalar equation for the stream function ψ. For non-Newtonian rheology this is a nonlinear elliptic partial differential equation because η depends on the derivatives of ψ. It takes the form:

$$(9) \qquad \left(\frac{\partial^2}{\partial z^2} - \frac{\partial^2}{\partial x^2}\right)\eta\left(\frac{\partial^2\psi}{\partial z^2} - \frac{\partial^2\psi}{\partial x^2}\right) + \frac{4\partial^2}{\partial x\partial z}\eta\frac{\partial^2\psi}{\partial x\partial z} = Ra\frac{\partial T}{\partial x}$$

In the case of constant Newtonian viscosity, we have the usual biharmonic equation

$$\nabla^4 \psi = Ra \frac{\partial T}{\partial x} \tag{10}$$

We note that the effective viscosity η depends on shear-deformation and is

$$\eta = \frac{1}{2} \left(\frac{1}{2} \left(\frac{\partial^2 \psi}{\partial x^2} - \frac{\partial^2 \psi}{\partial x^2} \right)^2 + 2 \left(\frac{\partial^2 \psi}{\partial z \partial x} \right)^2 \right)^{\frac{1-n}{2n}} \tag{11}$$

The energy equation is

$$\frac{\partial T}{\partial t} = \nabla^2 T + \frac{\partial \psi}{\partial x} \frac{\partial T}{\partial z} - \frac{\partial \psi}{\partial z} \frac{\partial T}{\partial x} \tag{12}$$

From eqn. (9) or (10), ψ can be solved in terms of T. Thus the nonlinearity of the temperature equation can be made more apparent by rewriting (12) as

$$\frac{\partial T}{\partial t} = \nabla^2 T + a(T) \frac{\partial T}{\partial z} + b(T) \frac{\partial T}{\partial x} \tag{13}$$

where $a(T)$ and $b(T)$ are derived from ψ in terms of T from the elliptic momentum equation.

The numerical methods for solving eqn. (9), (10) and (11) are based on bi-cubic splines and the details can be found in Malevsky and Yuen (1991 a, b). Basically in the case of nonlinear rheology ψ must be obtained by iterative techniques since the viscosity depends on ψ. Non-Newtonian computations require at least 5 times more time than Newtonian because of the subiterations needed in solving eqn. (9). Eqn. (11) is a nonlinear time-dependent advection-diffusion equation. Time-stepping is based on a second-order correct, predictor corrector method. We have employed a finite-element scheme based on the Lagrangian formulation of the total time derivative in the energy equation (Malevsky and Yuen, 1991b). This method is very efficient for strong time-dependent thermal convection, which is dominated by advection. Mesh sizes range from 120×28 grid points for the lowest Ra to 800×170 for the highest Ra cases. The grid is unevenly distributed along the vertical and equally spaced in the horizontal direction.

IV. Numerical Results. We will now display the differences in both the temperature and vorticity fields as a function of Ra, going from a soft turbulent (chaotic) state to the hard turbulent regime. The vorticity is given by the curl of the velocity vector and is a quantity, which measures the magnitude and direction of local rotation in the flow field. In Fig. 1 (see Plates section) are shown the temperature (left hand column) and vorticity fields (right hand column) for base-heated convection in an aspect-ratio 1.8 box for $Ra = 10^6, 10^8$ and 10^{10}. These snapshots are taken after several overturns. Regular cellular temperature fields are observed for $Ra = 10^6$, while the vorticity field is relatively smooth with a few blue (negative vorticity) peaks associated with incipient boundary-layer instabilities. The plumes are continuous at this Ra (see the red and blue boundary layers along the edges). As Ra is increased beyond 10^7, disconnected plumes begin to appear at Ra=10^8, but the vertical symmetry is still maintained by virtue of the accuracy of the code. Diapiric structures are formed readily and rise by themselves to the top (see orange patches). The vorticity field also changes form to one characterized by torn patches (red and blue colors represent respectively positive and negative values of vorticity). At Ra=10^8 the flow is now in the hard turbulent regime from the appearance of disconnected plumes and torn patches in the vorticity field. At a higher Ra of 10^{10} we observe thinner plumes with smaller diapirs. Such a high Ra may be characteristic of the Earth's mantle in the Archean about 4 billion years ago. Most of the heat transfer takes place by large-scale circulation. The interior becomes much more isothermal (green color) as a consequence of the strong convective mixing. Vorticity field becomes more torn up into positive (red) and negative (blue) patches. One is struck by the dramatic change in the vorticity pattern from Ra=10^6 to 10^{10}. The boundaries of these regions assume fractal-like appearance. We note that the asymptotic dependence of the vorticity spectra can give an indication of the way isovorticity lines fill the fluid volume (Benzi et al, 1990) and the ability of the system to mix efficiently.

Long-wavelength seismic velocity studies for the mantle indicate that the mantle structure is dominated by very large-scale heterogeneity (e.g. Masters et al., 1982), shown predominantly by the first two degrees for the whole mantle (Tanimoto, 1990a). Spectral amplitudes of mantle seismic heterogeneity as a function of horizontal wave number k (Tanimoto, 1990b) shows that the seismic anomaly spectra varies as 1/k. We have studied how the thermal anomaly spectra from convection would vary as a function of Ra. Most of the previous efforts on analyzing thermal spectra in mantle convection have been conducted for steady-state situations (e.g. Jarvis and Peltier, 1982). Machetel and Yuen (1987) called attention to the way time-dependent convection might influence the thermal spectra at the core-mantle boundary from bottom boundary-layer instabilities.

In Fig. 2 we show for Ra=3×10^6 the thermal anomaly spectra at four different depths in the top half of the layer. Most of the thermal energy is concentrated in the fundamental wavelengths. The asymptotic decay of the thermal spectra varies close to 1/k for wavenumbers greater than 30. The thermal spectra for Ra=3×10^7 are shown in Figure 3. The decay of the spectra shows a gradual change in slope for k around 100. The spectral amplitudes near the top are lower than for Ra= 3×10^6 on account of the greater convective vigor. As we increase Ra to 3×10^9, the spectra remain white until close to k=100 and then decay rapidly, faster than 1/k (Fig. 4). Thus the asymptotic decay rate of thermal spectra outside the thermal boundary layers may be used as a viable diagnostic tool for detecting hard turbulent convection (Malevsky and Yuen, 1991b). More theoretical work on this aspect is needed.

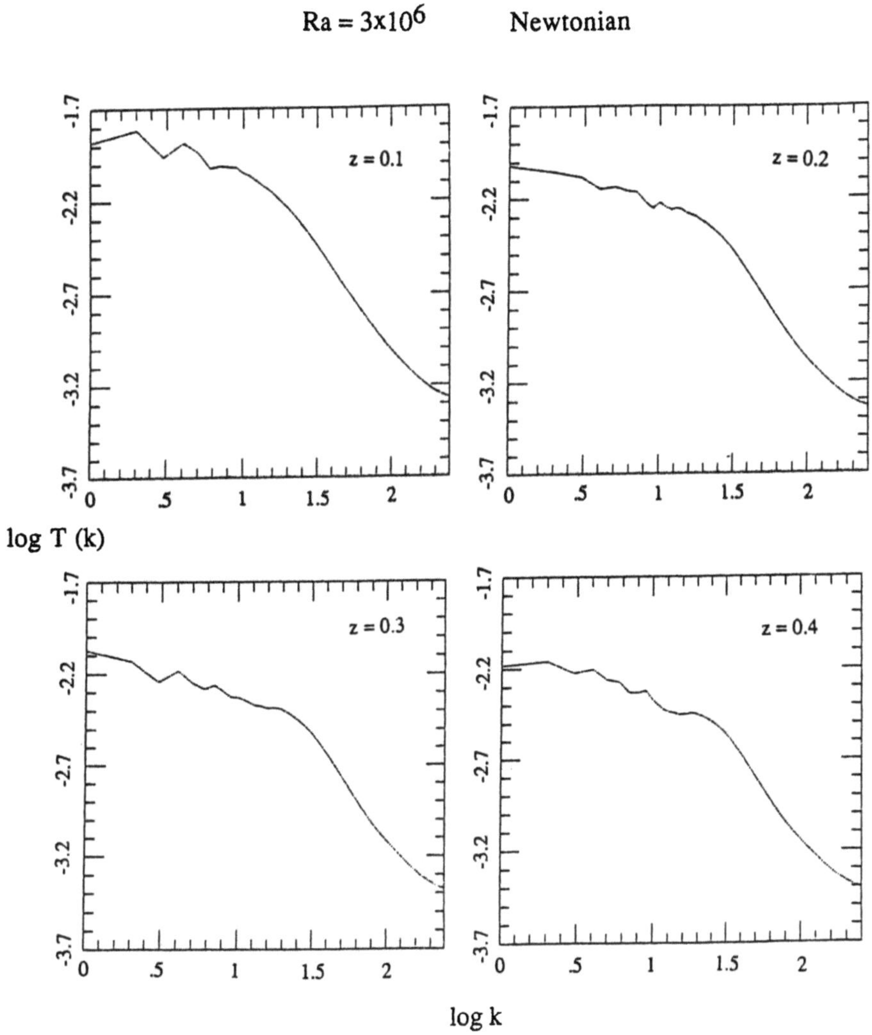

Fig. 2 Horizontal Fourier spectra of thermal anomalies for Ra=3×10^6. Depths are indicated in the figure with z=0.1 near the top. Spectra are taken at an instant after several overturns. 160×48 grid points are employed. An aspect-ratio of 1.8 is employed.

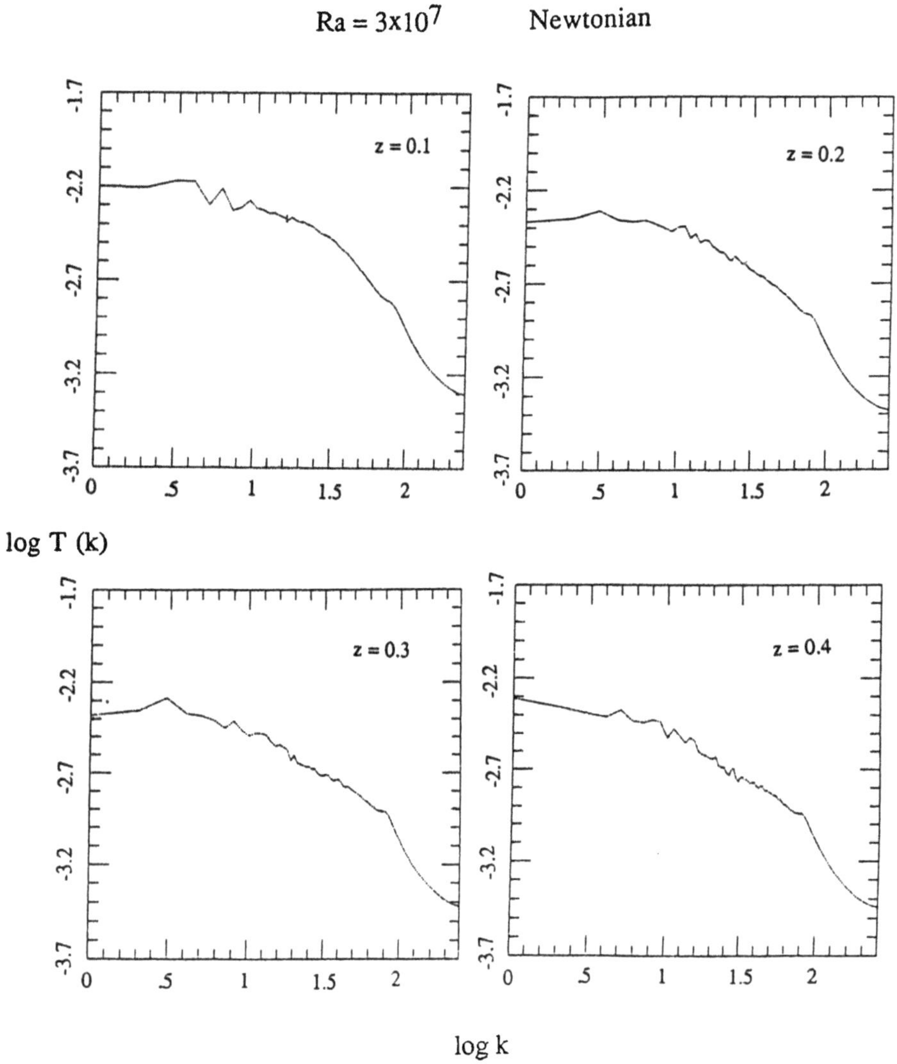

Fig. 3 Same as for Fig. 2 except Ra=3×10^7.160 × 48 points are used.

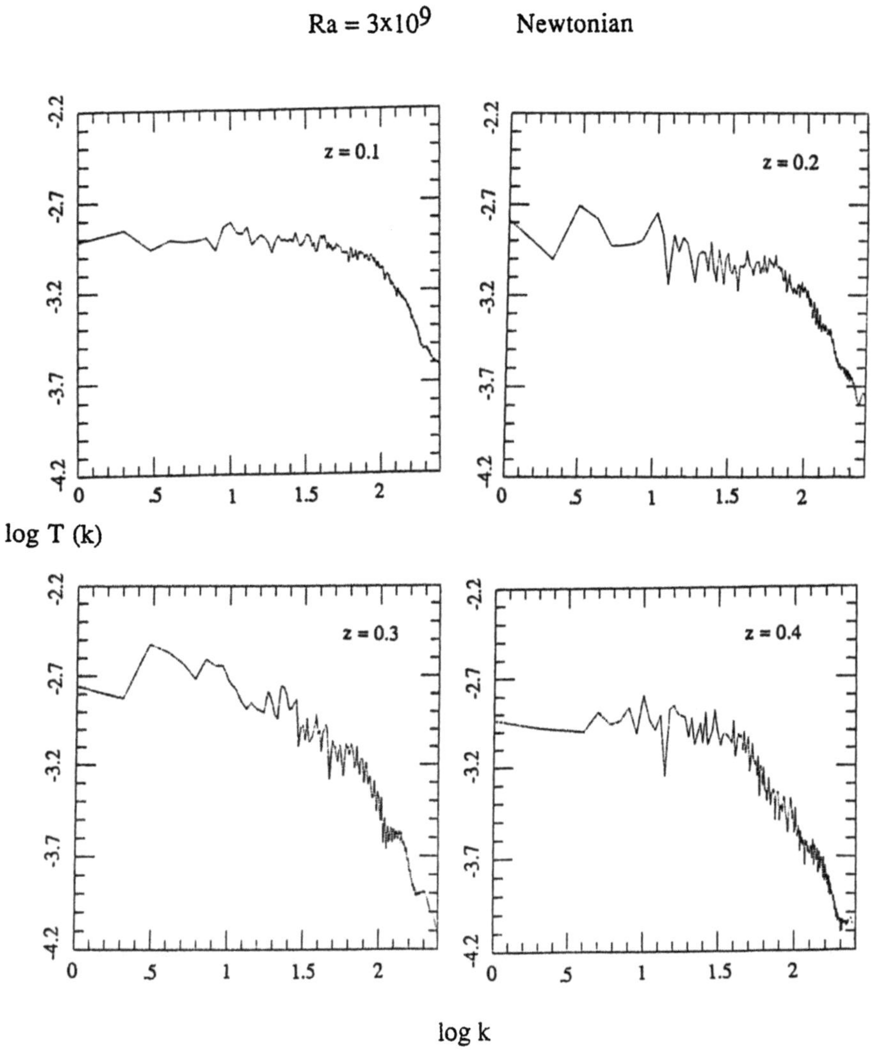

Fig. 4 Same as Fig. 2 except $Ra=3\times10^9$.600 $\times$ 140 points are used.

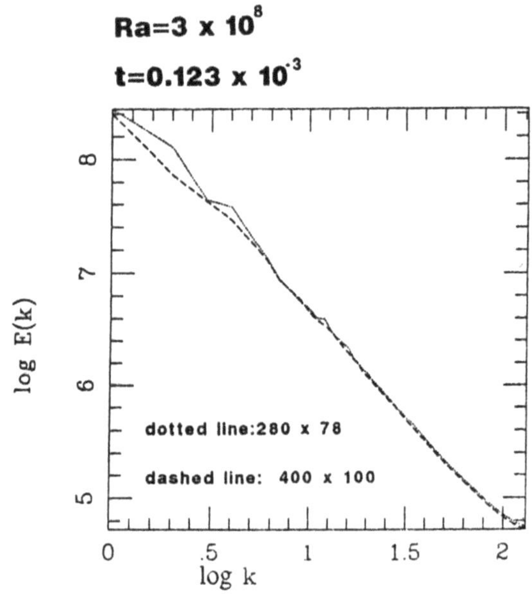

Fig. 5 Kinetic energy spectra for two different grids at Ra=3 $\times 10^8$. Aspect-ratio of the box is 1.8

We can define the kinetic energy spectrum E(k) as

$$E(k_r) = \frac{1}{2} \int_0^{\frac{\pi}{2}} \left(\overline{u}^2(k_r, k_\vartheta) + \overline{v}^2(k_r, k_\vartheta)\right) dk_\vartheta \tag{14}$$

where k_r and k_ϑ are the wavenumbers cast in polar coordinates and $\overline{u}_i$ denotes two-dimensional Fourier transform of the individual velocity component. For the sake of simplicity, we will henceforth denote k_r by k, thus $E(k) = E(k_r)$. Fig. 5 shows the kinetic energy spectra for Ra=3$\times 10^8$ as a function of k. Two different grid configurations, 280×78, and 400×100 points have been employed to demonstrate well the numerical convergence of this high Ra solution. E(k) decays as $1/\mathrm{k}^2$ (Malevsky and Yuen, 1991b) in accordance with the asymptotic theory by She (1989). Most of the previous efforts on non-Newtonian convection were conducted for n=3 with an aspect-ratio 2.5 box and a heated from below configuration (Christensen and Yuen, 1989). We have conducted numerical simulations with a power-law (n=3) rheology in a box with an aspect-ratio 5 also heated from below (Malevsky and Yuen, 1991a). Previous work (Christensen and Yuen, 1989) was restricted to mildly chaotic regime with time-averaged Nusselt numbers less than 10.

In Fig. 6 (see Plates section) are displayed color snapshots depicting the temporal evolution of the temperature T and viscosity fields for chaotic non-Newtonian convection with a time-averaged surface Nu of around 23. The appearance of disconnected plumes at this Nu marks the transition to hard turbulence for non-Newtonian convection. The threshold Nu for the transition to hard turbulence is smaller than for Newtonian convection by about 60% (Hansen et al., 1990). Sinking and rising plumes are shifting around continuously. Different diapirs are being shed constantly from the boundary layers. The vertical symmetry has been destroyed and the interior is filled with both continuous plumes and diapirs. Some of the blobs are being swept upward like passive tracers. Stable plumes can exist in relatively stagnant regions, while other diapirs are being ripped off by the vigorous convective circulations. The logarithmic viscosity field shows sharp variations in the viscosity with the white strips indicating low viscosity and black regions representing high viscosity. Viscosity variations span six orders in magnitude for this chaotic non-Newtonian flow. There exist low viscosity regions adjacent to descending and ascending boundary layers. The convecting region is filled with low (white) and high (black) viscosity patches. Low viscosity zones are not correlated with the boundaries of the convective cells but can also appear in the interior from abrupt boundary layer instabilities. There are no large long-lasting regions with extremely high or low values of the viscosity. Viscosity "grains" are constantly moving around in turbulent non-Newtonian convection. The viscosity field of strongly chaotic non-Newtonian convection appears to be more uniform for processes averaged over long wavelengths than for lower Ra. This is borne out by Fourier spectra of $1/\eta$ which show that the deviations of the components from the average value do not exceed an order of magnitude most of the time (Malevsky and Yuen, 1991a). The iso-viscosity contour levels in strongly chaotic non-Newtonian convection may be amenable to fractal-analysis, as in the case of gradients of convected scalar fields (Varosi et al., 1991). They may give further clues about the nonlinear nature of non-Newtonian convection.

A comparison of the vorticity fields between a soft-turbulent flow (Nu around 13) and a hard-turbulent circulation (Nu around 23) is displayed in the color panels of Fig. 7 (see Plates section). The dramatic differences between the two regimes are brought out very clearly by the vorticity signatures. It is important to compare the similarity in the disconnected vorticity fields between the Newtonian (Fig. 1) and the non-Newtonian cases. The evolution of the vorticity fields indicate that the flow lines are very chaotic and can be very efficient in mixing chemical heterogeneities in convection (Leong and Ottino, 1989, Rom-Kedar et al., 1990). Visualization of the vorticity structure may prove to be a powerful diagnostic tool for detecting hard turbulent convection. Analysis of vortex properties has shed much light on two-dimensional turbulence in the high Reynolds number regime (e.g. McWilliams, 1990).

non-Newtonian convection

Ra' = 1500, 4500

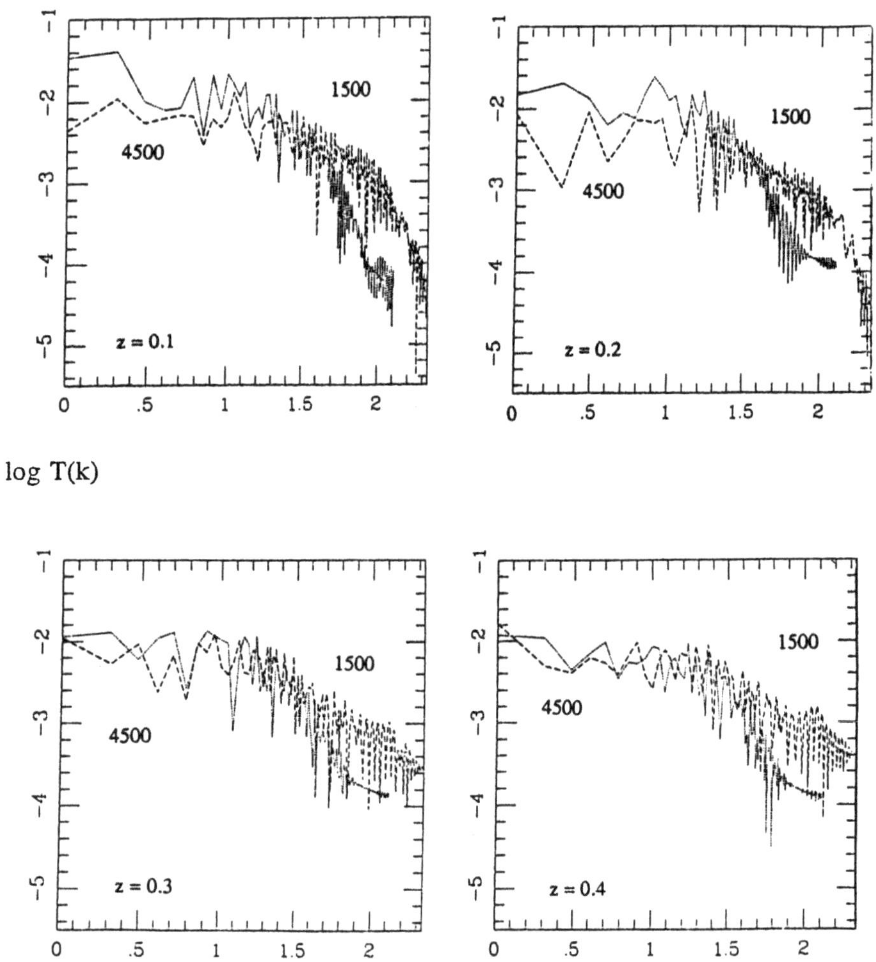

Fig. 8 Horizontal Fourier spectra for non-Newtonian convection. Time is taken from the last frame in Fig. 7. Dashed and dotted lines represent Ra'=4500 and 1500 respectively. An aspect-ratio five box is used.

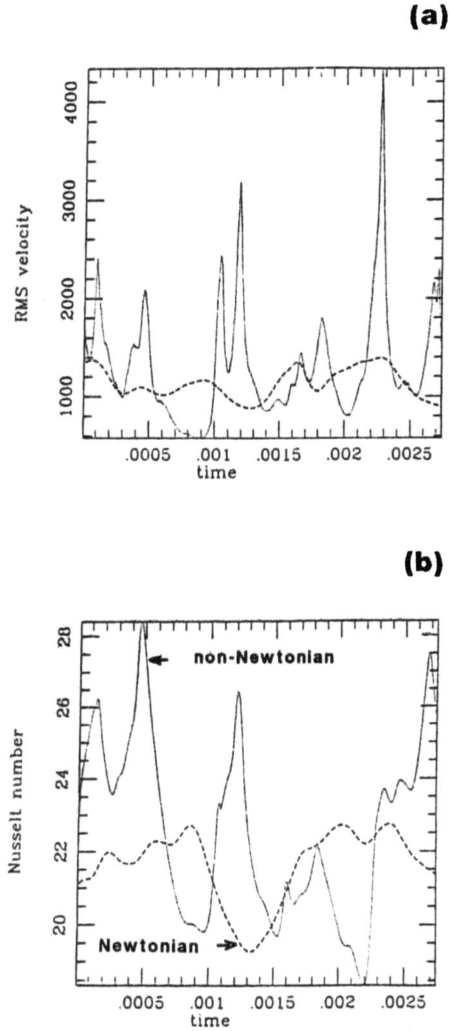

Fig. 9 Comparison of the time histories between Newtonian and non-Newtonian convection. Ra' is 4500 for non-Newtonian and Ra=2×10^6 for Newtonian rheology. An aspect-ratio five box is used.

The spectral characteristics of the thermal anomalies in non-Newtonian flows are also quite different from Newtonian signature. This is illustrated in Fig. 8 where the horizontal Fourier spectra for the two solutions shown in Fig. 7 are displayed for different depths. The spectral decay with k is different between the two rheologies. Near the surface the spectra decay faster than $1/k^2$. Away from the boundary layer the spectra decrease as $1/k^2$, as in the Newtonian hard turbulent case (Fig. 4). The use of spectra has always been an important aspect of turbulence study from Kolmogorov's theory (Kolmogorov, 1941). In geophysics there exist also some classical spectral scaling laws, such as decay of the Earth's gravity anomaly as $1/k^2$, (e.g. Kaula, 1980).

It is well known that the time-dependence of non-Newtonian flows is much more vigorous than the Newtonian solutions (Christensen, 1984). These differences are shown in Fig. 9 where the time-histories of the rms (root mean-squared) velocity and the surface Nu are provided for the two rheologies. The average values of the surface Nu are nearly the same for both cases (Nu around 23). These results would mean that mixing would be much more efficient in non-Newtonian convection. If the rheology of the upper mantle is non-Newtonian and the lower mantle is Newtonian, as suggested by recent laboratory experiments (Li and Karato, 1991), then the vertical variations in the power-law index of the flow laws can account for some of the differences in geochemical signatures from upper and lower mantle substances (DePaolo and Wasserburg, 1979). Non-Newtonian convection (dotted curves) is characterized by short intense periods with sharply peaked flows followed by an interval of quiescence. The amplitudes of the fluctuations in rms velocity and Nu are greater than the Newtonian quantities with approximately the same average magnitude. Nu for non-Newtonian convection exhibits smaller oscillations in relative terms than those associated with rms velocity. The peaks and troughs in the oscillations of Nu and rms velocity are correlated because of the nature of infinite Prandtl number fluids which force the velocity to follow instantaneously changes in the temperature field.

V. Closing Remarks. Our understanding of strongly time-dependent mantle convection is greatly dependent on our understanding of fundamental aspects of hard turbulent convection. To understand the dynamics of hard turbulent convection and their significance in the early thermal history of Earth, we need to know the topological and statistical details of various flow quantities, such as disconnected plumes and vorticity patches. While the approach presented here is based primarily on flow visualization, a discussion of the dynamical role of diapirs in turbulent convection must involve some type of statistical theory (e.g. Loewen et al, 1986). In a general conceptual framework we are concerned with many issues related to these so-called "coherent large-scale structures" in hard turbulent convection. We must address issues pertaining to the circumstances of their occurrence, their characteristic spatial scale, the statistical distribution of their inherent characteristics, such as size and shape, their temporal evolution and the aggregate dynamical behavior from interaction between small and large-scale flows.

The concepts of hard turbulent convection are something new for geoscientists.

This field is still evolving with many of the issues, such as kinks in Nu versus Ra relationship (e.g., Solomon and Gollub, 1991, Goldstein et al., 1990) and the particular shape of the probability distribution function (Castaing et al. 1989, Sano, et al., 1989), remaining unresolved. There is also the interesting possibility of a new transition occurring at Ra around 10^{11} (Wu et al., 1990). These new developments in hard turbulent convection will have a direct bearing on improving our understanding of the turbulent magma ocean (Tonks and Melosh, 1990). From a geophysical perspective, the presence of nonlinear rheology in mantle convection may be sufficient to bring about turbulent convection in the present day mantle, as the threshold Nu is around 0(20) and can further be lowered by other physical mechanisms capable of enhancing dramatically the local Rayleigh number, such as phase-transitions and temperature-dependent viscosity. The presence of an immobile lithosphere, as in the case of Venus, would also lower the threshold Nu. We hope this paper will motivate deeper examination of some of the questions raised here which have both fluid-dynamical and geophysical interests.

Acknowledgements. We are grateful to our colleagues Ulli Hansen and Alain P. Vincent for stimulating conversations and their contributions to works related to this research. We thank M. Lundgren and L. Weyer for help in preparing this manuscript. This research has been supported by the Innovative Research Program of N.A.S.A., Minnesota Supercomputer Institute and Army High Performance Computing and Research Center.

REFERENCES

[1] BALACHANDAR, S., MAXEY, M.R. AND L. SIROVICH, *Numerical simulation of high Rayleigh number convection*, J. Sci. Comp., 4 (1989), 219–235.

[2] BENZI, R., PALADIN, G. AND A. VULPIANI, *Power spectra in two-dimensional turbulence*, Phys. Rev. A., 42 (1990), 3654–3656.

[3] CASTAING, B., GUNARATNE, G., HESLOT, F., KADANOFF, L., LIBCHABER, A., THOMAE, S., WU, X.Z., ZALESKI, S. AND ZANETTI, G., *Scaling of hard thermal turbulence in Rayleigh-Bénard convection*, J. Fluid Mech., 204 (1989), 1–30.

[4] CHRISTENSEN, U.R., *Convection with pressure- and temperature-dependent non-Newtonian rheology*, Geophys. J.R. Astr. Soc., 77 (1984), 343–384.

[5] CHRISTENSEN, U.R. AND D.A. YUEN, *Time-dependent convection with non-Newtonian viscosity*, J. Geophys. Res., 94 (1989), 814–820.

[6] DAVIES, G.F., *Ocean Bathymetry and mantle convection 1. Large-scale flow and hotspots*, J. Geophys. Res., 93 (1988), 10408–10420.

[7] DEPAOLO, D.J. AND G.J. WASSERBURG, *Neodymium isotopes in flood basalts from the Siberian Platform and inferences about their mantle sources*, Proc. Nat. Acad. Sci. U.S.A., 76 (1979), 3056–3060.

[8] DZIEWONSKI, A.M., *Mapping the lower mantle: Determination of lateral heterogeneities in P velocity up to degree and order 6*, J. Geophys. Res., 89 (1984), 5929–5952.

[9] GOETZE, C. AND D.L. KOHLSTEDT, *Laboratory study of dislocation climb and diffusion in olivine*, J. Geophys. Res., 78 (1973), 5961–5971.

[10] GOLDREICH, P. AND A. TOOMRE, *Some remarks on polar wandering*, J. Geophys. Res., 74 (1968), 2555–2567.

[11] GOLDSTEIN, R.J., CHIANG, H.D. AND D.L. SEE, *High-Rayleigh number convection in a horizontal enclosure*, J. Fluid Mech., 213 (1990), 111–126.

[12] GORDON, R.G., *Late Cretaceous apparent polar wander of the pacific plate: Evidence for a rapid shift of the Pacific hotspots with respect to the spin axis*, Geophys. Res. Lett. 10 (1983), 709–712.

[13] GUDMUNDSSON, O., DAVIES, J.H., AND CLAYTON, R.W., *Stochastic analysis of global traveltime data: mantle heterogeneity and random errors in the ISC data*, Geophys. J. Int. 102 (1990), 25–43.

[14] GURNIS, M., *Bounds on global dynamic topography from Phanerozoic flooding of continental platforms*, Nature, 344 (1990), 754–756.

[15] HANSEN, U., YUEN, D.A. AND S.E. KROENING, *Transition to hard turbulence in thermal convection at infinite Prandtl number*, Phys. Fluids, A2(12) (1990), 2157–2163.

[16] HANSEN, U., YUEN, D.A. AND S.E. KROENING, *Mass and heat transport in strongly time-dependent thermal convection at infinite Prandtl number*, Geophys. Astrophys. Fluid Dyn., in press (1991).

[17] HESLOT, F., CASTAING, B. AND A. LIBCHABER, *Transitions to turbulence in helium gas*, Phys. Rev. A. 36 (1987), 5870–5873.

[18] HOUSEMAN, G.A., *The dependence of convection planform on mode of heating*, Nature, 332 (1988), 346–349.

[19] INOUE, H., FUKAO, Y., TANABE, K. AND Y. OGATA, *Whole mantle P-wave travel time tomography*, Phys. Earth Planet, Int., 59 (1990), 294–328.

[20] JARVIS, G.T., *Time-dependent convection in the Earth's mantle*, Phys. Earth Planet. Inter., 36 (1984), 305–327.

[21] JARVIS, G.T. AND W.R. PELTIER, *Mantle convection as a boundary-layer phenomenon*, Geophys. J.R. astr. Soc., 68 (1982), 385–424.

[22] JURDY, D.M., *Ridges, trenches and polar wander excitation*, J. Geophys. Res., 83 (1978), 4889–4894.

[23] KARIG, D.E., *Evolution of arc systems in the western Pacific*, Ann. Rev. Earth Planet. Sci., 2 (1974), 51–75.

[24] KAULA, W.M., *Material properties for mantle convection consistent with observed surface fields*, J. Geophys. Res., 85 (1980), 7031–7044.

[25] KOLMOGOROV, A.N., *The local structure of turbulence in incompressible viscous fluid for very large Reynolds number*, Dokl Akad. Nauk. USSR, 30 (1941), 301–307.

[26] LEONG, C.W. AND J.M. OTTINO, *Experiments on mixing due to chaotic advection in a cavity*, J. Fluid Mech., 209 (1989), 463–499.

[27] LI, P. AND S. KARATO, *Superplasticity in CaTi0₃ perovskite*, E.O.S., Transactions, A.G.U., vol. 72, no. 17 (1991), 287.

[28] LIU, M., YUEN, D.A., ZHAO, W. AND S. HONDA, *Development of diapiric structures in the upper mantle due to phase transitions*, Science, 252 (1991), 1836–1839.

[29] LOEWEN, S., AHLBORN, B. AND A.B. FILUK, *Statistics of surface flow structures on decaying grid turbulence*, Phys. Fluids, 29 (1986), 2388–2397.

[30] MACHETEL, P. AND D.A. YUEN, *Chaotic axisymmetrical spherical convection and large-scale mantle circulation*, Earth Planet. Sci. Lett., 86 (1987), 93–104.

[31] MACHETEL, P. AND D.A. YUEN, *Penetrative convective flows induced by internal heating and mantle compressibility*, J. Geophys. Res., 94 (1989), 10, 609–10, 662.

[32] MALEVSKY, A.V. AND D.A. YUEN, *Strongly chaotic non-Newtonian mantle convection*, Geophys. Astro. Fluid. Dyn., in press, (1991a).

[33] MALEVSKY, A.V. AND D.A. YUEN, *Characteristics-based methods applied to infinite Prandtl number thermal convection in the hard turbulent regime*, Phys. Fluids A3, 2105–2115, 1991b.

[34] MASTERS, G., JORDAN, T.H., SILVER, P.G., AND F. GILBERT, *Aspherical earth structure from fundamental spheroidal mode data*, Nature, 298 (1982), 609–613.

[35] MC WILLIAMS, J.C., *The vortices of two-dimensional turbulence*, J. Fluid Mech., 219 (1990), 361–385.

[36] MILLER, G.H., STOLPER, E.M., AND T.J. AHRENS, *The equation of state of a molten komatiite, 2. Application to komatiite petrogenesis and the hadean mantle*, J. Geophys. Res., 96, 11 (1991), 849–11.

[37] NISBET, E.G., *The Young Earth*, Allen and Unwin Inc., Boston, 1987, pp. 402.

[38] OHTANI, E., *The primordial terrestrial magma ocean and its implications for the stratification of the mantle*, Phys. Earth Planet. Inter., 38 (1985), 70–80.

[39] OLSON, P.L., AND G.M. CORCOS, *A boundary layer model for mantle convection with surface plates*, Geophys. J.R. astr. soc., 62 (1980), 195–219.

[40] PARMENTIER, E.M., TURCOTTE, D.L. AND K.E. TORRANCE, *Studies of finite amplitude non-Newtonian convection with applications to convection in the Earth's mantle*, J. Geophys. Res., 81 (1976), 1839–1846.

[41] RANALLI, G., *Rheology of the Earth: Deformation and flow processes in geophysics and geodynamics*, Chapter 10, Allen and Unwin, Boston, 1987.

[42] ROM-KEDAR, V., LEONARD A. AND S. WIGGINS, *An analytical study of transport, mixing and chaos in an unsteady vortical flow*, J. Fluid Mech., 214 (1990), 347–394.

[43] SANO, M., WU, X. Z AND L. LIBCHABER, *Turbulence in helium-gas free convection*, Phys. Rev. A. 40, (1989), 6421–6430.

[44] SHE, Z.S., *On the scaling laws of thermal turbulent convection*, Phys. Fluids A1 (1989), 911–913.

[45] SIROVICH, L., BALACHANDAR, S., AND M.R. MAXEY, *Simulations of turbulent thermal convection*, Phys. Fluids, A1 (1989), 1911–1913.

[46] SOLOMON, T.H. AND J.P. GOLLUB, *Sheared boundary layers in turbulent Rayleigh–Bénard convection*, Phys. Rev. Lett., 64 (1990), 2382–2385.

[47] SOLOMON, T.H. AND J.P. GOLLUB, *Thermal boundary layers and heat flux in turbulent convection: the role of recirculating flows*, Phys. Rev. A. 43 (1991), 6683–6693.

[48] SU, W.J. AND A.M. DZIEWONSKI, *Predominance of long-wavelength heterogeneity in the mantle*, Nature, 352 (1991), 121–126.

[49] TANIMOTO, T., *Long wavelength S-wave velocity structure throughout the mantle*, Geophys. J., 100 (1990a), 327–336.

[50] TANIMOTO, T., *Predominance of large-scale heterogeneity and the shift of velocity anomalies between the upper and lower mantle*, J. Phys. Earth, 38 (1990b), 493–509.

[51] TONKS, W.B. AND H.J. MELOSH, *The physics of crystal settling and suspension in a turbulent magma ocean*, in *Origin of the Earth*, ed. by H.E. Newsom and J.H. Jones (1990), 151–174.

[52] TURCOTTE, D.L. AND E.R. OXBURGH, *Finite amplitude convective cells and continental drift*, J. Fluid Mech., 28 (1967), 29–42.

[53] VAROSI, F., ANTONSEN, T.M. AND E. OTT, *The spectrum of fractal dimensions of passively convected scalar gradients in chaotic fluid flows*, Phys. Fluids A 3 (1991), 1017–1028.

[54] VINCENT, A.P., HANSEN, U., YUEN, D.A. MALEVSKY, A.V. AND S.E. KROENING, *On the origin of a characteristic frequency in hard thermal turbulence*, Phys. Fluids, A(3) (1991), 2222–2226.

[55] WEINSTEIN, S.A., OLSON, P.L., AND D.A. YUEN, *Time-dependent large aspect-ratio thermal convection in the Earth's mantle*, Geophys. Astrophys. Fluid Dyn., 47 (1989), 157–197.

[56] WHITEHEAD, J.A. AND D.S. LUTHER, *Dynamics of Laboratory diapir and plume models*, J. Geo. Res., 80 (1975), 705–717.

[57] WHITEHEAD, J.A. AND K.R. HELFRICH, *Magma waves and diapiric dynamics*, in *Magma Transport and Storage*, ed. by M.P. Ryan, John Wiley and Sons, Ltd., 1990, pp. 53–76,.

[58] WOODHOUSE, J.H. AND A.M. DZIEWONSKI, *Mapping the upper mantle: three-dimensional modelling of earth structure by inversion of seismic waveforms*, J. Geophys. Res., 89 (1984), 5953–5986.

[59] WOODWARD, R.L. AND G. MASTERS, *Lower mantle structure from SCS–S differential travel times*, Nature, 352 (1991), 231–233.

[60] WU, X.Z. KADANOFF L.,, LIBCHABER, A. AND M. SANO, *Frequency power spectrum of temperature fluctuations in free convection*, Phys. Rev. Lett., 64 (1990), 2140–2143.

ROUTES TO CHAOS IN THE SOLID EARTH*

DONALD L. TURCOTTE†, CHERYL A. STEWART‡, AND JIE HUANG†

Abstract. Two problems in solid earth dynamics are considered in this paper: mantle convection and brittle deformation of the crust. The equations for mantle convection can be specified and an essential feature is the very large Prandtl number. We examine routes to chaos in two-dimensional thermal convection in a fluid layer heated from below. Twelve, twenty-four and forty-mode expansions are considered. The twelve and forty-mode expansions evolve to chaotic flows. These problems are of particular interest because the model truncations are reasonable approximations to thermal turbulence at high Prandtl numbers. We also consider a low-order analog model for the interaction of faults in the earth's crust. Two slider-blocks are considered and are shown to experience a period-doubling route to chaos as long as the symmetry is broken. One implication is that seismicity is chaotic and earthquake prediction must be treated as a probabilistic problem.

1. Introduction. In this paper we consider two problems in solid-earth dynamics. The first problem is mantle convection. Solid-state thermal convection in the earth's mantle is responsible for plate tectonics, mountain building, seismicity, and volcanism. An important characteristic of mantle convection is that the Prandtl number is essentially infinite. We address the question whether infinite Prandtl number thermal convection is chaotic?

The second problem we address is crustal deformation. Deformation of the brittle upper crust occurs primarily on pre-existing faults and takes place during earthquakes. It is recognized that the frequency-size distribution of faults is also fractal. There are several important questions that must be addressed. One is the behavior of individual faults. It is generally hypothesized that velocity-weakening friction on faults is responsible for their stick-slip behavior. A second question is the importance of interactions between faults. Bak et al. [1] suggest that the crust is in a state of self-organized criticality.

It has long been recognized that slider-block models are low-order models for the behavior of faults [2]. We will show the conditions under which slider-block models behave chaotically and will show that they follow the Feigenbaum, period-doubling route to chaos.

2. Chaotic Mantle Convection. Thermal convection in the earth's mantle transports the heat generated by the radioactive isotopes of uranium, thorium, and potassium and also cools the mantle. The surface plates of plate tectonics are the thermal boundary layers of mantle convection cells. The plates are created at ocean ridges, they cool and become gravitationally unstable with respect to the mantle below, and they founder and sink at ocean trenches. The plates transport about 80% of the heat associated with mantle convection but there is evidence for other mechanisms of mantle convection.

---

*This research was supported under grant EAR-8915992 from the National Science Foundation and grants NAG 5-319 and NGT-50203 from the National Aeronautics and Space Administration.

†Department of Geological Sciences, Cornell University, Ithaca, NY 14853 U.S.A.

‡Now at: Laboratoire de Mathematiques, Universite de Nice, Parc Valrose 060345, Nice, France

Intraplate hot spots such as Hawaii are attributed to hot plumes ascending from hot, unstable boundary layers at the base of the convecting mantle. The Hawaiian plume track extends to the Aleutian islands indicating that this plume has been active for some 80 million years. This is an exception as most plume tracks are considerably shorter. Mantle convection is certainly unsteady, but is it "turbulent"? In order to approach this question it is necessary to understand thermal convection at very large Prandtl numbers. The first approach is to consider a layer of Boussinesq fluid; this is a bad approximation for mantle convection for a number of reasons. The most important is that the viscosity of the mantle has an exponential dependence on temperature and pressure. In addition the mantle is compressible, the rheology may not be Newtonian, and a spherical shell should be considered. Nevertheless, the basic question concerning chaotic behavior can be addressed in the simple layered configuration with a Boussinesq fluid.

The governing equations that we consider are

$$0 = \nabla^4 \psi - Ra\theta_x \tag{2.1}$$

$$\frac{D\theta}{Dt} = \nabla^2 \theta \tag{2.2}$$

These are the applicable equations for thermal convection in an infinite Prandtl number fluid. The variables are the nondimensional stream function ψ and temperature θ. Lengths have been nondimensionalized with the layer thickness h, time with $h^2\kappa$, and temperatures with the temperature difference across the layer $T_1 - T_2$. The nondimensional temperature is referenced to the conduction profile. The only parameter in this limit is the Rayleigh number $Ra = \rho\alpha g(T_2 - T_1)h^3/\kappa\mu$.

Following Saltzman [3], the stream function and the temperature are expanded in double Fourier series in x and y with coefficients that are functions of time only:

$$\psi(x,y,t) = \sum_{m=-\infty}^{\infty} \sum_{n=-\infty}^{\infty} \psi_{m,n}(t) \exp\left[2\pi i \left(\frac{mx}{\lambda} + \frac{ny}{2}\right)\right] \tag{2.3}$$

$$\theta(x,y,t) = \sum_{m=-\infty}^{\infty} \sum_{n=-\infty}^{\infty} \theta_{m,n}(t) \exp\left[2\pi \left(\frac{mx}{\lambda} + \frac{ny}{2}\right)\right] \tag{2.4}$$

where λ is the nondimensional wavelength of the fundamental mode. Lorenz [4] truncated these equations to a single term in the stream-function expansion and a pair of terms in the temperature expansion. The Lorenz expansion is

$$\psi = \frac{1}{2^{1/2}\lambda}(4+\lambda^2)A(\tau)\sin\left(\frac{2\pi x}{\lambda}\right)\sin \pi y \tag{2.5}$$

$$\theta = \frac{1}{\pi r}\left[C(\tau)\sin 2\pi y - 2^{1/2}B(\tau)\cos\left(\frac{2\pi x}{\lambda}\right)\sin \pi y\right] \tag{2.6}$$

where

(2.7)
$$\tau = \pi^2 \left[1 + \left(\frac{2}{\lambda}\right)^2\right] t$$

(2.8)
$$r = \frac{Ra}{Ra_c}$$

where Ra_c is the critical Rayleigh number for the onset of convection. These equations satisfy the boundary conditions $\psi = \nabla^2\psi = \theta = 0$ at $y = 0, 1$. In the limit of infinite Prandtl number the Lorenz equations become

(2.9)
$$A = B$$

(2.10)
$$\frac{dB}{d\tau} = (r-1)B - BC$$

(2.11)
$$\frac{dC}{d\tau} = bC + B^2$$

where $b = 4/(1 + [2/\lambda]^2)$. The steady state solution $A = B = C$ corresponds to heat conduction without flow. An additional pair of allowed solutions are

(2.12)
$$A = B = \pm[b(r-1)]^{1/2}$$

(2.13)
$$C = r - 1$$

These solutions correspond to cellular rolls either rotating clockwise or counterclockwise. A stability analysis for the conduction solution shows that it is stable for $r < 1$ and unstable for $r > 1$, a pitchfork bifurcation occurs at $r = 1$ ($Ra = Ra_c = 27\pi^2/4$). The steady state solution consisting of cellular rolls is stable for the entire range $r > 1$. In the limit of infinite Prandtl number the Lorenz equations do not yield chaotic solutions. This is expected since in this limit they reduce to a pair of equations and two degree of freedom systems generally do not yield chaotic solutions.

In order to consider this problem further Stewart and Turcotte [5] considered a higher order expansion of the Lorenz equations. In the limit of infinite Prandtl number the Fourier coefficients of the stream function $\psi_{m,n}$ can be related to the Fourer coefficients of the temperature $\theta_{m,n}$ using (2.1) with the result

(2.14)
$$\psi_{m,n} = \left(\frac{2\pi i m}{\lambda}\right) \frac{\theta_{m,n}}{\left(\frac{4\pi^2 m^2}{\lambda^2} + \pi^2 n^2\right)^2}$$

This result is then substituted into (2.2). The lowest consistent order of truncation beyond that used by Lorenz is $m = 2$ for the expansion in $x(m = 0, 1, 2)$ and $n = 4$ for the expansion in $y(n = 1, 2, 3, 4)$. This truncation yields a set of 12 ordinary differential equations for the time dependence of the temperature coefficients

$$\frac{d\theta_{m,n}}{dt} = -\sum_{p=-2}^{2}\sum_{q=-4}^{4} \frac{2\pi^2}{\lambda}(mq - np)\left(\frac{2\pi p}{\lambda}\right) \cdot \frac{\theta_{p,q}\theta_{m-p,n-q}}{[(4\pi^2 p^2/\lambda^2) + \pi^2 q^2]^2} \tag{2.15}$$

$$+Ra\left(\frac{2\pi m}{\lambda}\right)^2 \cdot \frac{\theta_{m,n}}{[(4\pi^2 m^2 \lambda^2) + \pi^2 n^2]^2} - \pi^2\left(\frac{4m^2}{\lambda^2} + n^2\right)\theta_{m,n}$$

It is necessary to take twice the resolution in the vertical direction as in the horizontal direction in order to resolve the $u\partial\theta/\partial x$ term in the energy equation.

The time evolution of the twelve coefficients $\theta_{0,1}, \theta_{0,2}, \theta_{0,3}, \theta_{0,4}, \theta_{1,1}, \theta_{1,2}, \theta_{1,3}, \theta_{1,4}, \theta_{2,1}, \theta_{2,2}, \theta_{2,3}, \theta_{2,4}$ is found by integrating the twelve equations given by (2.15) numerically. The time evolution can be thought of as trajectories in a twelve dimensional space. It is convenient to project the 12-dimensional trajectories onto the two-dimensional phase-space consisting of $\theta_{1,1}$ and $\theta_{2,1}$. These correspond to the fundamental mode and the first subharmonic. There are two parameters in this problem, the Rayleigh number, Ra or r, and the wavelength λ. In this discussion solutions are given only for the critical value of the wavelength $\lambda = 2^{3/2}$.

At the subcritical Rayleigh numbers $0 < r < 1 (0 < Ra < 657.512)$ the only fixed point of the solution is at the origin and it is stable, there is no flow. For higher Rayleigh numbers, the two fixed points corresponding to clockwise and counterclockwise rotations in the fundamental model $\theta_{1,1}$ become stable. The steady state solution for $Ra = 10^4 (r = 15.21)$ is given in Table 1. It is seen that only six of the 12 coefficients are nonzero: $\theta_{0,2}, \theta_{0,4}, \theta_{1,1}, \theta_{1,3}, \theta_{2,2}$, and $\theta_{2,4}$. This solution was obtained by specifying an initial condition near the origin and studying the time evolution of the 12 coefficients using (2.15). This time evolution projected onto the $\theta_{1,1} - \theta_{2,1}$, plane is given in Figure 1a. Although the subharmonic coefficient $\theta_{2,1}$ is zero at the fixed point, it is nonzero during the time evolution.

The steady-state solution for $Ra = 3 \times 10^4$ $(r = 45.62)$ is also given in Table 1. It is seen that only four of the twelve coefficients are nonzero: $\theta_{0,2}, \theta_{0,4}, \theta_{2,1}, \theta_{2,3}$. At this Rayleigh number the fundamental mode and its associated coefficients $\theta_{1,1}, \theta_{1,3}, \theta_{2,2}$, and $\theta_{2,4}$, are zero at the stable fixed point. The time evolution of the solution projected onto the $\theta_{1,1} - \theta_{2,1}$ plane is given in Figure 1b. Finally, the steady-state solution for $Ra = 4.3 \times 10^4$ $(r = 65.39)$ is given in Table 1. It is seen that eight of the twelve coefficients are now nonzero: $\theta_{0,1}, \theta_{0,2}, \theta_{0,4}, \theta_{2,1}, \theta_{2,3}, \theta_{2,4}$. All of the $\theta_{1,n}$, coefficients are zero including the fundamental mode $\theta_{1,1}$. The time evolution of the solution projected onto the $\theta_{2,1} - \theta_{1,1}$ plane is given in Figure 1c. It is seen that the evolution prior to entering the stable fixed point is much more complex; the solution oscillates in the positive $\theta_{2,1}$ quadrants before entering the negative quadrants.

Table 1

NUMERICAL VALUES OF THE FOURIER COEFFICIENTS OF THE FIXED POINTS OF THE 12-MODE EQUATIONS (2.15)

	Ra		
	10^4	3×10^4	4.3×10^4
$r = Ra/R_c$	15.21	45.62	65.39
$\theta_{0,1}$	0.000	0.000	-2120.222
$\theta_{0,2}$	-1506.097	-4571.719	-6663.231
$\theta_{0,3}$	0.000	0.000	-366.705
$\theta_{0,4}$	-286.133	-2191.940	-3355.777
$\theta_{1,1}$	506.978	0.000	0.000
$\theta_{1,2}$	0.000	0.000	0.000
$\theta_{1,3}$	66.170	0.000	0.000
$\theta_{1,4}$	0.000	0.000	0.000
$\theta_{2,1}$	0.000	1117.832	-1415.752
$\theta_{2,2}$	426.445	0.000	-398.378
$\theta_{2,3}$	0.000	509.313	-623.561
$\theta_{2,4}$	105.531	0.000	-490.614

The time evolution of the solution for $Ra = 5.6 \times 10^4$ ($r = 68.44$) is given in Figure 1d; it is fully chaotic and no fixed points are stable. The flow alternates between aperiodic oscillations about the two fundamental modes (clockwise and counterclockwise) and the two subharmonic modes (clockwise and counterclockwise). All 12 coefficients are nonzero and are time-dependent. The time dependence of the $\theta_{2,1}$ and $\theta_{1,1}$ coefficients are shown in Figure 2. The resemblance to the time behavior of the Lorenz attractor is striking. Oscillatory behavior of the $\theta_{2,1}$ mode amplifies until the flow undergoes a burst in the fundamental $\theta_{1,1}$, mode, where it is briefly trapped before flipping into the $\theta_{2,1}$ mode with the opposite sense of rotation.

Figure 1. Numerical solutions of the 12-mode infinite Prandtl number equations projected onto the $\theta_{1,1}, \theta_{2,1}$ plane of the 12-dimensional phase space. (a) $Ra = 10,000$ ($r = 15.21$), (b) $Ra = 30,000$ ($r = 45.62$), (c) $Ra = 43,000$ ($r = 65.39$), (d) $Ra = 45,000$ ($r = 65.39$).

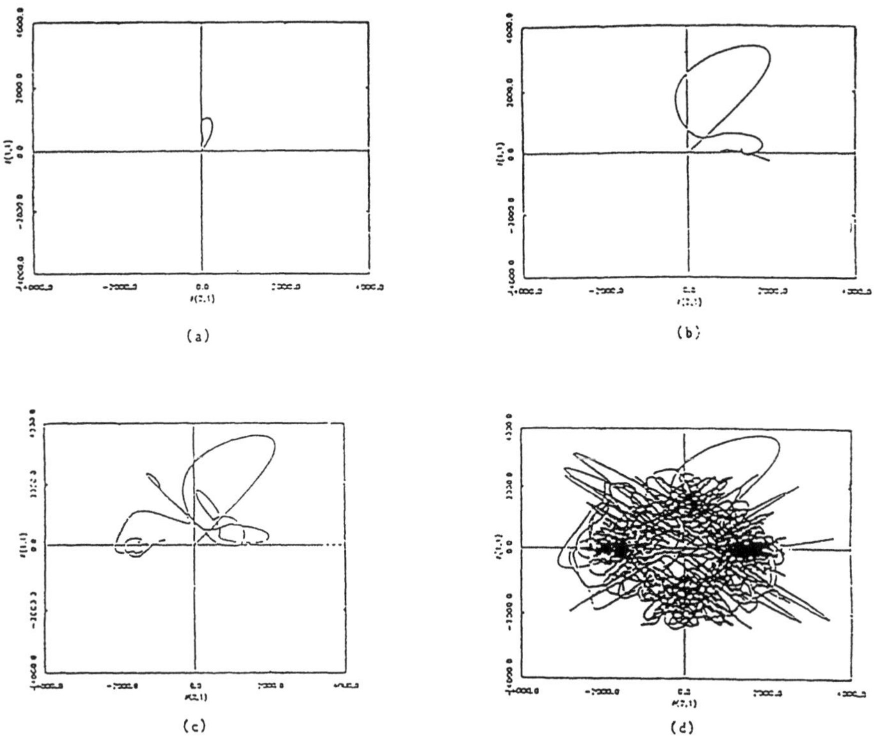

In order to better understand the transitions in the behavior of the time-dependent solutions, we calculated the location of the fixed points as a function the Rayleigh number. Figure 3 gives two projections of the loci of fixed points as a function of Rayleigh number of this system. The solid lines denote stable fixed points, the dashed lines unstable fixed points, and the open circles Hopf bifurcations. Figure 3a shows a projection on to the $Ra, \theta_{1,1}$ plane. The original system of equations is symmetric under the transformations $\theta_{1,1} \to -\theta_{1,1}$ and $\theta_{2,1} \to -\theta_{2,1}$; therefore we only show the positive quadrant.

Figure 2. Time dependence of the coefficients (a) $\theta_{1,1}$ and (b) $\theta_{2,1}$ for the solution given in Figure 1d, $Ra = 45,000$.

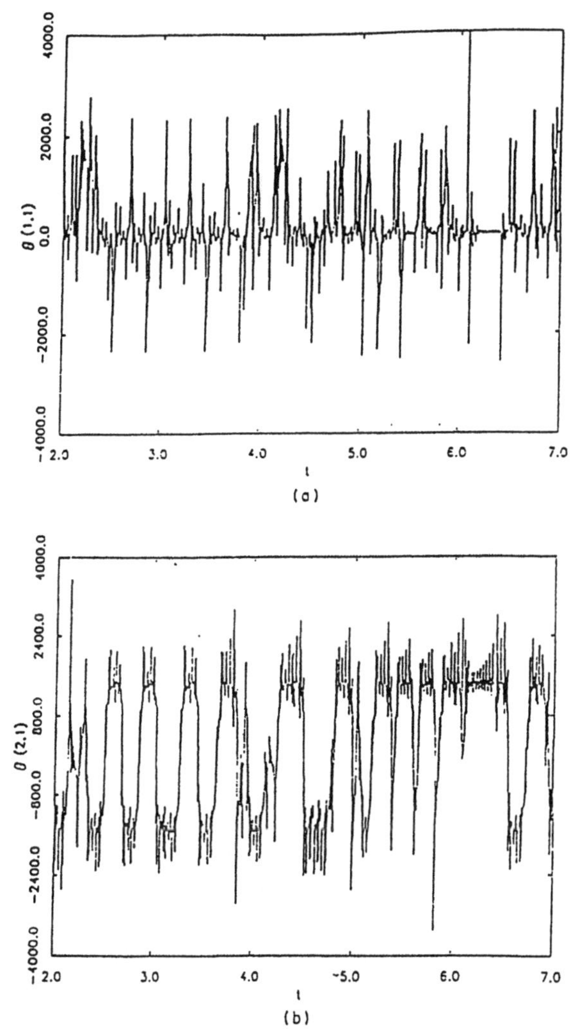

The origin representing the conduction solution becomes unstable and bifurcates at Rayleigh number $Ra = 657.512$, spawning two stable symmetric solutions which contain no component in the $\theta_{2,1}$ mode. One branch of this solution appears in the positive quadrant and is labelled "$\theta_{1,1}$ pure" in Figure 3a to distinguish it from the mixed-mode solution where there is some contribution from the $\theta_{2,1}$ mode.

Each branch becomes unstable and undergoes a subcritical pitchfork bifurcation at $Ra = 3.802 \times 10^4$, producing four unstable mixed mode solutions, labeled "$\theta_{1,1}$ mixed". Each $\theta_{1,1}$ mixed-mode branch sweeps back to a saddle bifurcation at $Ra = 1.909 \times 10^4$. This type of bifurcation configuration (subcritical pitchfork plus two saddles) typically produces hysterisis effects when the saddle has one stable branch and one unstable branch. Here the "$\theta_{1,1}$ mixed" solution has one unstable manifold (out of 12) on one side of the saddle, and two unstable manifolds on the other.

Figure 3. Bifurcation diagram for the 12-mode infinite Prandtl number equations. The fixed points of (2.5) are projected onto the (a) $\theta_{1,1}, Ra$ plane (b) onto the $\theta_{2,1}, Ra$ plane. Stable branches are shown as solid lines, instable branches as dashed lines, pitchfork bifurcations as solid circles, and Hopf bifurcations as open circles.

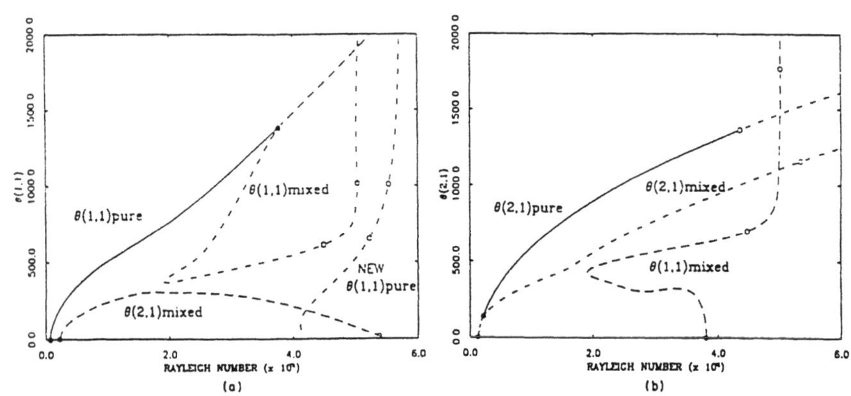

The second bifurcation of the conduction solution is at $Ra = 1315.023$, where two unstable symmetric fixed points dominant in the subharmonic $\theta_{2,1}$ mode appear. Since these fixed points contain no component in the $\theta_{1,1}$ mode, we call these "$\theta_{2,1}$ pure" unstable solution branches. Each of these branches becomes stable and undergoes a pitchfork bifurcation at $Ra = 2041.918$, resulting in the branches solution labeled "$\theta_{2,1}$ mixed" in Figures 3a and 3b. Then $\theta_{2,1}$ mixed mode branches nearly connect with the $\theta_{1,1}$ mixed mode branches.

Both the fundamental and the subharmonic pure mode solutions are stable between $Ra = 2369$ and $Ra = 3.802 \times 10^4$. The trajectories in Figures 1a and 1b have the same initial condition, yet in Figure 1a it converges to the fundamental subharmonic stable fixed point. Presumably this is because the unstable mixed-mode branches disrupt the seperatix between the basin of attraction of the fundamental and subharmonic pure mode solutions. Note that the transition from Figure 1a to Figure 1b occurs at a Rayleigh number above the stability limit of the fundamental mode.

The third bifurcation of the conduction solution is at $Ra = 4.140 \times 10^4$, where two unstable symmetric solutions in the fundamental mode $\theta_{1,1}$ appear. These are labelled "new $\theta_{1,1}$ pure" in Figure 3a. Each of these undergo Hopf bifurcations at $Ra = 5.23 \times 10^4$ and $Ra = 5.53 \times 10^4$. At no point does the origin itself undergo a Hopf bifurcation, nor does the conduction solution bifurcate to mixed mode solutions.

Figure 4 Loci of the fixed points from Figure 3 projected onto the $\theta_{1,1}, \theta_{2,1}$ plane. Superimposed as a dotted line is the time evolution of chaotic solution from Figure 1d.

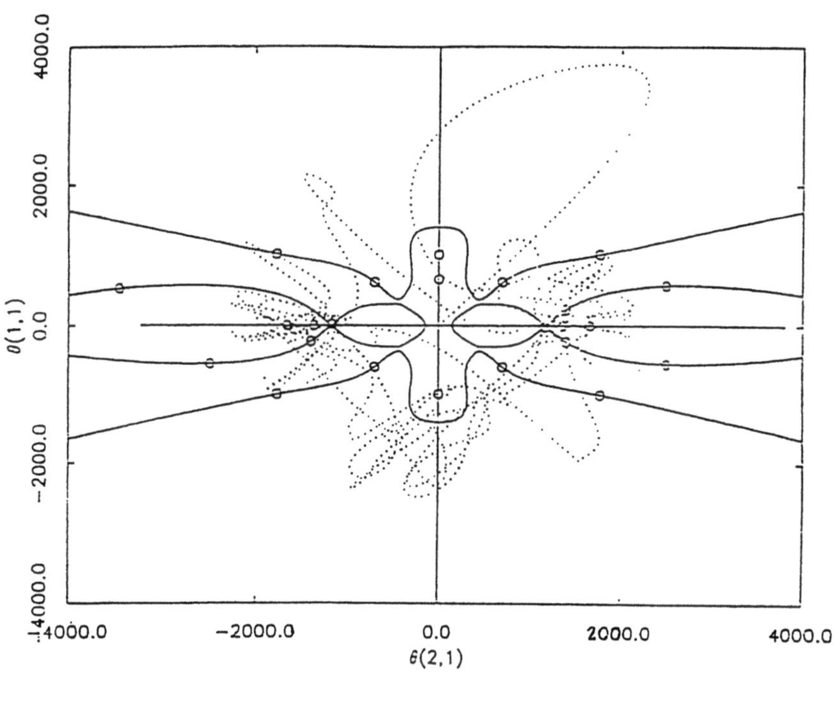

We detected no Hopf bifurcations for the conduction or fundamental harmonic solutions; however, the stable subharmonic branch undergoes two Hopf bifurcations (Figure 3b), one at $Ra = 4.37 \times 10^4$ and one at $Ra = 6.36 \times 10^4$. Each mixed-mode saddle ($\theta_{1,1}$ mixed) undergoes two Hopf bifurcations, at $Ra = 4.491 \times 10^4$

and $Ra = 5.039 \times 10^4$. Each mixed-mode saddle ($\theta_{2,1}$ mixed) undergoes six Hopf bifurcations. Each of these Hopf bifurcations sheds stable or unstable periodic orbits which are responsible for the oscillations of the trajectory at $Ra = 4.5 \times 10^4$ shown in Figure 1d. In Figure 4, the first 7000 points of the trajectory at $Ra = 4.5 \times 10^4$ are projected onto the $\theta_{1,1}, \theta_{2,1}$ plane (dotted line) superimposed on the central portion of the branches of the fixed points (solid lines). Note that the trajectory weaves aperiodically around several Hopf bifurcations (circles).

Physically, infinite Prandtl number, high Rayleigh number convection becomes time-dependent through boundary layer instabilities that generate thermal plumes. In terms of spectral expansions, these instabilities result form the nonlinear coupling in the convective terms of the heat equation.

The 12-mode expansion gives solutions that are clearly chaotic. A valid question is whether higher order expansions also yield chaotic solutions. In order to examine this question 24-and 40- mode expansions have been considered. At $Ra = 10^4$, both models converge to fixed points, and the energy-distribution among the Fourier modes is qualitatively correct; modes which should be near-zero are near-zero; while large-amplitude modes have comparable amplitudes, within 10%. At $Ra = 5 \times 10^4$, however, the temporal behavior of the 24-model and the 40-mode model diverge significantly; the 40-mode model goes to a fixed-point while the 24-mode model goes to a periodic orbit. At higher Rayleigh numbers, this relationship inverts, the 24-mode model going to a stable fixed-point, and the 40-mode model to a periodic orbit. Therefore, it is not always true that adding more modes in the expansion will reduce the time-dependent behavior of the solutions - or the dimension of the attractor; sometimes the opposite is true.

Figure 5. Periodic orbit of the 40-mode system at $Ra = 1.9 \times 10^5$.

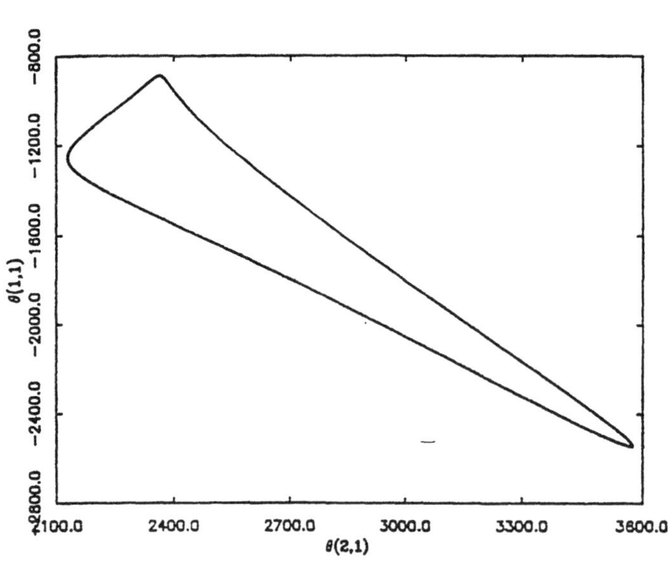

In Figure 5 we show the periodic orbit of the 40 mode system projected onto the $\theta_{1,1} - \theta_{2,1}$ plane at Rayleigh number $Ra = 1.9 \times 10^5$. Because the system is symmetric under the transformations $\theta_{m,n} \rightarrow -\theta_{m,n}$ this periodic orbit also occurs in the other quadrants of the system. Figure 6 shows the periodic orbit after bifurcation to a 2-Torus. This occurs near Rayleigh number $Ra = 1.95 \times 10^5$. A second temporal frequency has been introduced that causes the periodic orbit to wind around in two dimensions, remaining on the surface of a torus. The ratio between the two frequencies is approximately 50.

Figure 6 Periodic orbit after the transition to a 2-torus in the 40-mode system at $Ra = 1.19 \times 10^5$.

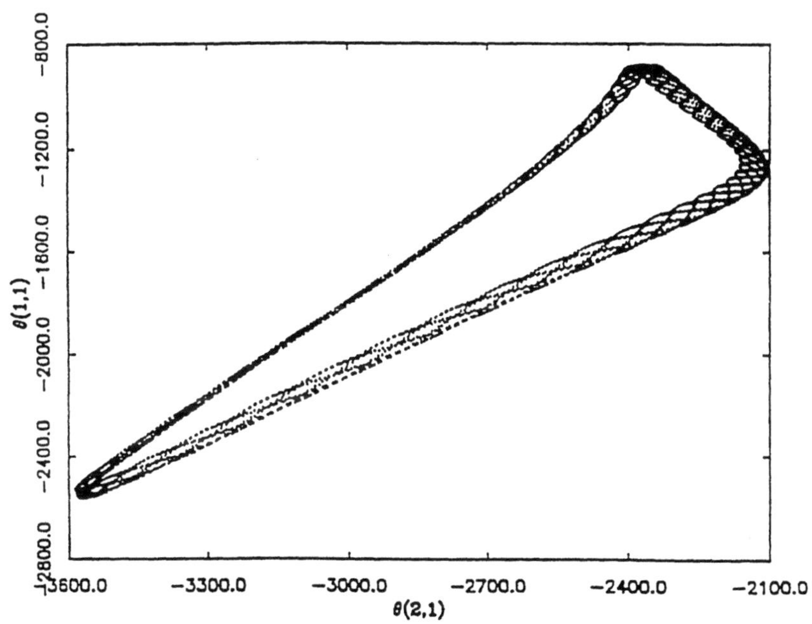

The study of thermal convection at infinite Prandtl number may be the best way to relate low-order chaotic systems to high-order turbulent systems. At infinite Prandtl number the viscous damping occurs in the same wavelength as the buoyancy input of energy. The 40-mode expansion considered here approaches the resolution of published numerical calculations that exhibit fluid turbulence [6-11].

3. Chaotic Tectonics. Displacements on preexisting faults are the primary mechanism for deformation of the brittle upper crust. These displacements occur during earthquakes. Faults exhibit stick-slip behavior. The stick-slip behavior is explained by velocity-weakening friction on the faults.

The behavior of faults can be modeled using a spring-block model [2]. Others have applied this model to simulate earthquake features including foreshocks, mainshocks, and aftershocks, pre- and post-seismic slip and earthquake statistics [12-17]. Most studies involve numerical simulations of multi-block models with spatial variations and a variety of friction laws.

There are at least three sources of complexity which are important to seismicity: nonlinear friction, fault heterogeneities, and fault interactions. Dynamic instabilities associated with complicated friction laws are well known from studies using single-block models [18-22]. Under certain conditions, the motion of a system subject to rate and state dependent friction laws demonstrates unstable and chaotic bounded oscillations [21]. The contribution of fault dynamics was investigated by Nussbaum and Ruina [23]. They used a two-block model with spatial symmetry to simulate a strike-slip fault. They found spatially asymmetric events but no chaotic behavior was observed, i.e. all solutions went to either equilibrium points or period-two limit cycles. Perez Pascual and Lomnitz-Adler [24] have examined the related problem of coupled relaxation oscillators.

Figure 7 Illustration of the slider block model. Two blocks are connected to a constant velocity driver and to each other with springs. Stick-slip behavior occurs if the surface friction is velocity weakening. If the blocks are asymmetric, for example different friction forces, chaotic behavior can occur.

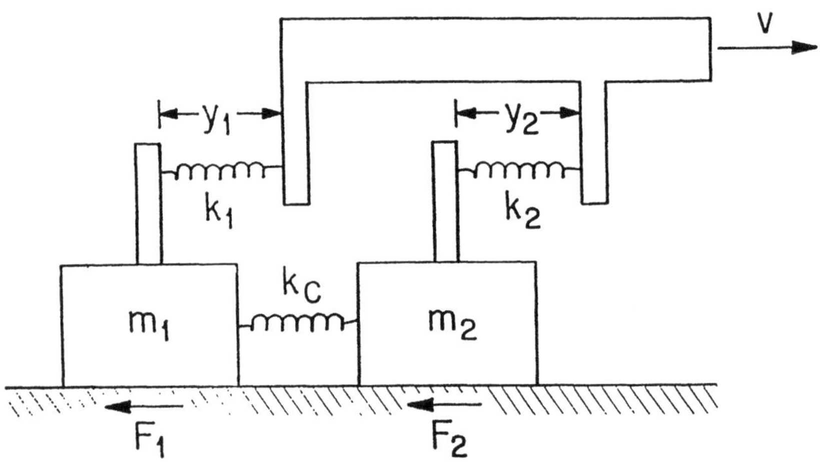

Slider-block models are low-dimension analogs to tectonic deformation. We consider the two-block system illustrated in Figure 7. Two sliding blocks are coupled to each other and to a constant velocity driver by linear elastic springs of stiffness

k_c, k_1, and k_2 respectively. Other model parameters include the block masses m_1 and m_2, and the frictional forces F_1 and F_2. The position coordinates for the blocks, referred to the constant velocity driver, are y_1 and y_2. The two-block model is a fourth order system, y_1, y_2, v_1, v_2. It is the simplest system for which chaotic behavior might be expected.

The equations of motion for this two-block system are

$$m_1\ddot{y}_1 + (k_1 + k_c)y_1 - k_c y_2 = F_1 \tag{3.1}$$

$$m_2\ddot{y}_2 + (k_2 + k_c)y_2 - k_c y_1 = F_2 \tag{3.2}$$

Several additional assumptions are made to simplify the model. First, symmetry of the model is assumed except for the frictional forces. Accordingly, we take $m_1 = m_2 = m$, and $k_1 = k_2 = k$. In addition, the coupling spring constant is defined in terms of $k : k_c = \alpha k$. A second assumption is that the loading velocity of the driver is sufficiently slow that we may consider it to be zero during the sliding of the blocks. The ratio of the frictional forces of the two blocks β is introduced such that $F_2 = \beta F_1 = \beta F$. With these assumptions incorporated into the model, (1) and (2) can be written

$$m\ddot{y}_1 + (1 + \alpha)ky_1 - \alpha k y_2 = F \tag{3.3}$$

$$m\ddot{y}_2 + (1 + \alpha)ky_2 - \alpha k y_1 = \beta F \tag{3.4}$$

In this study, we choose a velocity-weakening friction law of the form

$$F = \frac{F_0}{1 + |\dot{y}/v_f|} \tag{3.5}$$

The frictional forces decrease monotonically as velocity increases. In essence, this is a coupled oscillator system with damping from the velocity-weakening friction. Velocity-weakening phenomena have been observed in laboratory experiments [25].

It is convenient to introduce the nondimensional variables $Y_i = y_i k/F_0 (i = 1, 2)$, and a dimensionless time $\tau = t\sqrt{k}/m$. The failure criteria on Y_1 and Y_2 are now given by

$$Y_1 + \alpha(Y_1 - Y_2) = 1 \tag{3.6}$$

$$Y_2 + \alpha(Y_2 - Y_1) = \beta \tag{3.7}$$

and the equations of motion during slip are

$$\ddot{Y}_1 + Y_1 + \alpha(Y_1 - Y_2) = \frac{1}{1 + \gamma|\dot{Y}_1|} \tag{3.8}$$

$$\ddot{Y}_2 + Y_2 + \alpha(Y_2 - Y_1) = \frac{1}{1+\gamma|\dot{Y}_2|} \tag{3.9}$$

where $\gamma = F_0/v_f\sqrt{mk}$ is a measure of the amount of velocity weakening. These equations are numerically integrated using a fourth order Runge–Kutta scheme. The dynamic evolution of the system is recorded only after the system reaches a statistically steady state. Results illustrating the chaotic behavior of a spring-block system using static-dynamic friction have been given by Huang and Turcotte [26].

Figure 8 Phase portrait for a symmetrical two-block system. The phase plane coordinates are the position coordinates of the two blocks. Two examples are shown of the infinite number of allowed cycles. The particular cycle found depends upon the initial conditions for the run.

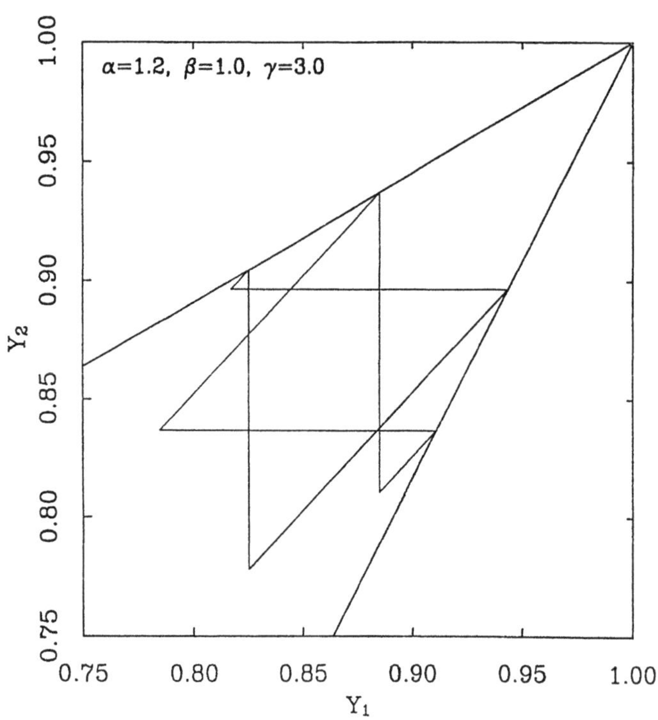

We first consider the behavior of the system when it is symmetric $\beta = 1$. After an initial transient the system exhibits cyclic behavior. Two examples are illustrated in Figure 8. The diagonal lines represent the failure envelopes for the two blocks given by (3.6) and (3.7). The diagonal lines with unit slope represent the strain accumulation phases, the blocks stick and the springs extend at equal rates. Vertical lines represent the slip of block 2 (Y_1 constant) and horizontal lines represent the slip of block 1 (Y_2 constant). Slip occurs when the diagonal accumulation lines intersect the failure envelope. Slip continues until the block comes to rest, it then sticks. For the parameters chosen the blocks slip alternately in a cyclic manner. There are an infinite number of allowed solutions in which the slips are equal but the lengths of the strain accumulation phases vary. The particular limiting solution depends on the initial conditions.

When the system is asymmetric, i.e. the ratio of frictional forces β is not equal to one, the dynamical picture changes completely. An example of a chaotic orbit with 100 events plotted in the phase plane (Y_1, Y_2) is shown in Figure 9. We note that there are single-block failures at some times, while at other times both blocks slide together. If an initial slip event from one failure line encounters the second failure line the second block also slips. Failure curves outside the failure envelope represent a multiple event.

Figure 9 Phase portrait for an asymmetric two-block system. For these parameter values the evolution of the system is chaotic.

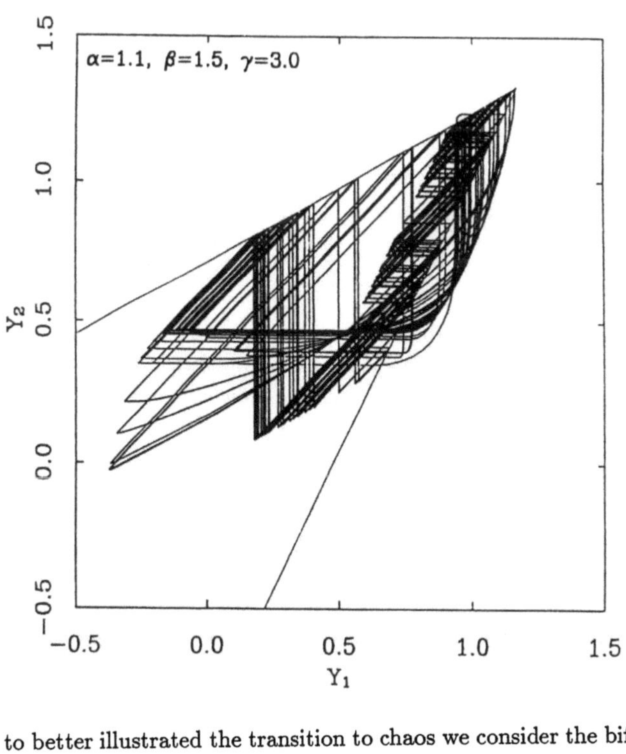

In order to better illustrated the transition to chaos we consider the bifurcation diagram shown in Figure 10. For this example the velocity softening parameter γ is varied while holding the parameters α and β fixed. Similar results are obtained if the other parameters are varied. In Figure 10 the values $Y_2 - Y_1$ after a slip event are given for various values of γ. There appears to be a window of limit cycle behavior near $\gamma = 3$. Before examining this window in some detail we will consider the variation of the Lyapunov exponent under these conditions.

Figure 10 Bifurcation diagram giving the values of $Y_2 - Y_1$ after a large number of slip events for various values of the velocity-weakening parameter γ.

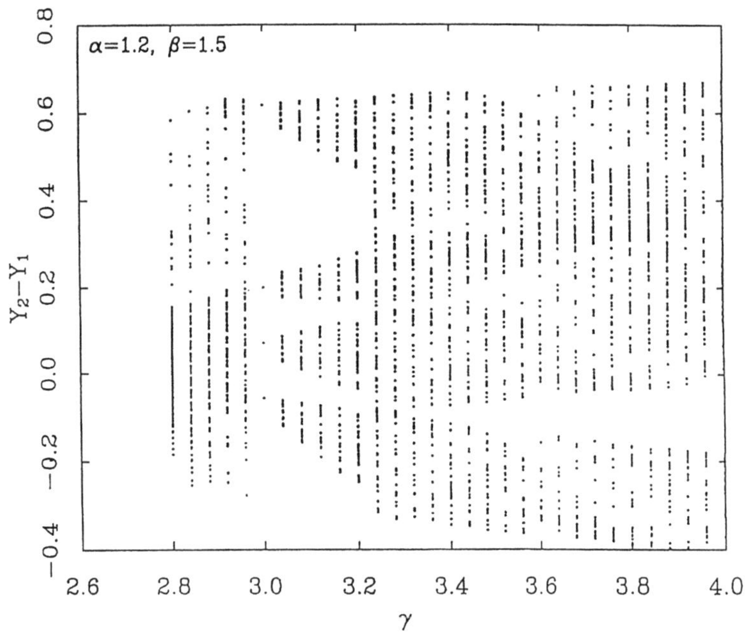

In a chaotic system, orbits that arise from nearby initial conditions diverge exponentially on the average. Lyapunov exponents measure the rates of divergence or convergence. Positive Lyapunov exponents indicate chaos while negative exponents indicate stability.

For a one-dimensional iteration map of known functional form $x_{n+1} = f'(x_n)$, the Lyapunov exponent exponent λ is given by

$$\lambda = \lim_{n \to \infty} \frac{1}{n} \sum_{i=0}^{n-1} \log_2 |f'(x_i)| \tag{3.10}$$

A nonzero λ means that a pair of infinitesimally separated points initially a distance dx_0 apart diverge or converge exponentially on the average according to $dx_n = dx_0 2^{n\lambda}$. The derivation of (3.10) is straight forward if we notice the fact that $dx_{n+1} = f'(x_n)dx_n$, i.e. the derivative $f'(x_n)$ represents relative local stretching or contraction for each iteration. Then we have

$$2^{n\lambda} = \frac{dx_n}{dx_0} = \frac{dx_n}{dx_{n-1}} \cdots \frac{dx_1}{dx_0} = |f'(x_{n-1})|\,|f'(x_{n-2})| \cdots |f'(x_0)| \tag{3.11}$$

Values of the Lyapunov exponent λ corresponding to the bifurcation diagram, Figure 10, are given in Figure 11. The window of stable behavior near $\gamma = 3$ is clearly shown.

Figure 11 Dependence of the Lyapunov exponent λ on the velocity-weakening parameter γ for the same parameter range given in Figure 10.

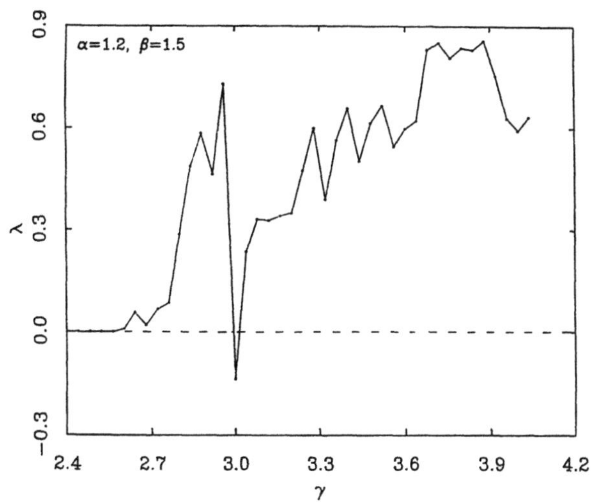

Figure 12 Details of the bifurcation diagram given in Figure 10 in the vicinity of the stable window at $\gamma = 3$.

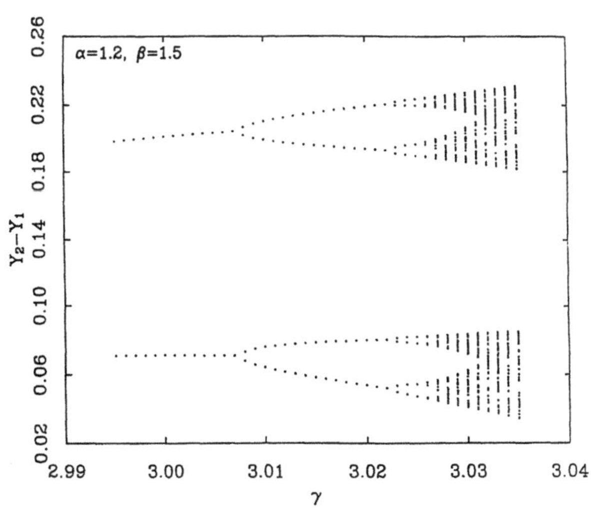

In order to examine the route to chaos in greater detail we illustrate a portion of the bifurcation diagram given in Figure 10 in the vicinity of $\gamma = 3$, this is done in Figure 12. This is a classic example of the period doubling route to chaos. Values of the Lyapunov exponent in this region are given in Figure 13. The phase portrait for the limit cycle at $\gamma = 3.000$ is given in Figure 14. Four slip events occur in this limit cycle. Two events involve the weaker block, one larger event involves the stronger block, and one large event involves both blocks. All initial conditions for the system evolve to this limit cycle for these parameter values.

At $\gamma_1 = 3.00700$ the system experiences a flip bifurcation and the period doubles. This period doubling is clearly illustrated in Figure 12 and the Lyapunov exponent goes to zero at this value as expected. We find that we can determine the bifurcation point more accurately from the calculation of the Lyapunov exponent than from the bifurcation diagram. The phase portrait for the period-doubled limit cycle at $\gamma = 3.008$ is given in Figure 15.

Figure 13 Values of the Lyapunov exponent from Figure 11 in the vicinity of the stable window at $\gamma = 3$.

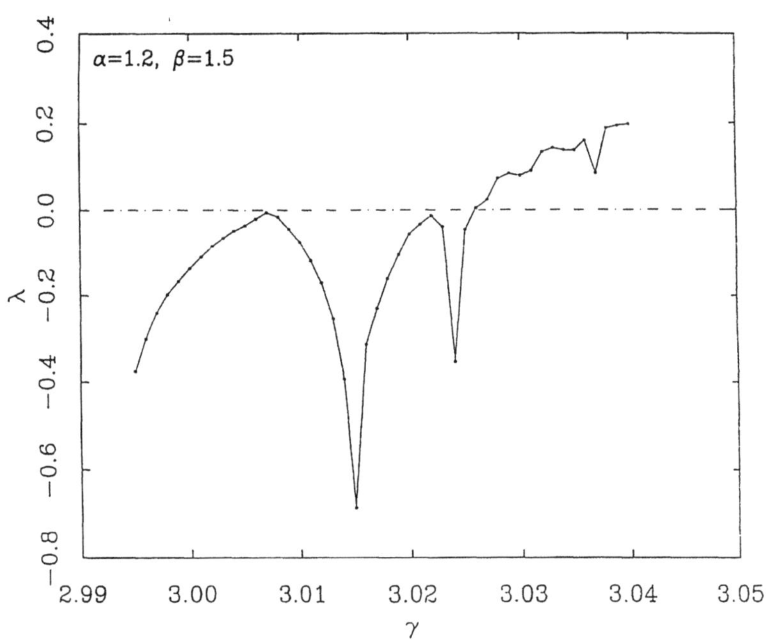

Figure 14 Phase portrait for the four event limit cycle at $\gamma = 3.000$.

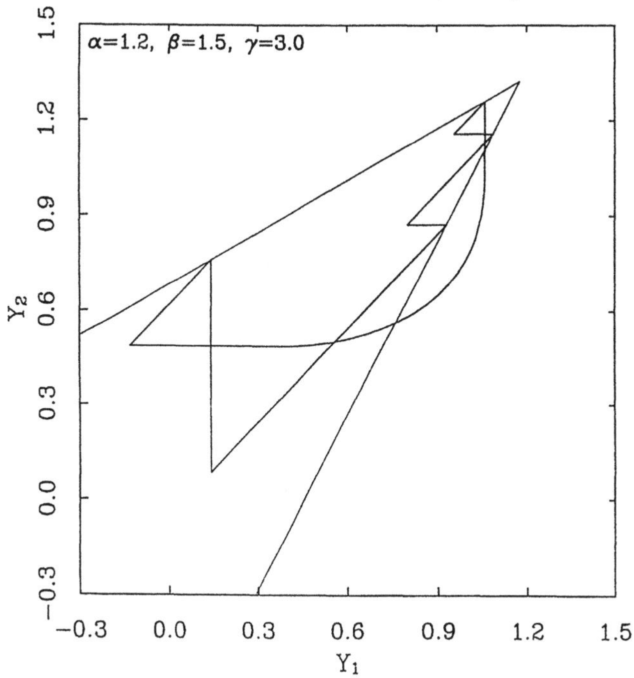

Figure 15 Phase portrait for the eight event period-doubled limit cycle at $\gamma = 3.008$.

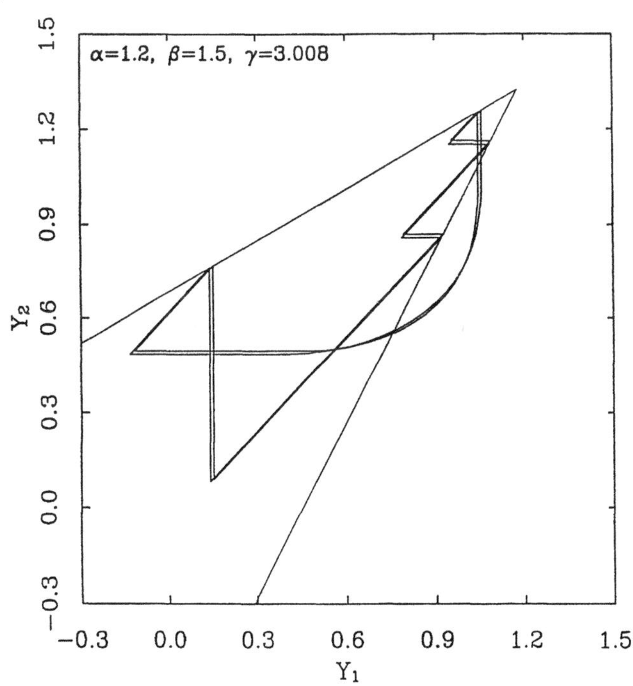

In order to determine the further examples of period-aoubling the detailed behavior of the Lyapunov exponent in the vicinity of $\gamma = 3.025$ is given in Figure 16. The subsequent examples of period doubling occur at $\gamma_2 = 3.02240, \gamma_3 = 3.02590$, and $\gamma_4 = 3.02654$. It is of interest to apply the Feigenbaum criteria for the period-doubling route to chaos for these results. This criteria is given by

$$\gamma_c = \frac{F\gamma_{k+1} - \gamma_k}{F - 1} \tag{3.12}$$

where the Feigenbaum number $F = 4.669202$ and γ_c is the onset of chaotic behavior. Taking $\gamma_2 = 3.02240$ and $\gamma_1 = 3.00700$ we find $\gamma_{c1} = 3.02660$. Taking $\gamma_3 = 3.02590$ and $\gamma_2 = 3.02240$ we find $\gamma_{c2} = 3.02685$. Taking $\gamma_4 = 3.02654$ and $\gamma_3 = 3.02590$ we find $\gamma_{c3} = 3.02671$. Using the Fergenbaum criteria the predicted transition to chaos is in excellent agreement with the first positive Lyapunov exponent in Figure 16.

4. Conclusions. We have examined two problems in solid earth dynamics. We have considered truncated expansions of the equations governing thermal convection at infinite Prandtl number. Our studies indicate chaotic behavior and are a first attempt to bridge the gap between low-order truncations and numerical calculations indicating "turbulent" solutions.

Our studies of slider-block models show classical chaotic behavior. It remains for future studies to develop the relations between these analog models and distributed seismisity in the crust.

Figure 16 Values of the Lyapunov exponent from Figure 11 in the vicinity of the onset of chaotic behavior at $\gamma = 3.0267$.

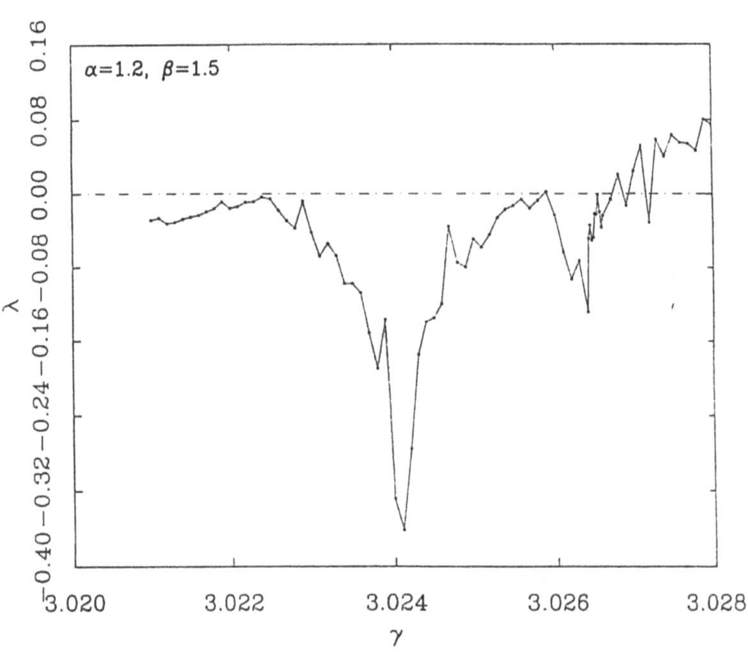

REFERENCES

[1] P. Bak, C. Tang and K. Wiesenfeld, *Self-organized criticality*, Phys. Rev., A38 (1988), pp. 364–374.

[2] R. Burridge and L. Knopoff, *Model and theoretical seismicity*, Seis. Soc. Am. Bull, 57 (1967), pp. 341–371.

[3] B. Saltzman, *Finite amplitude free convection as an initial value problem*, J. Atmos. Sci. 19 (1962), pp. 329–341.

[4] E.N. Lorenz, *Deterministic nonperiodic flow*, J. Atmos. Sci. 20 (1963), pp. 130–141.

[5] C.A. Stewart and D.L. Turcotte, *The route to chaos in thermal convection at infinite Prandtl number 1. Some trajectories and bifurcations*, J. Geophys. Res. 94 (1989), pp. 13,707–13,717.

[6] U.R. Christensen, *Time-dependent convection in elongated Rayleigh-Benard cells*, Geophys. Res. Let. 14 (1987), pp. 220–223.

[7] T.B. Lennie, D.P. McKenzie, D.R. Moore and N.O. Weiss, *The breakdown of steady convection*, J. Fluid Mech. 188 (1988), pp. 47–85.

[8] U. Hansen and A. Ebel, *Time-dependent thermal convection for multiscale flow in the earth's mantle*, Geophys. J. 94 (1988), pp. 181–191.

[9] A.P. Vincent and D.A. Yuen, *Thermal attractor in chaotic convection with high-Prandtl-number fluids*, Phys. Rev. A38 (1988), pp. 328–334.

[10] ———, *Multimode mean-field filtering experiments for chaotic thermal convection*, Geophys. Astrophys. Fluid Dyn. 47 (1989), pp. 131–156.

[11] U.R. Christensen and D.A. Yuen, *Time-dependent convection with non-Newtonian viscosity*, J. Geophys. Res. 94 (1989), pp. 814–820.

[12] J.H. Dieterich, *Time-dependent friction as a possible mechanism for aftershocks*, J. Geophys. Res. 77 (1972), pp. 3371–3781.

[13] J.B. Rundle and D.D. Jackson, *Numerical simulation of earthquake sequences*, Seis. Am. Bull. 67 (1977), pp. 1363–1377.

[14] S. Cohen, *Computer simulation of earthquakes*, J. Geophys. Res. 82 (1977), pp. 3781–3796.

[15] T. Cao and K. Aki, *Seismicity with mass-spring model and a displacement hardening-softening friction law*, Pure Ap. Geophys. 122 (1984), pp. 10–24.

[16] ———, *Seismicity simulation with rate- and state-dependent friction law*, Pure Ap. Geophys. 124 (1986), pp. 487–513.

[17] J.M. Carlson and J.S. Langer, *Mechanical model of an earthquake fault*, Phys. Rev. A40 (1989), pp. 6470–6484.

[18] J.D. Byerlee, *The mechanics of stick-slip*, Tectonophys. 9 (1970), pp. 575–486.

[19] J.H. Dieterich, *Constitutive properties of faults with simulated gouge*, in Mechanical Behavior of Crustal Rocks, N.L. Carter, M. Friedman, J.M. Morgan, and D.W. Steans, eds., American Geophysical Union, Geophys. Mono. 24 (1981), pp. 103–120.

[20] A. Ruina, *Slip instability and state variable friction laws*, J. Geophys. Res. 88 (1983), pp. 10,359–10.370.

[21] J.C. Gu, J.R. Rice, A.L. Ruina and S.T. Tse, *Slip motion and stability of a single degree of freedom elastic system with rate and state dependent friction*, J. Mech. Phys. Solids 32 (1984), 167–196.

[22] J.R. Rice and S.T. Tse, *Dynamic motion of a single degree of freedom system following a rate and state dependent friction law*, J. Geophys. Res. 91 (1986), pp. 521–530.

[23] A. Nussbaum and A. Ruina, *A two degree-of-freedom earthquake model with static/dynamic friction*, Pure Ap. Geophys. 124 (1987), pp. 629–656.

[24] R. Perez–Pascual and J. Lomnitz–Adler, *Coupled relaxation oscillators and circle maps*, Physica D30 (1988), p. 61–82.

[25] T.E. Tullis and J.D. Weeks, *Constitutive behavior and stability of frictional sliding of granite*, Pure Ap. Geophys. 124 (1986), pp. 383–414.

[26] J. Huang and D.L. Turcotte, *Are earthquakes an example of deterministic chaos*, Geophys. Res. Let. 17 (1990), pp. 223–226.

MANTLE PHASE TRANSITIONS, LAYERED CHAOTIC CONVECTION, AND THE VISCOSITY OF THE DEEP MANTLE

W.R. PELTIER* AND L.P. SOLHEIM*

Abstract. Recent high pressure measurements of the phase diagram of mantle material (Ito and Takahashi, 1989) have demonstrated rather clearly that the phase-loop for the divariant Spinel-post Spinel transition is extremely narrow in pressure and that the Clapeyron slope of this transition is near $-2.8 MPa/^{o}K$. Extremely high resolution calculations of the nature of convective mixing in a flow in which both this and the Olivine-Spinel transition are present, demonstrate that the endothermic transition suffices to episodically enforce a high degree of layering upon the circulation, since mass flux through the phase boundary is on occasion strongly inhibited. The presence of the internal thermal boundary layer that develops across the phase boundary when such layering is present has important implications for the expected radial variation of mantle viscosity. We invoke recent inferences of the radial variation of mantle viscosity, based upon mantle tomography and non-hydrostatic geoid anomalies, to suggest that the observed variation of this transport property is in accord with expectations based upon the layered convection scenario. Inferences of mantle viscosity based upon glacial rebound data are also in accord with this view, since these data also require that the increase of viscosity across the 670 km discontinuity is small.

1. Introduction. The issue of the radial variation of the viscosity of the Earth's mantle is very clearly connected to the issue of the style of the convective circulation that fills this region of the planetary interior. Because viscosity is a strong function of temperature and because temperature, in turn, is a strong function of depth, we expect viscosity to vary strongly with radius. We do not find it surprising, therefore, that the surface lithosphere which comprises the cold upper thermal boundary layer of the convective circulation is a region of extremely high viscosity. Similarly, if the D'' layer at the base of the mantle were to comprise the hot lower thermal boundary layer of a mantle general circulation that was to some significant degree heated from below, then we could be reasonably sure that this region would be one of anomalously low viscosity (e.g. Peltier, 1980). The issue as to whether the style of mantle mixing is "whole mantle" or "layered" is synonymous with the issue as to whether at some depth in the mantle there might exist an interior thermal boundary layer across which the radial heat transport is entirely diffusive. Various arguments have been advanced in the recent literature in support of both of these scenarios (e.g. Peltier and Jarvis, 1982; Jeanloz and Morris, 1987). A detailed recent review of this issue in the form of a series of extended essays by different authors from a variety of different perspectives (Peltier [ed.], 1989) ended with the conclusion that none of the various arguments that had been advanced at that time could be regarded as definitive.

The history of this issue is of course "long" in the geophysical literature, and dates from the earliest writings from the "mobilist" perspective. A line of argument that was seen by some to be especially telling, as of a decade ago, was that

*Department of Physics, University of Toronto, Toronto, Ontario, Canada M5S 1A7

based upon the trace element geochemistry of erupted basalts (O'Nions et al., 1977; Wasserburg and DePaulo, 1979; Allegre, 1982). On the basis of these data, from the Sm/Nd and Rb/Sr isotopic systems, it was suggested that the observations showed a clear indication of the existence of two mantle reservoirs, one which was essentially undifferentiated and one which had been differentiated to produce the crust of the present day continents. Because the ratio of the volumes of these two reservoirs was calculated to be close to the ratio of the volume of the upper mantle to the lower mantle, these authors concluded that the convective circulation must be layered with the interface separating the two reservoirs coincident with the seismic discontinuity at 670 km depth. With the increasing number of isotopic measurements that came available in the subsequent decade, the early clarity of the scenario implied by these data became progressively more obscured. The most recent detailed analysis of this argument (Hart and Zindler, 1989) concludes that the error in the inference of the volumes of the "pristine" and "depleted" reservoirs is of the order of 50%, so that no particularly compelling argument is possible on the basis of these data alone.

Earlier arguments in favour of the layered convection scenario focused upon the observation that seismicity ceased at a depth near 670 km (e.g. Isaaks and Molnar, 1971). If the cold downgoing slab did not penetrate below this depth this observation would obviously be simply explained. One of the earliest explanations as to how mixing across the 670 km horizon might be impeded invoked the notion that this boundary was due to a chemical discontinuity. Richter and Johnson (1971), for example, showed that even a very modest increase in mean atomic weight across this level would suffice to strongly inhibit (in fact to essentially entirely prevent, at least in small amplitude theory) radial mass flux. D.L. Anderson is one who has long argued on the basis of the observed radial variation of elastic properties with depth that the 670 km boundary was to a significant degree due to a chemical discontinuity. An additional perspective on the basis of which this interpretation of the 670 km horizon has been advocated has been based upon the observed sharpness of the increase of the body wave velocities that occurs across it. Jeanloz (e.g. 1981, 1989), for example, has argued that the boundary was too sharp to be caused by an equilibrium phase transition and so must be to a significant degree chemical. This argument has also been strongly undermined by the recent publication of the very important paper by Ito and Takahashi (1989) on the high pressure phase diagram of the dominant mineralogy of the mantle. Using the large volume high pressure apparatus they have demonstrated not only that the phase transition from Spinel to a mixture of Perovskite and Magnesiowüstite takes place at a pressure corresponding to a depth of 670 km but also that the phase loop for this divariant phase transition is extremely thin, sufficiently so to explain, of itself, the strong reflections of elastic waves that occur at this depth. There can be no dispute therefore that the 670 km discontinuity is caused by this equilibrium phase transition. Whether a chemical discontinuity might also exist is of course not resolved by these measurements. Rather, what is demonstrated is that such contrast is not in fact required by the data that has usually been invoked to support its existence.

In seeking to further understand the basis on which one might be obliged to conclude either in favour of the "whole mantle" or "layered" hypotheses it seems to us that it is important to understand dynamically how layering might have come to be established in the course of the thermal evolution of the planet since its initial accretion out of the primitive solar nebula (e.g. Safronov, 1972). Given that the planet most probably formed sufficiently hot to have differentiated very early in its evolution, it is hard for us to imagine how layering of the mantle convection process could have been established by a chemical discontinuity. In this hot initial state, mantle mixing would have been much more vigorous than today and therefore, in the absence of a specific non-chemical process that would enforce layering, the establishment of any degree of chemical layering would have been highly improbable. When viewed from the perspective of the thermal history it therefore seems clear that in order for layering of the circulation to develop an isochemical mechanism is required. It is equally clear that the only such mechanism available is one that would be associated with the mantle phase transitions themselves.

In order to address the issue as to whether mantle phase transitions are capable of enforcing a layered style of mixing it is clearly necessary to solve the dynamical equations for thermal convection, with phase boundaries present, in the regime of parameter space in which the earth lives. It was concluded in Peltier (1989), repeating arguments previously presented, that the mantle general circulation must be assumed to be whole mantle in style unless it can be shown that dynamical effects associated with the Spinel-post Spinel transition itself could be shown to be strongly inhibiting of the radial flux of mass. If such inhibition were to occur as a consequence of this phase transition then a chemical discontinuity could become established as the earth continued to evolve, a consequence of which would most probably be that the circulation would become even more strongly layered. In this layered convection scenario a chemical contrast across 670 km depth would therefore be a consequence rather than the cause of the layering of the circulation.

Our intention in this paper is to explore the plausibility of the above described scenario. In the next section we will describe a sequence of very high resolution numerical experiments that have recently been performed to investigate the impact that mantle phase transitions might be expected to have upon the radial structure of the circulation, using the best available laboratory data to constrain the properties of the phase boundaries. As we will show, when the Rayleigh number of the flow is fixed to a value near that expected to be characteristic of the present day earth then the circulation is, episodically, quite strongly layered. These calculations demonstrate that an expected consequence of layering is the establishment of a thermal boundary layer at 670 km depth across which the temperature increases by about $300°K$. In sections 3 and 4 we proceed to investigate the reasonableness of this scenario from the perspective of the radial profile of mantle viscosity that has been inferred on the basis of two different lines of argument. These inferences are shown to be entirely consistent with the scenario in which an internal thermal boundary layer exists. Our conclusions are summarized in section 5.

2. An axi-symmetric numerical model of mantle convection with phase transitions. The very high resolution analyses of the mantle convection process that we will describe here have been performed with a modified version of the model recently described in detail in Solheim and Peltier (1990), the main alterations of the dynamical system being those required to incorporate the full dynamical impact of multiple equilibrium phase transformations and the introduction of a new depth dependence for the coefficient of thermal expansion α. The necessity of the latter adjustment is clear on the basis of the recent experimental measurements of Chopelas and Boehler (1989) who demonstrate that the following empirical expression provides an excellent fit to the observations:

$$\frac{\alpha}{\alpha_o} = \left(\frac{\rho}{\rho_o}\right)^{-\delta_T} \tag{1}$$

in which $\delta_T \approx 5 (4 \leq \delta_T \leq 6)$ is the Anderson-Gruneissen parameter. This very useful new constraint on an earth property that is extremely important to our understanding of the thermal convection process has been further discussed by Anderson et al. (1990) and Reynard and Price (1990).

The axi-symmetric convection model that we have developed to analyse the influence of phase transitions and other physical effects is governed by the following non-dimensional equations that have been written in streamfunction – vorticity form:

$$\eta\frac{\partial^2 w}{\partial r^2} + \left(2\frac{\partial\eta}{\partial r} + \frac{g\eta\rho_r}{K_T}\right)\frac{\partial w}{\partial r} + \frac{\eta}{r^2}\frac{\partial^2 w}{\partial\theta^2} - \frac{\eta\cot\theta}{r^2}\frac{\partial w}{\partial\theta}$$

$$+\left(\frac{\partial^2\eta}{\partial r^2} + \frac{g\rho_r}{K_T}\frac{\partial\eta}{\partial r}\right) w = \rho_r\alpha g\sin\theta\left\{\frac{\partial A}{\partial\theta} + \frac{2h_1\Gamma_1(1-\Gamma_1)}{\alpha_o T_c}\left(\frac{\tau_o\ell_1}{C_p} - \frac{\Delta_1}{\rho_r\alpha}\right)\frac{\partial r_{p_1}}{\partial\theta}\right.$$

(2a)

$$\left.+\frac{2h_2\Gamma_2(1-\Gamma_2)}{\alpha_o T_c}\left(\frac{\tau_o\ell_2}{C_p} - \frac{\Delta_2}{\rho_r\alpha}\right)\frac{\partial r_{p_2}}{\partial\theta}\right\} + \frac{4g\eta}{3K_T\rho_r}\frac{\partial\rho_r}{\partial r}\left(\frac{1}{r^2}\frac{\partial^2\psi}{\partial\theta^2} - \frac{\cot\theta}{r^2}\frac{\partial\psi}{\partial\theta}\right),$$

$$\frac{\partial^2\psi}{\partial r^2} - \frac{1}{\rho_r}\frac{\partial\rho_r}{\partial r}\frac{\partial\psi}{\partial r} + \frac{1}{r^2}\frac{\partial^2\psi}{\partial\theta^2} - \frac{\cot\theta}{r^2}\frac{\partial\psi}{\partial\theta} = -w\rho_r, \tag{2b}$$

$$\frac{\partial A}{\partial t} = \frac{\kappa}{Ra}D^2 A - \frac{J(A,\psi)}{\rho_r r^2\sin\theta} + \frac{\mu}{C_p Ra} + \frac{\tau_o\Phi}{\rho_r C_p} - \frac{\tau(A+T_o)}{\rho_r r^2\sin\theta}\frac{\partial\psi}{\partial\theta} + \zeta, \tag{2c}$$

(2d)

$$\beta_1\left\{A + T_o + \frac{\tau_o}{2\alpha_o T_c C_p}[\ell_1 + \ell_2 + \ell_2\tanh((r_{p_2} - r_{p_1})h_2)]\right\} + p_o - p_r = 0,$$

(2e)

$$\beta_2\left\{A + T_o + \frac{\tau_o}{2\alpha_o T_c C_p}[\ell_1 + \ell_2 + \ell_1\tanh((r_{p_1} - r_{p_2})h_1)]\right\} + p_o - p_r = 0.$$

In the above equations w is the transformed vorticity ($w = \omega + \sin\theta$), ψ is the streamfunction, and $J(.,.) = \partial(.,.)/\partial(r,\theta)$ the Jacobean operator. The additional operator D^2 in the system (2) is:

$$D^2 = \frac{\partial^2}{\partial r^2} + \left(\frac{2}{r} + \frac{1}{k}\frac{\partial k}{\partial r}\right)\frac{\partial}{\partial r} + \frac{1}{r^2}\frac{\partial^2}{\partial\theta^2} + \frac{\cot\theta}{r^2}\frac{\partial}{\partial\theta}.$$

Our method of incorporating the dynamical impact of phase transitions with Clapeyron slopes and latent heats of reaction β_i and $\ell_i (i = 1, 2)$ and density jumps $\Delta_i = \rho_{i+1} - \rho_i$, is through the introduction of the transformed temperature A, defined as:

(3)
$$A = T - \frac{\rho_o}{\alpha_o T_c C_p} (\ell_1 \Gamma_1 + \ell_2 \Gamma_2),$$

in which the Γ_i are phase density functions defined as:

(4)
$$\Gamma_i = \frac{1}{2} \left\{ \frac{1}{2} + \tanh \left[(r_{p_i}(\theta, t) - r) \, h_i \right] \right\},$$

which represent the fraction of the $(i+1)^{\text{st}}$ phase that exists at position (r, θ) and time t. The parameter h_i determines the length scale over which the ith phase change takes place and should correspond, so far as is numerically possible, to the actual thicknesses of the divariant phase loops of the Olivine-Spinel and Spinel-post Spinel transitions.

The remaining variables that appear in the equations for the model system (2) include the $r_{p_i}(\theta, t)$ which represent the radial location of the ith phase boundary at latitude θ and time t. Further parameters include g the (radially varying – PREM) gravitational acceleration; η the molecular viscosity (we will present results only for $\eta(r)$= constant, although the system (2) includes the terms that arise when η is depth dependent); K_T the isothermal bulk modulus; ρ_r the depth dependent background density in the anelastic model (including the density jumps due to the phase transitions); α the coefficient of thermal expansion; κ the thermal diffusivity; k the thermal conductivity; C_p the heat capacity at constant pressure; and $\tau = g\alpha d / C_p$ the dissipation number. The main control variable in the model is the Rayleigh number $Ra = \rho_o g_o \alpha_o T_c d^3 / \kappa_o \eta_o$ while the nondimensional internal heating parameter is $\mu = \rho_o \chi d^2 / k_o T_c$ in which χ is the (assumed constant) internal heating rate per unit mass. The temperature scale $T_c = T_{cmb} - T_s$ and the viscous dissipation of motion into heat in this model is represented by the function Φ. The final field ζ that appears in (2) arises straightforwardly due to the transformation (3) from the absolute temperature T to the new variable A. No useful purpose will be served here by reproducing the rather complex mathematical expressions for either Φ or ζ. The system (2) constitutes a set of 5 simultaneous partial differential equation in the 5 unknown fields $\omega, \psi, A, r_{p_1}, r_{p_2}$. We have found that the transformation from absolute temperature T to the new variable A helps considerably in numerically stabilizing the computation of the crucial impact of phase boundary deflection.

The main feature of the dynamical system (2) upon which we will focus attention here is described by the phase function Γ_i as mentioned above. In these phase functions the parameters h_i determine the thicknesses of the divariant phase loops across which the ith phase change occurs. For the Olivine-Spinel transition the loop thickness is approximately 10 km (Akaogi and Ito, 1989; Katsura and Ito, 1989) whereas for the Spinel-Post Spinel transition it is closer to 5 km (Ito and Takahashi 1989). Clearly if we are to model the dynamical impact of these phase transitions accurately we shall have to ensure either that the phase loop thicknesses employed in the numerical computations equal these experimentally determined thicknesses or that the results are insensitive to the loop thicknesses employed. One outcome of the analyses to be reported here is the demonstration that failure to employ sufficiently small loop thicknesses for these dynamical calculations may lead to underestimates of the ability of the Spinel-Post Spinel transition to inhibit convection across it.

It is also important, of course, that the Clapeyron slopes assigned to these transitions be equal to the experimentally determined values. Based upon the laboratory experiments of Akaogi and Ito (1989) and Katsura and Ito (1989) we will therefore assign a Clapeyron slope for the exothermic Olivine-Spinel transition of $2MPa/°K$, while on the basis of the work of Ito and Takahashi (1989) we will take the Clapeyron slope for the endothermic Spinel-post Spinel transition to be $-2.8MPa/°K$. Our intention will be to demonstrate that with the loop thicknesses reasonably set and the Claperyron slopes equal to the above values, then convection through the 670 km phase discontinuity is strongly inhibited.

In fixing the Clapeyron slopes to these representative values we are conscious that our choice of $+2MPa/°K$ for the $\alpha \rightarrow \beta$ transition might be seen as rather low. However, for high temperatures ($> 1600°C$, see Figure 10 of Akaogi et al. 1989) the slope of the boundary of the mixed phase region on the low pressure side is near 2.4 $MPa/°K$ whereas on the high pressure side it is $< 1.7MPa/°K$. Our choice of $2MPa/°K$ is based upon the average of these two values. For lower temperatures ($< 1600°C$) the slope of the boundary of the mixed phase region is the same on the high pressure side but on the low pressure side it rises to $+4MPa/°K$. Averaging the slopes of the two boundaries of the mixed phase region we might therefore suppose that at low temperatures a representative number would be $(1.7+4.0)/2 = +2.85MPa/°K$ whereas at high temperature the equivalent number would be $+2.0MPa/°K$ as mentioned above. The average of these values ($+2.4MPa/°K$) would then appear to be the best single estimate possible for "the" Clapeyron slope of the (actually divariant) Olivine $\rightarrow$ Spinel ($\alpha \rightarrow \beta$) mantle phase transformation. This is considerably smaller than the value of $+4MPa/°K$ that has been assumed in much of the current literature on the influence of phase transitions on the mantle convection process.

Previous work on the ability of mantle phase transitions to inhibit convection through them has been considerably less than definitive. The earliest work based upon linear stability theory suggested that the Spinel-post Spinel transition should inhibit convection (Busse and Schubert, 1971; Schubert and Turcotte, 1971; Schubert et al., 1975). However, as pointed out in Peltier (1985) these analyses were improbably scaled. With appropriate rescaling the small amplitude analysis demonstrates that the impact of latent heat release (which is destabilizing for the Spinel-post Spinel transition) dominates the impact of the buoyancy force associated with the phase boundary deflection (which is stabilizing for this endothermic horizon). As the Rayleigh number increases, however, the magnitude of the former effect should increase like $Ra^{2/3}$ whereas the magnitude of the latter effect should increase like Ra. This demonstrates that at sufficiently high Rayleigh number the stabilizing influence of the phase boundary deflection should overcome the destabilizing effect of the latent heat release. Above some transition Rayleigh number we therefore expect the Spinel-post Spinel transition to be stabilizing. The Cartesian geometry analyses of Christensen (1982) and Christensen and Yuen (1984) have revealed a tendency for some weak layering to develop at very high Rayleigh number in models with a single transition (Spinel-post Spinel), but the layering was not suggested for Earth like values of the Rayleigh number. Recently, Machetal and Weber (1991) have obtained a similar result for a single phase transition in a simple model in axially symmetric spherical geometry but for a Rayleigh number considerably lower than that which characterizes the present-day earth. Yuen et al. (1990) have argued that when both the Olivine-Spinel and the Spinel-post Spinel transitions are included in the simulation then no layering is evident in the Earth-like regime. Although this is plausible, given that the Olivine-Spinel transition will enhance convection if the Spinel-post Spinel transition inhibits it, it seems unlikely that the impact of the former could ever annihilate the effect of the latter. The density jump across the shallower transition is considerably smaller than that across the deeper transition and since the degree of inhibition/enhancement of convection that is caused by a phase transition is determined to first order by the product of the phase boundary deflection and this density jump, the $\alpha \rightarrow \beta$ transition would have to have a very large Clapeyron slope to be equally effective as the deeper boundary. Numerically accurate computations are clearly required to decide the issue. Although we will describe here only results for a Clapeyron slope of the Olivine-Spinel transition of $+2MPa/^{\circ}K$, ongoing analyses using significantly higher slopes have shown that the impact of this parameter is not severe.

The results to be reported below were obtained on the CRAY Y-MP 2/64 computer system at the Los Alamos National Laboratory. The system (2) was integrated forward in time in the streamfunction - vorticity formulation for the axisymmetric fields and the in-core multigrid solver MUDPACK (from the National Centre for Atmospheric Research) was employed to invert the elliptic equations (2a) and (2b) prior to each advection step of the temperature field based upon (2c). Using these numerical methods we have been able to integrate (2) on grids with 769 mesh points in the radial direction ($\Delta r = 3.75$ km) and 1025 mesh points in the aximuthal direction. We believe these calculations to be the highest resolution simulations (in radius) of the mantle convection process that have been performed.

In Figure 1a (see Plates section) we show two cross-sectional views of the temperature field from this high resolution model sufficiently late in the simulation that the chaotic time dependent flow (Jarvis and Peltier 1982, Solheim and Peltier 1990) is approaching a statistically steady state. Both calculations are for heated from below flows for which the Rayleigh number $Ra = 10^7$ so that both are operating in the "Earth-like" regime. Inspection of this Figure demonstrates that in the no phase transitions case, both the hot upwelling and cold downwelling plumes traverse the radial extent of the mantle unimpeded. In the case with two phase transitions, however, no hot plume is observed (at this time) to be rising through the 670 km boundary, nor is any cold plume observed to be sinking through the boundary. Of special note in this Figure is that "puddles" of high temperature low seismic velocity material are accumulating on the underside of the 670 km phase boundary, a signature that might be resoluble using seismic tomography. In Figure 1b (see Plates section) we show the radial velocity fields that correspond to the temperature fields shown in Figure 1a. It is important to note, as stated on these Figures, that in the two phase boundary case the thicknesses of both phase loops have been set at 50 km, which is approximately an order of magnitude thicker than implied by the laboratory data. We might reasonably enquire as to whether the layering apparent in the Figure for the case of two phase boundaries depends upon this specified thickness.

This degree of layering has been investigated quantitatively through the use of a radial mass flux diagnostic F that we first introduced in Solheim and Peltier (1991). This is defined as

$$F = \frac{\langle \rho_r |w| \rangle}{\frac{1}{(r_s - r_{cmb})} \int_{r_{cmb}}^{r_s} \langle \rho_r |w| \rangle dr} \tag{5}$$

in which w is the radial velocity, $\rho_r(r)$ is the basic state density field in the anelastic model, and the angle brackets $\langle \ \rangle$ denote an azimuthal average. In Figure 2 we show a sequence of time averaged diagnostics $F(r)$ for a sequence of two phase transition models that differ from one another only in the widths that are assigned to the phase loops at 400 km and 670 km depth, which are taken equal for purposes of illustration. The mass flux diagnostic is shown for models in which the phase loop thicknesses are set at 100 km, 50 km, and 25 km. For comparison purposes we have also superimposed on this Figure the time averaged mass flux diagnostic for the flow at $Ra = 10^7$ with no phase transitions. Inspection of this Figure very clearly shows that the 670 km phase boundary has a strongly inhibiting influence upon the mixing of material through it, at least at this instant of time, and that this inhibition increases as the prescribed thickness of the phase loop decreases. The smallest loop thickness for which we have been able to compute (due to lack of CPU resources) has been 25 km, which is still a factor of 5 higher than the observed loop thickness for the Spinel-post Spinel phase boundary. It does appear, however, that the impact of decreasing loop thickness has begun to diminish somewhat by this point. We therefore believe that the more extensive analysis of this result that we are now pursuing will only be altered in detail by this ongoing refinement. The result clearly suggests that even in the absence of chemical layering across the 670 km phase transition, the mantle convective circulation in the modern earth might be, on occasion, substantially layered.

In attempting to establish whether convective layering of the phase transition induced kind discussed above is actually present in the real earth, we wish also to suggest here that seismologists involved in the application of tomographic techniques should focus upon attempting to resolve the depth dependent power spectrum of the inferred lateral heterogeneity. To illustrate the possible utility of this diagnostic we show on Figures 3 and 4 the depth dependent lateral heterogeneity (of temperature) spectrum from the same two simulations at $Ra = 10^7$ used to construct Figure 1. Figure 3 is for the simulation with no phase transitions and Figure 4 is for the simulation with two phase transitions and phase loop thicknesses of 50 km. On each Figure we provide three different presentations of the expansion $T(r,\theta) = \sum_{\ell=o}^{\infty} C_\ell(r) P_\ell(\cos\theta)$. In the top frame we show $|C_\ell|$ vs. ℓ as a function of radius r using a black dot rendering to indicate $|C_\ell|$. Superimposed upon this we show the integrated spectral power as a function of depth (solid line). Below the top frame we show power spectral histograms in the top thermal boundary layer (50 km depth), at 400 km depth (the Olivine-Spinel boundary), at 670 km depth (the Spinel-post Spinel boundary), and in the bottom thermal boundary layer (D'') at the base of the mantle.

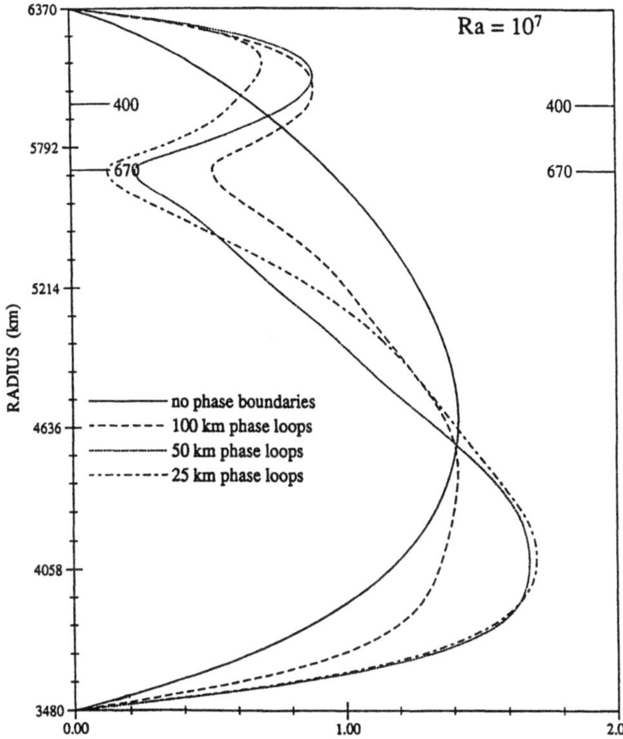

Figure 2: Radial mass flux diagnostic from equation (5) in the text for axi-symmetric simulations of the mantle convection process at a Rayleigh number of 10^7. The solid line (see Figure legend) is for the model with no phase transitions. The remaining curves are for models with two phase transitions and the simulations differ from one another only in the thicknesses that are assigned to the phase loops. Note that as the thickness of the phase loops decreases the flow goes more and more strongly layered. Since the observed thickness of the phase loop for the actual Spinel-post Spinel transition at 670 km depth is approximately 5 km we expect that the extent of the layering (if the above described trend should continue) might, on occasion, be extreme.

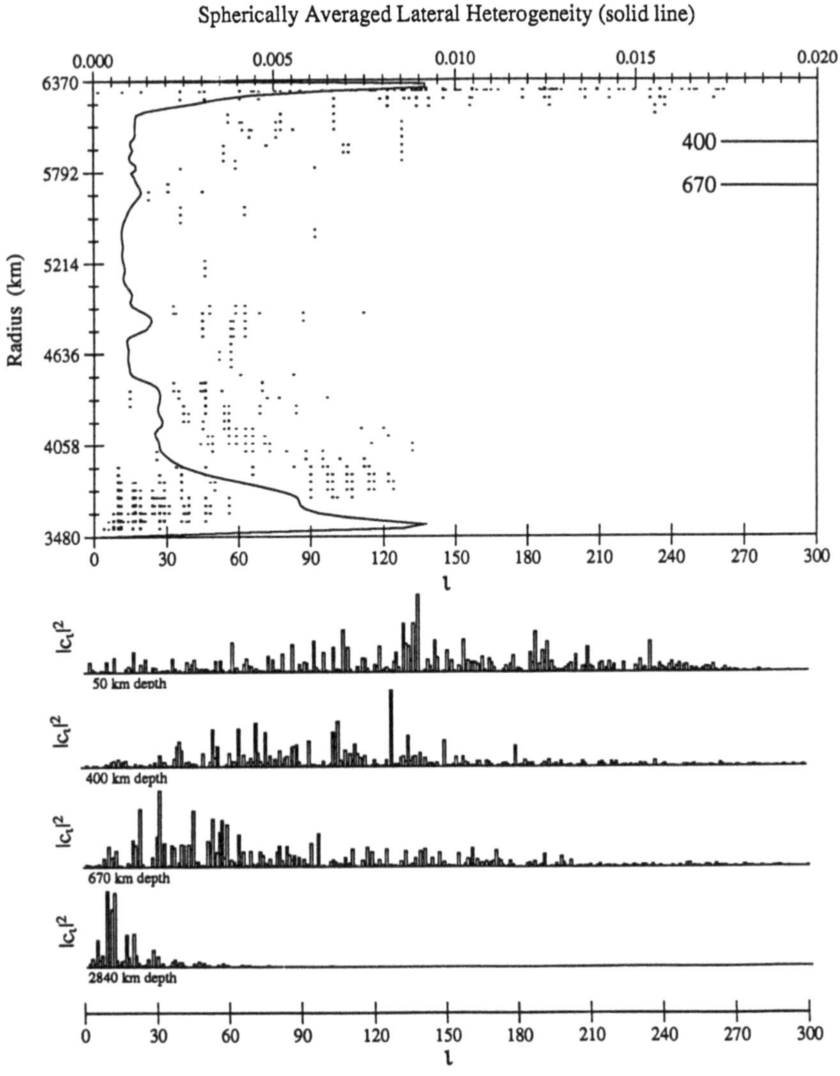

Figure 3: Depth dependent lateral heterogeneity spectra for the axi-symmetric convection model with $Ra = 10^7$ and no phase transitions. The black line on the upper plate shows the integrated power in the depth dependent lateral heterogeneity spectrum.

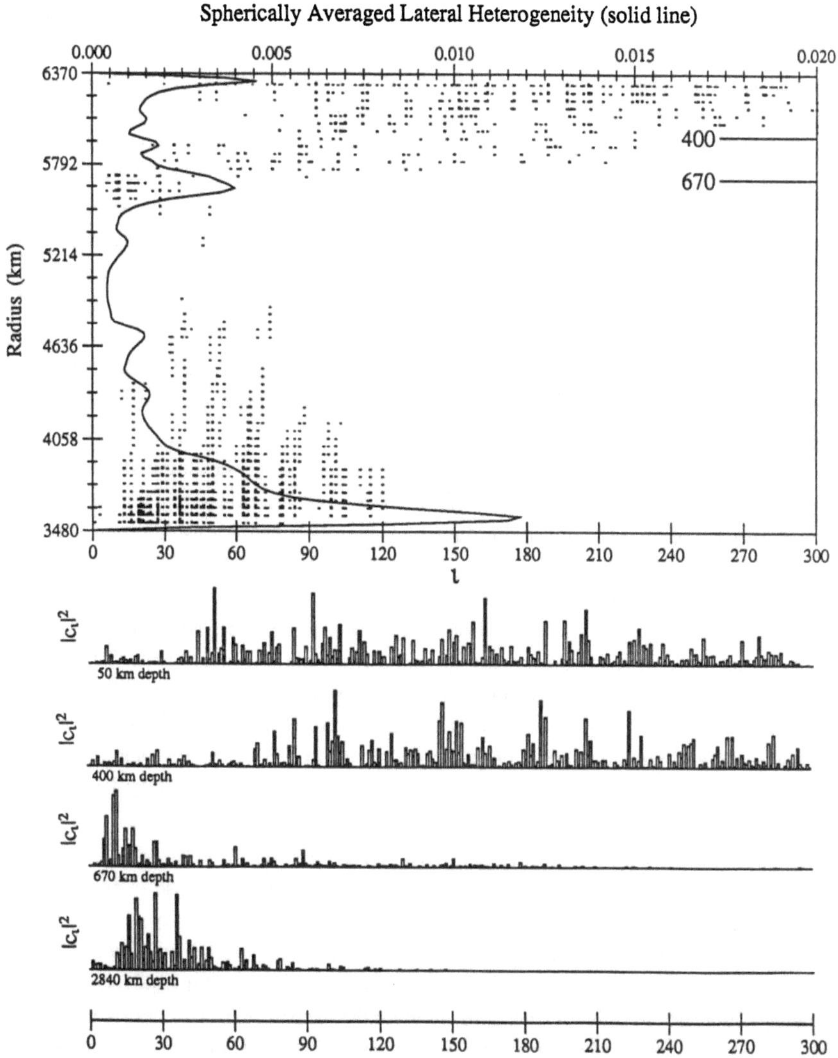

Figure 4: Same as Figure 3 but for the model at $Ra = 10^7$ with two phase transitions. Note the concentration of variance in the secondary thermal boundary layer that develops at 670 km depth.

Inspection of Figure 3 demonstrates that strong temperature variance is characteristic of both the top and bottom (cold and hot) thermal boundary layers. One should also note the strong systematic shift of power to lower spherical harmonic degree that is evident as one moves from the upper cold boundary layer to the lower hot boundary layer. Comparing the diagnostic on Figure 3 with that for the corresponding 2 phase transition case on Figure 4, it is clear that there is, in the latter case, a strong concentration of spectral variance also at the depth of the 670 km discontinuity across which mass flux is strongly inhibited by the dynamical influence of the endothermic Spinel-post Spinel phase transformation and an internal boundary layer develops. Noticeable also is the enhanced power at high spatial frequencies in the upper mantle and the dominance of low spatial frequencies in the lower mantle even at the depth of the 670 km discontinuity. These characteristics are similar over all to those revealed by global seismic tomographic imaging (e.g. see Forte and Peltier, 1987).

It should be clear that the depth dependent spectral power diagnostic advocated here as a "prospecting" tool for the transition zone boundary layer that must be characteristic of phase transition induced layering is considerably different from the spectral diagnostic suggested earlier in Jarvis and Peltier (1986). More important for our purposes in the present paper, however, is the magnitude of the temperature increase that develops across the boundary layer itself. In Figure 5 we illustrate this expected temperature increase by superimposing the geotherms delivered by the simulations in which phase loop thicknesses were fixed at 200 km, 100 km, and 50 km respectively. Inspection of this Figure demonstrates that if the flow existed in the layered style that our time dependent computations suggest should develop on occasion, then the temperature jump across the internal boundary layer should be near $300°K$.

This is a rather important number because, as demonstrated by detailed analysis presented previously in Peltier and Jarvis (1982), it is the increase in temperature that would be required to compensate for the increase in viscosity across the 670 km horizon (due to the change in atomic packing) that would be observed if the temperature change through this boundary were adiabatic. Our analyses therefore suggest that no significant increase of viscosity across 670 km depth should be observed if the flow were layered. In the following sections we will investigate the extent to which inferences of mantle viscosity based upon the use of seismic tomographic models of the interior density field to predict non-hydrostatic geoid anomalies and plate motion data, and upon glacial isostatic adjustment analyses, may be invoked to assess the veracity of this expectation.

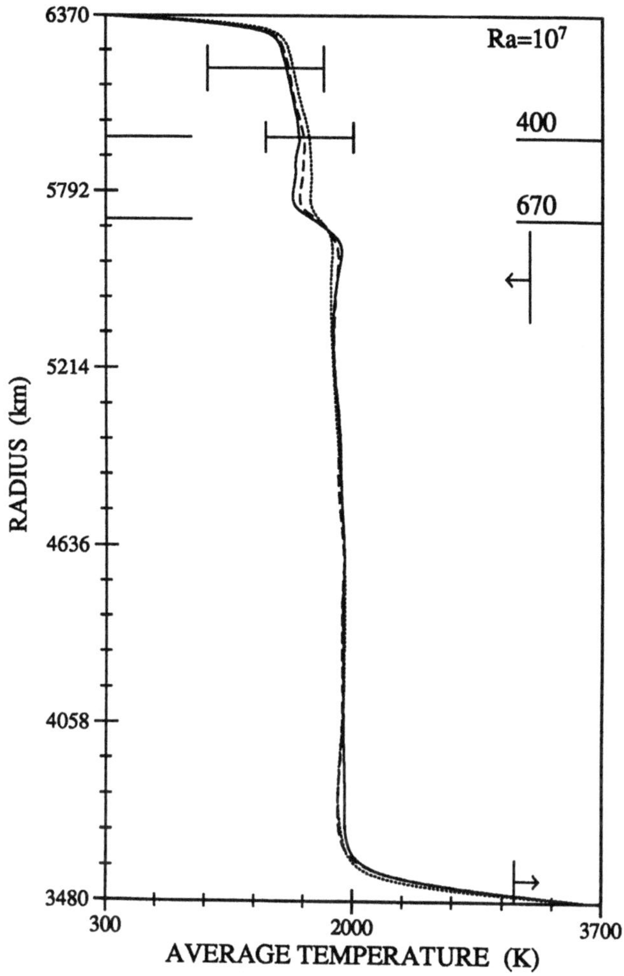

Figure 5: Time averaged geotherms for convection models with two phase transitions at Rayleigh number $R_a = 10^7$ over an interval of time through which the flow is layered. The solid, dashed and dotted curves are respectively for models in which the phase loop thicknesses are set at 50 km, 100 km, and 200 km. Note that the temperature increase that develops across the 670 km boundary layer is approximately $300°K$. The horizontal bars represent the constraints on the temperature field provided by observational data (Jeanloz and Morris, 1986).

3. Inferences of mantle viscosity from non-hydrostatic geoid and plate motion data. Of the variety of methods whereby one may hope to infer the variation of viscosity in the earth's mantle, the one most directly connected with the convection process itself is clearly that in which the lateral heterogeneity of density in this part of the earth is inferred on the basis of the techniques of body and surface wave seismic tomography. Given this density field the momentum and energy equations decouple, rendering it possible to predict the flow velocity everywhere, and also the anomalies in the gravitational field that are caused by this density heterogeneity and the induced flow, if the viscosity field is also known. Since the gravitational field and flow velocities are observed at the surface and since the longest wavelength components of the density heterogeneity may be inferred from seismic tomography, the observations may be construed to pose an inverse problem for mantle viscosity. All of the work on this problem that has been completed to date has been based upon the (perhaps inappropriate) assumption that the viscosity is a function of radius only. That this assumption is wrong in principle follows from the fact that the effective viscosity of a solid is determined by the thermally activated processes of dislocation climb or glide and/or diffusion, all of which have exponential temperature dependence. In order to infer the lateral density heterogeneity in the mantle from seismic tomographic reconstructions of the variations of seismic wave speed one assumes that these fluctuations are induced by variations of temperature and that the temperature derivatives of the body wave velocities are known on the basis of laboratory (+ theoretical) analyses. Given the temperature fluctuation one may therefore infer the associated density variation on the basis of the (laboratory and/or theoretically determined) coefficient of thermal expansion. In a flow in which density heterogeneity is determined by temperature heterogeneity, it is therefore clear that lateral viscosity heterogeneity must also exist that is perfectly correlated with the density heterogeneity. Although work is ongoing that seeks to understand the impact that this lateral viscosity heterogeneity must have upon the inference of its radial variation (Forte and Peltier, in preparation), all of the analyses that have been completed to date have been based upon the assumption that mantle viscosity is a function of depth only. It is the results from these analyses that we will proceed to review in this section.

The earliest work in this line (which basically elaborated upon arguments first presented by S.K. Runcorn) focused upon the problem of using observations of non-hydrostatic geoid anomalies to infer mantle viscosity contrast (Hager 1984, Hager et al. 1985, Hager and Richards 1989, Richards and Hagar 1989). To be fair, this work dealt for the most part with demonstrating that simple models of mantle flow could be employed to predict the observed geoid given a tomographic reconstruction of the interior density field, and an "appropriate" model of mantle viscosity. In fact, as discussed in detail in Forte and Peltier (1987), the geoid data alone cannot be employed to infer the absolute value of mantle viscosity but only the ratio (in a two layer parameterization, say) of the upper mantle viscosity to the lower mantle viscosity. In his most recent work (e.g. Hager and Clayton, 1989; Hager 1991) Hager has inferred on this basis contrasts as high as a factor of 300 between the upper and the lower mantle in parameterizations represented by a small number of layers and with the lower mantle having the higher viscosity. A model of the radial variation of viscosity such as this is clearly very difficult to reconcile with expectations based upon the phase transition induced episodically layered convection scenario that we advanced in the last section. The increase in temperature across the thermal boundary layer at 670 km depth would appear to very strongly mitigate against any significant increase of viscosity across this seismic horizon. In what follows we will very briefly review and discuss the recent reanalyses of this problem by Forte and Peltier (1991b) which demonstrate not only that the inferences of Hager are at best non-unique but also that the best fitting two layer parameterizations of the depth variation of mantle viscosity are characterized by increasing strength at depths far in excess of 670 km. These models of Forte and Peltier are in good accord with expectations based upon the previously discussed convection scenario.

The results that we will summarize are based upon calculations with a model based upon the same anelastic compressibility approximation as was employed for purposes of the dynamical analyses presented in the last section. Here, however, the full impact of self-gravitation is included. In Cartesian tensor notation the equations of the model are:

(6a) $$u_{k,k} + \frac{\dot{\rho}_o}{\rho_o} u_r = 0$$

(6b) $$\sigma_{ij,j} + \rho_o(\phi_1)_{,i} - \rho_1 g_o \hat{r} = 0$$

(6c) $$\sigma_{ij} = -P_1\delta_{ij} + \eta \cdot \left(u_{i,j} + u_{j,i} - \frac{2}{3}\delta_{ij} u_{k,k} \right)$$

(6d) $$(\phi_1)_{,kk} = -4\pi G \rho_1$$

in which the overdot indicates differentiation with respect to $r, u_r = \hat{r} \bullet \mathbf{u}$, g_o and $\rho_o(r)$ are the gravitational acceleration and density in the hydrostatic equilibrium state and ρ_1 and ϕ_1 are the perturbations of density and gravitational potential from this state that arise in consequences of the velocity field $\mathbf{u}$. When η is a function of r only then (6) can be solved to deliver predictions of the spherical harmonic spectrum of the horizontal divergence of the flow velocity at the earth's surface $(\nabla_H \bullet \mathbf{u})_\ell^m(r = a)$, of the geoid anomaly $(Ge)_\ell^m$, and of the topography on the core-mantle boundary δb_ℓ^m in the forms, respectively:

(7a) $$(\nabla_H \bullet \mathbf{u})_\ell^m(r = a) = \frac{g_o}{\eta_o} \int_b^a S_\ell(r)(\rho_1)_\ell^m(r)dr$$

(7b) $$(Ge)_\ell^m = \frac{3}{\bar{\rho}(2\ell + 1)} \int_b^a G_\ell(r)(\rho_1)_\ell^m(r)dr$$

(7c) $$\delta b_\ell^m = \frac{1}{\Delta\rho_{cmb}} \int_b^a B_\ell(r)(\rho_1)_\ell^m(r)dr$$

Hager and colleagues were the first to employ (7b) to infer earth properties whereas Forte and Peltier (1987, 1989, 1991a,b) have introduced the additional constraints (7a) and (7c). As previously discussed, seismic tomography delivers the low degree and order coefficients of the lateral density heterogeneity spectra $(\rho_1)_\ell^m(r)$. $S_\ell(r)$,

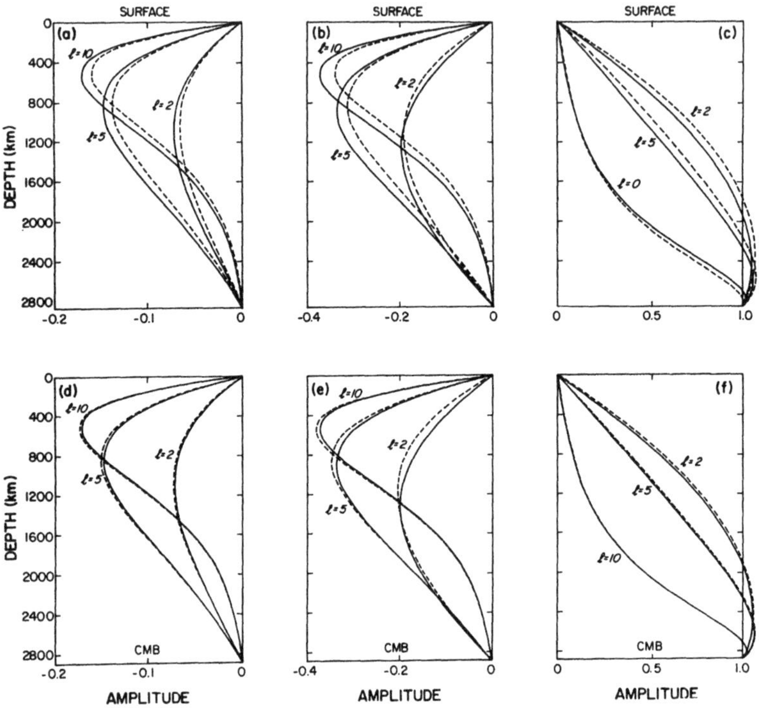

Figure 6: In (a)-(f) the solid lines are the kernels for a radially compressible (PREM) isoviscous mantle. In (a)-(c) the dashed lines are kernels for an incompressible isoviscous mantle and in (d)-(f) the dashed lines are kernels for a compressible isoviscous mantle in which the internal self-gravitational loading, $\dot{\rho}_o\phi_1$, is ignored. (a) and (d) Horizontal divergence kernels. (b) and (e) Nonhydrostatic geoid kernels. (c) and (f) CMB topography kernels.

$G_\ell(r)$ and $B_\ell(r)$ are kernels that depend upon the radial viscosity and density structure in the reference state, examples of which are presented in Figure 6.

Inspection of this Figure, plates (a)-(c) and (d)-(f) of which present the kernels (7a)-(7c) respectively for a radially compressible (PREM) isoviscous mantle (solid lines) demonstrate the manner in which these three different geophysical observables sample the internal density field. The dashed lines on plates (a)-(c) compare kernels for a incompressible isoviscous mantle to demonstrate the influence of compressibility whereas the dashed lines on plates (d)-(f) compare kernels of an compressible isoviscous mantle in which the influence of self-gravitational loading is ignored. In Figure 7 we present the same sequence of comparisons for a two layer viscosity model that has a factor of 8 increase in viscosity at a depth of 1200 km. Forte and Peltier (1991b) present a detailed discussion of the differential sensitivities revealed by these comparisons, the most important points of which concern the demonstration of the importance of *both* compressibility and self-gravitation on the longest wavelength components of the geoid.

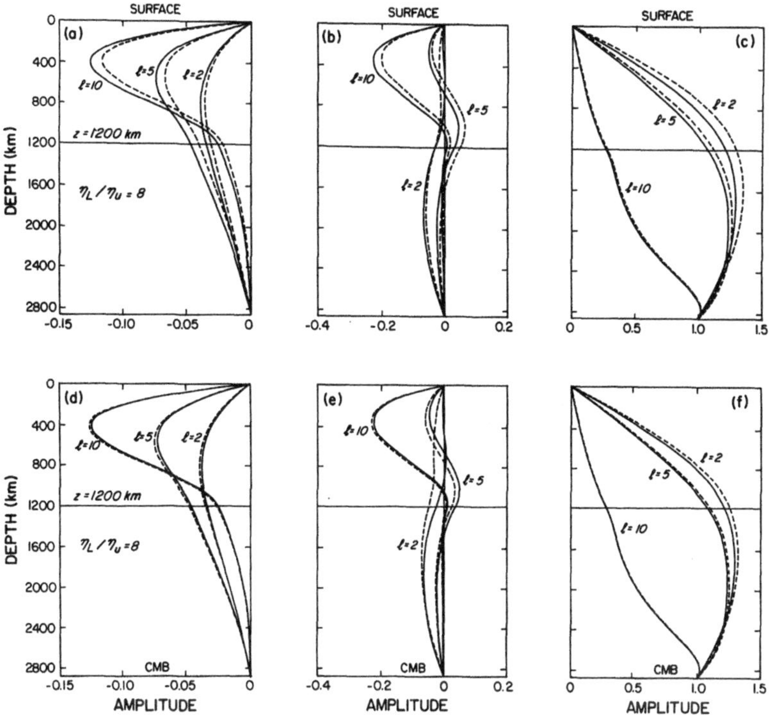

Figure 7: All kernels in this figure are for a two-layer viscosity profile with a factor of 8 viscosity jump at 1200 km depth. In (a)-(f) the solid lines are the kernels for a radially compressible (PREM) mantle. In (a)-(c) the dashed lines are for an incompressible mantle and in (d)-(f) the dashed lines are kernels for a compressible (PREM) mantle in which the internal self-gravitational loading, $\dot{\rho}_o\phi_1$, is ignored. (a) and (d) Horizontal divergence kernels. (b) and (e) Nonhydrostatic geoid kernels. (c) and (f) CMB topography kernels.

Recent analyses with the theoretical structure embodied in (7) have led to a number of conclusions relevant to the issue of convection timescale viscosity stratification that is of interest to us here. Even using the early, and relatively crude, tomographic models of Woodhouse and Dziewonski (1984) for the upper mantle (corrected for the presence of slabs as in Forte and Peltier (1989) and of Dziewonski (1984) for the lower mantle [models M84C and L02.45 respectively], surprisingly good fits to the observed geoid, plate motion, and core-mantle-boundary (CMB) topography fields have been obtained. Of interest from the present perspective is that the optimal two layer viscosity parameterization (in the sense of maximum variance reduction) which fits these data is one with the boundary separating the layers at 1200 km depth and a viscosity increase across the interface of approximately a factor of 10. When the interface is placed at 670 km depth the inferred viscosity contrast is significantly higher but the variance reduction is significantly lower.

Examples of the fits to the data that have been achieved with these models are shown firstly in Figure 8 for geoid height anomalies in the degree range $0 \leq \ell \leq 5$. Plate (a) in this sequence shows the observed field based upon GEM10B data (Lerch et al. 1979) while plate (b) shows the best prediction of the anelastically compressible model and plate (c) that for the inferior incompressible model. The misfit in plate (d) demonstrates that this is almost entirely a consequence of an erroneous degree 2 and order 0 coefficient which current work on the tomographic inversions at Harvard University (A.M. Forte, personal communication) is attempting to address.

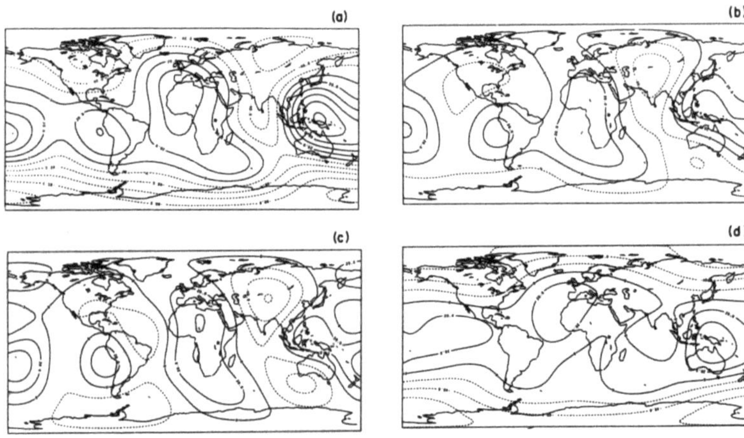

Figure 8: (a) the GEM10B nonhydrostatic geoid of Lerch et al. (1979) (filtered by removal of the hydrostatic flattening calculated by Nakiboglu [1982]) in the degree range $\ell = 2 - 5$. The contour interval is 20 m. (b) The predicted nonhydrostatic geoid (in the range $\ell = 2 - 5$) for a compressible mantle with a factor of 9 viscosity jump at 1200 km depth. The internal density heterogeneity is obtained from the 'SLAB' densities extended to 1100 km depth, M84R in the upper mantle, and L02.45 in the lower mantle (see Forte and Peltier, 1991b for details). The contour interval is 20 m. (c) The predicted nonhydrostatic geoid (in the range $\ell = 2 - 5$) for an incompressible mantle. The viscosity and lateral density heterogeneity model is the same as in (b). The contour interval is 20 m. (d) The residual geoid obtained by subtracting the predicted field in (b) from the observed geoid in (a). The contour interval is 20 m.

In Figure 9 we compare predicted and observed surface velocities using a slightly modified form of the theory described in Forte and Peltier (1991b) that incorporates the impact upon the calculation of the fact that the horizontal velocity field at the earth's surface is not only *not* entirely poloidal, as is assumed in the construction of (7), but in fact is such that fully 50% of the kinetic energy is carried by the toroidal constituent. Plate (a) of Figure (9) shows the horizontal divergence of the observed tectonic plate velocities computed from the model of Minster and Jordan (1978) in the degree range $\ell = 1 - 15$. The prediction of this field according to the modified convection model is shown in plate (b), inspection of which demonstrates the extremely high quality fit to the data that is delivered by our theory. Plates (c) and (d) similarly compare the observed and predicted radial vorticity and demonstrate the equally encouraging fit of the theory to the data for the toroidal component of the flow.

Figure 10 shows a similar sequence of comparisons between seismologically observed (plate a) and theoretically predicted (plates b and c) core mantle boundary topography in the degree range $\ell = 2 - 4$. The observed field is that inferred by Morelli and Dziewonski (1987). In plate (b) the predicted *cmb* topography is that for a two layer model with an order of magnitude increase in viscosity across 1200 km depth whereas in plate (c) the predicted *cmb* deflection incorporates the influence of a thin layer adjacent to the *cmb* in which the mantle viscosity is sharply reduced which shows that the physical effect of this feature is modest. Excepting for the North Atlantic and Africa regions the theoretical predictions of the simple theory embodied in (7) provides an extremely satisfactory reconciliation of the observations.

It is important to understand that all of the above described fits of the theory to observations have been achieved with simple two layer models in which the only substantial viscosity variation occurs across the rather deep horizon 1200 km beneath the earth's surface, not across the 670 km seismic discontinuity. Above this level the viscosity is everywhere $10^{21} Pa\ s$. As discussed in Peltier and Jarvis (1982), if the radial variation of temperature through the 670 km boundary were adiabatic then we would expect the viscosity to increase across it by about one order of magnitude. Our review of the evidence from seismic tomography and the observables connected with the convection process suggest that no such increase of viscosity across this level is required by the data and that best fitting models have no such contrast. The reasons why the results of our analyses differ so strikingly from the previous results of Hager (referenced above) are discussed at length in Forte and Peltier (1991b). More detailed discussion of the results presented in this section of the present paper will also be found therein.

As it happens there is an additional perspective on the basis of which this inference of weak viscosity contrast across 670 km depth may be further reinforced. This perspective is rather more "classical" than those from which we have viewed this issue in the present section.

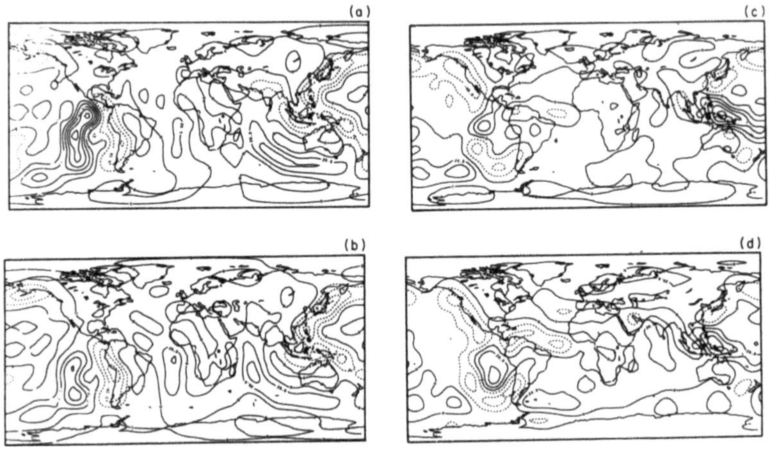

Figure 9: (a) The horizontal divergence of the observed tectonic plate velocities (Minster and Jordan, 1978) in the degree range $\ell = 1-15$. The contour interval is 20×10^{-9} rad/yr. (b) The predicted horizontal divergence of the surface plate motions (in the range $\ell = 1-15$) obtain by filtering the predicted divergence in Figure 8b according to the poloidal-toroidal coupling theory discussed in detail in Forte and Peltier (1991b). The contour interval is 20×10^{-9} rad/yr. (c) The observed radial vorticity of the observed tectonic plate velocities (Minster and Jordan, 1978) in the degree range $\ell = 1-15$. The contour interval is 20×10^{-9} rad/yr. (d) The predicted radial vorticity of the surface plate motions (in the range $\ell = 1-15$) obtained by filtering the predicted divergence in Figure 8b according to equation (49) in the text. The contour interval is 20×10^{-9} rad/yr. *Note*: In calculating the predicted fields in (b) and (d) we use the reference viscosity $\eta_o = 4.2 \times 10^{21}\, Pa\ s$, obtained on the basis of our kinetic-energy matching calculation (see Forte and Peltier, 1991b for details).

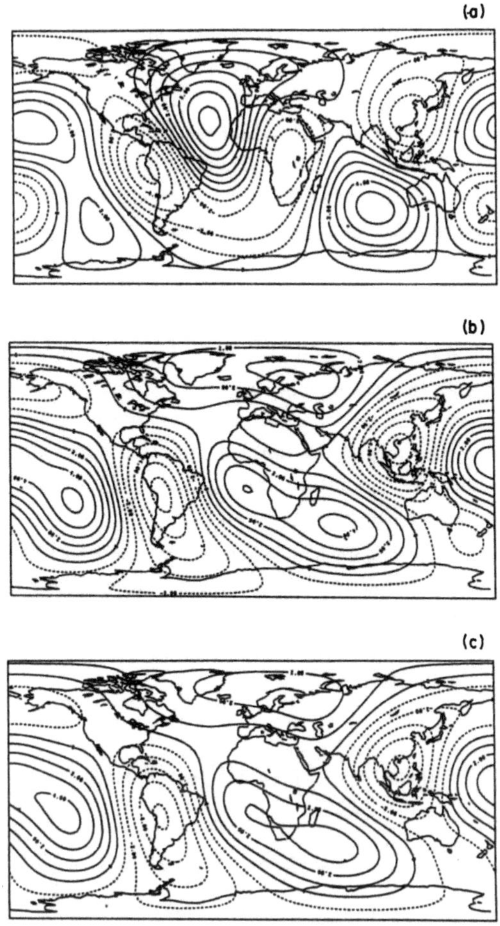

Figure 10: (a) The seismically observed CMB topography (Morelli and Dziewonski, 1987) in the degree range $\ell = 2-4$. The contour interval is 1 km. (b) The flow-induced CMB deflections predicted with a compressible flow model with a factor of 9 viscosity jump at 1200 km depth. The lateral density heterogeneity model is the same as in Figure 4b. The contour interval is 1 km. (c) The flow-induced CMB deflections in a mantle which also includes the radial, exponential viscosity decrease in D'' so that the mantle viscosity at the CMB is 30,000 times smaller than the mantle viscosity at the top of D''.

4. Inferences of mantle viscosity from the observables of glacial isostatic adjustment. The use of observations of the adjustment of the earth's shape following large scale deglaciation to infer the viscosity of the planetary mantle has constituted something of a geophysical "industry" since the earliest efforts of Haskell (1935, 1936). Although the theoretical model that Haskell employed to infer a value of $10^{21} Pa\ s$ for the molecular viscosity of the mantle, based upon observations of the postglacial rebound of Fennoscandia, consisted of a simple half space of constant density and viscosity, his value for the viscosity of the upper mantle has not been significantly refined by later analysis. Considerably more elaborate analyses have unfortunately been necessary to quantify the radial resolving power of the Fennoscandia data and to incorporate the data from the North American continent which are controlled by the melting of this much larger ice sheet than that which covered Fennoscandia at the last glacial maximum of the present ice age. Models suitable for this purpose must be spherical and self-gravitating and if sea level data are to be employed as the target observables then the model should also include the capability to predict the nonuniform redistribution of meltwater over the ocean basins that occurs when ice sheets melt.

A recent review of the theoretical structure required to make predictions of glacial isostatic adjustment effects will be found in Peltier (1989). In this structure the relative sea level history $S(\theta, \psi, t)$, a function of latitude θ, longitude ψ, and time t, that is induced by a deglaciation history $I(\theta, \psi, t)$ is obtained as the solution of the following integral equation:

$$
\begin{aligned}
S(\theta, \psi, t) = C(\theta, \psi) \Bigg[\int_{-\infty}^{t} \iint_{\Omega} & \{\rho_I I(\theta', \psi', t') + \rho_w S(\theta', \psi', t')\} \cdot \\
& \cdot \left\{ \frac{\Phi(\gamma, t-t')}{g} - \Pi(\gamma, t-t') \right\} d\Omega' dt' + \frac{\Delta\Phi(t)}{g} \Bigg].
\end{aligned} \tag{8}
$$

In this equation $C(\theta, \psi)$ is the ocean function ($C = 1$ where there is ocean and $C = 0$ where there is land), γ is the angular separation between source point (θ', ψ') and field point (θ, ψ) over the surface of the earth Ω, ρ_I and ρ_w are the densities of ice and water respectively, $\phi(\gamma, t-t')$ is the viscoelastic Green function for the gravitational potential perturbation, g is the surface gravitational acceleration, $\Pi(\gamma, t-t')$ is the radial displacement Green function and $\Delta\Phi(t)$ is a time dependent function required to ensure that (8) conserves mass. When predictions of the associated variations in the surface gravitational field are important, expressed for example by a free air gravity anomaly $\Delta g(\theta, \psi, t)$, these may be obtained by evaluation of the following convolution integral:

$$
\begin{aligned}
\Delta g(\theta, \psi, t) = & \int_{-\infty}^{t} \iint_{\Omega} L(\theta', \psi, t') G(\gamma, t-t') d\Omega' dt \\
& - \frac{2\Delta\Phi(t)}{2}.
\end{aligned} \tag{9}
$$

In this expression $L(\theta', \psi', t') = \rho_I I + \rho_w S$ is the total load (ice + water), $G(\gamma, t-t')$ is the Green function for the free air gravity anomaly (Mitrovica and Peltier, 1989), and "a" is the earth's radius.

For present purposes we will employ this structure to predict both relative sea level histories and the free air gravity anomaly over the Hudson Bay Region. In order to implement the theory we need a deglaciation history $I(\theta, \psi, t)$ that is as accurate as possible and so will employ the recently published ICE-3G model of Tushingham and Peltier (1991a). The spatial distribution of the northern hemisphere component of this ice model at the last glacial maximum ($18,000^{14}C$ years ago) is shown in the polar projection in the frame at the apex of Figure 11 (see Plates section). Based upon the ICE-3G melting history we have inverted (8) on a basis of spherical harmonics to degree and order 128 for a variety of simple models of the radial viscoelastic structure that differ from one another only in the viscosity contrast across 670 km depth. The models all have PREM radial elastic structure and 120 km thick lithospheres. On Figure 12 we compare the predicted relative sea level histories to the observed histories at six sites around Hudson Bay (see inset for locations) that were once covered by the Laurentide ice sheet. Predictions are shown for an isoviscous mantle model with viscosity equal to $10^{21} Pa\ s$, and for models which differ from this only in the lower mantle where their viscosities are set equal to $2 \times 10^{21} Pa\ s$, $4.5 \times 10^{21} Pa\ s$, and $10^{22} Pa\ s$ respectively. The relative sea level observations, from the extensive compilation recently published by Tushingham and Peltier (1991b), are shown as the vertical bars. Inspection of this Figure demonstrates that only the isoviscous model or that with the weakest viscosity contrast across 670 km depth ($\times 2$) are able to reconcile the observations.

Given the rather accurate new deglaciation model ICE-3G it is interesting to enquire as to the magnitude of the free air gravity anomaly that is expected over the Hudson Bay region in association with the existing degree of isostatic disequilibrium that remains there, especially since the rather large amplitude of the observed anomaly has often been invoked in the past (e.g. Walcott 1980) to argue that the viscosity of the lower mantle must be significantly in excess of the viscosity of the upper mantle. On Figure 13 we show predictions of the present day free air gravity anomaly for each of the viscosity models that were employed to make the relative sea level predictions shown on Figure 11. Inspection of this Figure shows that the magnitude of the free air anomaly that should exist over present day Hudson Bay, if the mantle is isoviscous with viscosity $10^{21} Pa\ s$, is only about 2 $mgal$. Even for the model with an increase of viscosity across 670 km depth by a factor of 10 the predicted present day anomaly is only 30 $mgal$. This must be compared to the observed anomaly shown on Figure 14, plates a and b of which show reconstructions of the free air anomaly in spherical harmonic expansions truncated at degree and order 32, based upon surface and satellite observation respectively. Since the peak observed anomaly is near $-47 mgal$ (in the satellite field - GEMT2) it is clear that the radial viscoelastic models that are required to fit the relative sea level data cannot simultaneously reconcile the observed free air gravity anomaly.

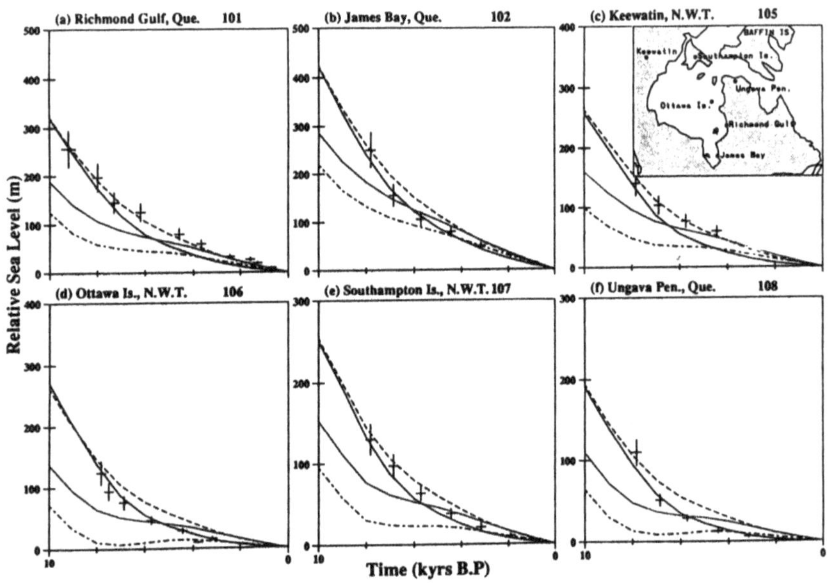

Figure 12: Observed relative sea level histories for the sites shown on the inset location map are represented by the vertical bars. At each site we show predictions of the relative sea level history for four different viscosity models, each of which has a 120 km thick lithosphere in which the viscosity is assumed infinite and an upper mantle extending to 670 km depth in which the viscosity is assumed to be equal to $10^{21} Pa\ s$. The solid, dashed, dotted, and dash-dotted curves are for models in which the lower mantle viscosity is $10^{21} Pa\ s$, $2 \times 10^{21} Pa\ s$, $4.5 \times 10^{21} Pa\ s$, and $10^{22} Pa\ s$ respectively.

The requirement for a large increase of mantle viscosity with depth in order to explain the observed free air gravity anomaly over Hudson Bay led Walcott (e.g. 1980) to argue that the deep earth must be very much more viscous than the upper mantle. However, as we have seen, models that postulate a large increase of viscosity across 670 km depth are unable to reconcile relative sea level histories. The explanation of this enigma is in fact to be found in terms of the mantle convection process discussed in the last section. If we employ a "best available" model from seismic tomography (e.g. Su and Dziewonski, 1991) to predict the surface gravitational field over the North American continent using the simple convection theory discussed in the last section, then we obtain the prediction for the northern hemisphere labelled CONV8 in Figure 11 (the discussion to be presented in this section is a precis of that to appear in Peltier et al. 1992). This prediction is carried out to degree and order 8 in the spherical harmonic expansion. The observed northern hemisphere field to the same degree and order is also shown on Figure 11 where it is labelled GEMT2. Although the ability of the convection model to explain the free air anomaly over Hudson Bay is clear even at this scale, the fact is made more apparent on Figure 15 (see Plates section) where the predicted geoid and free air anomalies are shown for the North American region along with the satellite observed free air anomaly. The excellence of the fit of the theoretical prediction to the observation is such as to constitute rather convincing evidence that virtually the entire gravity anomaly in this region is caused by the mantle convection process and has nothing to do with glacial isostatic adjustment.

To summarize the results of this section we therefore note that the radial viscosity structure required to fit the sea level data is characterized by very weak or non existent viscosity contrast across the seismic discontinuity at 670 km depth. The free air gravity anomaly that has been invoked to argue for high contrast is not in fact associated with glacial isostatic disequilibrium at all, but rather is caused by convection within the mantle.

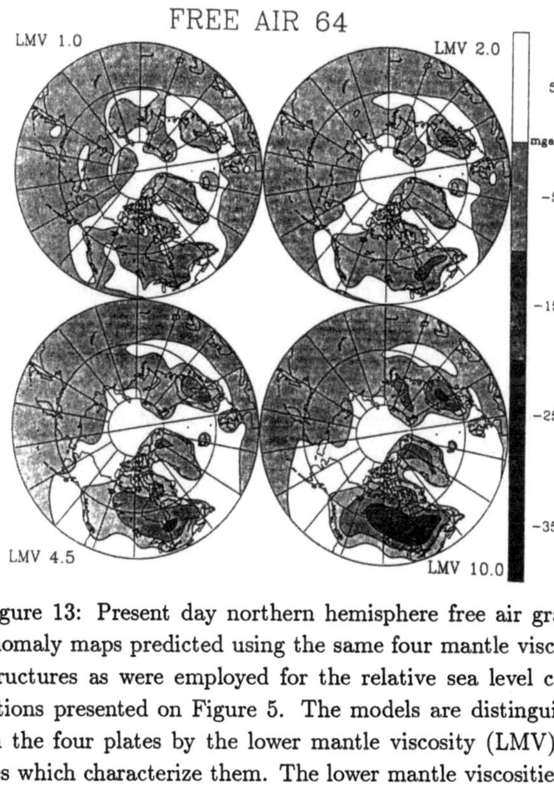

Figure 13: Present day northern hemisphere free air gravity anomaly maps predicted using the same four mantle viscosity structures as were employed for the relative sea level calculations presented on Figure 5. The models are distinguished on the four plates by the lower mantle viscosity (LMV) values which characterize them. The lower mantle viscosities are indicated in multiples of $10^{21} Pa\ s$.

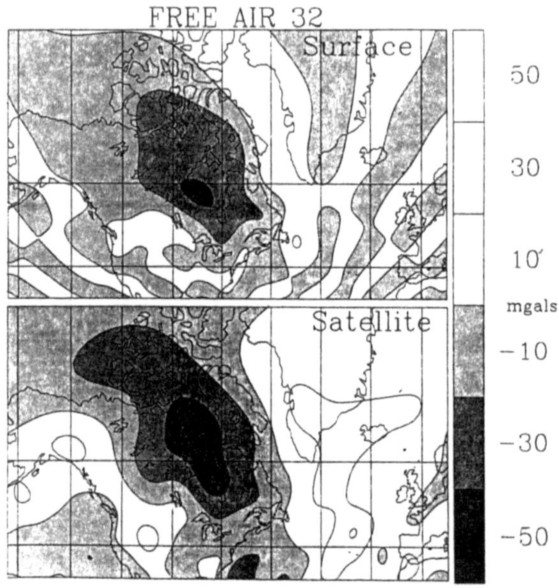

Figure 14: The free air gravity anomaly over Canada as determined on the basis of surface pendulum observations (plate a) and on the basis of analyses of the orbits of artificial Earth satellites (plate b). Both maps have been constructed on the basis of spherical harmonic expansions of the field truncated at degree and order 32. The Gibbs oscillations evident in the surface field off the east and west coasts of North America are an artifact of the fact that only data from the continental interior were employed to determine the spherical harmonic coefficients.

5. Conclusions. On the basis of direct integrations of the field equations for thermal convection in a fluid with multiple internal phase transitions we have computed spherical axisymmetric solutions in the earth-like chaotic regime. With the properties of the exothermic Olivine-Spinel transition and the endothermic Spinel-post Spinel transition set to values compatible with recent laboratory measurements, we found the flow to be quite strongly layered (episodically) when this characteristic of the circulation is measured by a time averaged radial mass flux diagnostic. This result clearly has important implications for our understanding of the thermal and chemical evolution of the planet. Since layering would have been even more pronounced in the early earth, it seems clear that this could have lead to differential chemical differentiation of the upper and lower mantle regions as hypothesized previously on the basis of trace element geochemistry in the Sm-Nd and Rb-Sm systems. A chemical discontinuity could therefore conceivably develop across the boundary that would further reinforce the layering, perhaps involving a relative depletion of Fe in the upper mantle compared to the lower mantle. In this scenario the chemical discontinuity at 670 km is a consequence rather than the cause of layered convection, given the nature of the geotherm that our dynamical models predict.

We find that the observed weak variation of mantle viscosity across the 670 km seismic discontinuity is most easily understood in terms of (at least episodically) layered convection. Our convection models suggest that the thermal boundary layer that develops across the 670 km phase transition should be characterized by a temperature rise of about $300° K$. Analysis in Peltier and Jarvis (1982) shows that this is the order of the temperature increase that would be required in order to cancel the viscosity increase that would be expected due to the change in atomic packing that occurs in association with the phase transition if the temperature variation through the transition were adiabatic. It is our contention that both postglacial rebound data and data such as plate velocities and nonhydrostatic geoid anomalies that are related to the convection process agree that the viscosity increase across 670 km depth is small. The increase of viscosity with increasing depth in the lower mantle that is required by the non-hydrostatic geoid data would be an expected consequence of a difference in slope of the melting curve in the lower mantle and the lower mantle adiabat.

Acknowledgements. We are indebted to Dr. Charles Keller of the Los Alamos National Laboratory for allowing us access to the CRAY Y-MP 2/64 system. This research was supported by NSERC grant A9627.

REFERENCES

AKAOGI, M. AND E. ITO, *Olivine-modified spinel-spinel transitions in the system* $Mg_2SiO_4 - Fe_2 SCO_4$*: Calorimetric measurements, Thermochemical Calculation, and Geophysical Application*, J. Geophys. Res., 94, no. B11 (1989), pp. 15671–15685.

ALLEGRE, C.J., *Chemical geodynamics*, Tectonophysics, 81 (1982), pp. 109–132.

ANDERSON, O.L., A. CHOPELAS, AND R. BOEHLER, *Thermal expansion vs. pressure at constant temperature: a re-examination*, in Geophys. Res. Lett., 17 (1990), pp. 685–688.

BUSSE, F.H. AND G. SCHUBERT, *Convection in a fluid with two phases*, J. Fluid Mech., 46 (1971), pp. 801–812.

CHOPELAS, A., AND R. BOEHLER, *Thermal expansion measurements at very high pressure, systematics, and a case for a chemically homogeneous mantle*, Geophys. Res. Lett., 16 (1989), pp. 1347–1350.

CHRISTENSEN, U., *Phase boundaries in finite amplitude mantle convection*, Geophys. J.R. Astron. Soc., 68 (1982), pp. 487–497.

CHRISTENSEN, U.R., AND D.A. YUEN, *Layered convection induced by phase transitions*, J. Geophys. Res., 90, no. B12 (1984), pp. 10291–10300.

DZIEWONSKI, A.M., *Mapping the lower mantle: determination of lateral heterogeneity in P velocity up to degree and order 6*, J. Geophys. Res., 89 (1984), pp. 5929–5952.

FORTE, A.M. AND W.R. PELTIER, *Plate tectonics and aspherical earth structure: the importance of poloidal-toroidal coupling*, J. Geophys. Res., 92 (1987), pp. 3645–3679.

FORTE, A.M. AND W.R. PELTIER, *Core mantle boundary topography and whole-mantle convection*, Geophys. Res. Lett., 16 (1989), pp. 621–624.

FORTE, A.M. AND W.R. PELTIER, *Mantle convection and core-mantle boundary topography: explanations and implications*, Tectonophysics, 187 (1991a), pp. 91–116.

FORTE, A.M. AND W.R. PELTIER, *Viscous flow models of global geophysical observables. I. Forward Problems*, J. Geophys. Res., in press 1991b.

HAGER, B.H., *Subducted slabs and the geoid: constraints on mantle rheology and flow*, J. Geophys. Res., 89 (1984), pp. 6003–6015.

HAGER, B.H., *Mantle viscosity: a comparison of models from postglacial rebound and from the geoid, plate driving forces, and advoctive heat flux*, in Glacial Isostasy, Sea Level and Mantle Rheology, R. Sabadini ed., Kluwer Academic Publishers, Dordrecht (1991), pp. 493–513.

HAGER, B.H. AND R.W. CLAYTON, *Constraints on the structure of mantle convection using seismic observations, flow models, and the geoid*, in Mantle Convection, edited by W.R. Peltier, Gordan and Breach Science Publishers, New York, (1989), pp. 657–763.

HAGER, B.H., R.W. CLAYTON, M.H. RICHARDS, R.P. COMER, AND A.M. DZIEWONSKI, *Lower mantle heterogeneity, dynamic topography and the geoid*, Nature, 313 (1985), pp. 541–545.

HAGER, B.H. AND M.A. RICHARDS, *Long wavelength variations in Earth's geoid: physical models and dynamical implications*, Phil. Trans. R. Soc. Lond., A328 (1989), pp. 309–327.

HART, S. AND A. ZINDLER, *Constraints on the nature and development of chemical heterogeneities in the mantle*, in Mantle Convection (W.R. Peltier ed.), Gordan and Breach Science Publishers, New York (1989), pp. 261–388.

HASKELL, N.A., *The motion of a viscous fluid under a surface load. 1*, Physics (N.Y.), 6 (1935), pp. 265–269.

HASKELL, N.A., *The motion of a viscous fluid under a surface load . 2*, Physics (N.Y.), 7 (1936), pp. 56–61.

ISAAKS, B., AND MOLNAR, P., *Distribution of stresses in the descending lithosphere from a global survey of focal mechanism solutions of mantle earthquakes*, Rev. Geophys. Space Phys., 9, pp. 103–174.

ITO, E. AND E. TAKAHASHI, *Postspinel transformations in the system* $Mg_2SiO_4 - Fe_2SiO_4$ *and some geophysical implications*, J. Geophys. Res., 94, no. B8 (1989), pp. 10637–10646.

JARVIS, G.T. AND W.R. PELTIER, *Mantle convection as a boundary layer phenomenon*, Geophys. J.R. Astron. Soc., 68 (1982), pp. 389–427.

JEANLOZ, R. AND S. MORRIS, *Temperature distribution in the crust and mantle*, Ann. Rev. Earth Planet. Sci., 14 (1986), pp. 377–415.

JEANLOZ, R. AND A.B. THOMPSON, *Phase transitions and mantle discontinuities*, Rev. Geophys. Space Phys., 21 (1983), pp. 51–74.

KATSURA, T. AND E. ITO, *The system* $Mg_2SiO_4 - Fe_2SiO_4$ *at high pressures and temperatures: precise determination of stabilities of Olivine, modified Spinel, Spinel*, J. Geophys. Res., 94, no. B11 (1989), pp. 15663–15670.

LERCH, F.S., S.M. KLOSKO, R.E. LAUBSCHER AND C.A. WAGNER, GRAVITY MODEL IMPROVEMENT USING GEOS3 (GEM9 AND GEM10), J. Geophys. Res., 84 (1979), pp. 3897–3916.

MACHETAL, P., AND WEBER, PATRICE, *Intermittent layered convection in a model mantle with an endothermic phase change at 670 km*, Nature, 350 (1991), pp. 55–57.

MINSTER, J.B. AND T.H. JORDAN, *Present-day plate motions*, J. Geophys. Res., 83 (1978), pp. 5331–5354.

MITROVICA, J.X. AND W.R. PELTIER, *Pleistocene deglaciation and the global gravity field*, J. Geophys. Res., 94 (1989), pp. 13651–13671.

MORELLI, A., AND A.M. DZIEWONSKI, *Topography of the core-mantle boundary and lateral homogeneity of the liquid core*, Nature, 325 (1987), pp. 678–683.

OLSON, P., P.G. SILVER, AND R.W. CARLSON, Nature, 344 (1990), pp. 209–215.

O'NIONS, R.K., N.M. EVENSON, AND P.J. HAMILTON, *Geochemical modelling of mantle differentiation and crustal growth*, J. Geophys. Res., 84 (1979), pp. 6091–6101.

PELTIER, W.R., *Mantle convection and viscosity*, In: A. Dziewonski and E. Boschi (Editors), Physics of the Earth's Interior, North Holland, Amsterdam (1980), pp. 362–431.

PELTIER, W.R., *Ice age geodynamics*, Annu. Rev. Earth Planet. Sci., 9 (1981), pp. 199–225.

PELTIER, W.R., *Mantle convection and viscoelasticity*, Ann. Rev. Fluid Mech., 17 (1985), pp. 561–608.

PELTIER, W.R., *Mantle viscosity*, in W.R. Peltier (ed.), Mantle Convection: Plate Tectonics and Global Dynamics, Gordan and Breach Sci. Publ., New York (1989), pp. 389–478.

PELTIER, W.R., A.M. FORTE, J.X. MITROVICA, AND A.M. DZIEWONSKI, *Earth's gravitational field: seismic tomography resolves the enigma of the Laurentian anomaly*, Nature, in press, 1991.

PELTIER, W.R., AND G.T. JARVIS, *Whole mantle convection and the thermal evolution of the earth*, Phys. Earth Planet. Int., 29 (1982), pp. 281–304.

PELTIER, W.R., G.T. JARVIS, A.M. FORTE AND L.P. SOLHEIM, *Radial structure of the mantle general circulation*, in W.R. Peltier ed. (op. cit.) (1989), pp. 765–815.

REYNARD, BRUNO AND GEOFFREY D. PRICE, *Thermal expansion of mantle minerals at high pressures – a theoretical study*, Geophys. Res. Lett., 17 (1990), pp. 689–692.

RICHARDS, M.A. AND B.H. HAGER, *Geoid anomalies in a dynamic earth*, J. Geophys. Res., 89 (1984), pp. 5987–6002.

RICHTER, F.M., AND C. JOHNSON, *Stability of a chemically layered mantle*, J. Geophys. Res., 79 (1974), pp. 1635–1639.

SAFRONOV, V.S., *Evolution of the Protoplanetary Cloud and Formation of the Earth and Planets*, Moscow: Nauka. In Russian (1969). English transl: NASA TT-F-677, 1972.

SCHUBERT, G. AND D.L. TURCOTTE, *Phase changes and mantle convection*, J. Geophys. Res., 76 (1971), pp. 1424–1432.

SCHUBERT, G., D.A. YUEN, AND D.L. TURCOTTE, *Role of phase transitions in a dynamic mantle*, Geophys. J.R. Astron. Soc., 42 (1975), pp. 705–735.

SOLHEIM, L.P. AND W.R. PELTIER, *Heat transfer and the onset of chaos in a spherical, axisymmetric, anelastic model of whole mantle convection*, Geophys. Astrophys. Fluid Dyn., 53 (1990), pp. 205–255.

SOLHEIM, L.P. AND W.R. PELTIER, *The influence of solid-solid phase transformations on the radial mixing length in mantle convection*, Can. J. Earth Sci., in press, 1991.

SU, WEI-JA, AND A. M. DZIEWONSKI, *Predominance of long-wavelength heterogeneity in the mantle*, Nature, 352 (1991), pp. 121–126.

TUSHINGHAM, A.M. AND W.R. PELTIER, *Ice-3G: A new global model of late Pleistocene deglaciation based upon geophysical predictions of post-glacial relative sea level change*, J. Geophys. Res., 96 (1991a), pp. 4499–4523.

TUSHINGHAM, A.M. AND W.R. PELTIER, *Validation of the ICE-3G model of Würm-Wisconsin deglaciation using a global data base of relative sea level histories*, J. Geophys. Res., in press (1991b).

WALCOTT, R.L., *Rheological models and observational data of glacio-isostatic rebound*, In "Earth Rheology, Isostasy, and Eustasy" (N.-A. Mörner ed.), 3–10, Wiley (New York), 1980.

WASSERBURG, G.J. AND D.J. DEPAULO, *Models of Earth structure inferred from Neodymium and Strontium isotopic abundances*, Proc. Natl. Acad. Sci. U.S.A., 76, (1979), pp. 3594–3598.

WOODHOUSE, J.H. AND A.M. DZIEWONSKI, *Mapping the lower mantle: Three dimensional model of earth structure by inversion of seismic waveforms*, J. Geophys. Res., 89 (1984), pp. 5953–5986.

YUEN, D.A., W. ZHAO, AND M. LIU, *Effects of phase transitions on mantle convection*, EOS, 71, no. 43 (1990), p. 1293.

VIGOROUS MOTIONS IN MAGMA CHAMBERS AND LAVA LAKES

HERBERT E. HUPPERT* AND M. GRAE WORSTER†

Abstract. Magma chambers and lava lakes are storage reservoirs of molten rock, which gradually cools and solidifies after emplacement. The molten rock is a multicomponent silicate melt that crystallizes to form a solid whose components differ locally from those of the original melt. This difference in composition can drive strong convective motions in the melt and cause large-scale zonation of minerals.

This paper considers a series of general, one-dimensional problems in which a layer of a one- or two-component melt is cooled from above. We describe the fundamental fluid-mechanical processes involved, which include: the influence of compositional effects on the solidification; the formation of mushy layers — regions in which fluid and solid co-exist; and the kinetic effects associated with disequilibrium thermodynamics. The solutions of the resultant mathematical models are compared with data from laboratory experiments using aqueous solutions. The agreement between the two sets of results is shown to be very good. Calculations using the simple, geologically oriented, two-component anorthite-diopside system are presented and their geological interpretation discussed briefly.

1. Introduction. Many, if not most, investigations in geophysics involve solving inverse problems, though they are not always formulated as such. The continual appearance of inverse problems occurs because our measurements of the earth are mainly constrained to be taken on the surface, with only a few measurements, such as the gravity field and the magnetic field, taken from satellites. This means that our knowledge of the interior of the earth is gained almost entirely by inference rather than by pure deduction. However, before the concepts and techniques of inverse theory can be applied to any problem, it is necessary to have a complete understanding of solutions related to the direct problem.

This general statement is true for the particular investigations of the physical and chemical processes that occur in magma chambers, which are large storage reservoirs of molten rock that are supposed to exist a few kilometres beneath all volcanoes, of both the subaerial and submarine variety. No measurements have ever been made within a molten magma chamber, and are unlikely to be made in the foreseeable future. Our knowledge is thus based on measurements from the occasional outpourings from such chambers, as seen, for example, in figure 1 (see Plates section), and from the theoretical concepts and calculations of the processes that are thought to occur within the chambers. One process of considerable importance in the temporal evolution of magma chambers is the slow transfer of heat from the chamber to the surrounding country rocks and the resulting solidification that occurs. In order to investigate some aspects of this evolution, the current paper presents the solutions to a series of one-dimensional problems describing the solidification of an initially fluid

*Institute of Theoretical Geophysics, University of Cambridge, 20 Silver Street, Cambridge CB3 9EW, UK. The work of the first author is partially supported by a grant from Venture Research International.

†Department of Engineering Sciences & Applied Mathematics, *and* Department of Chemical Engineering, Northwestern University, Evanston Illinois 60208, USA

layer due to cooling that takes place at the top of the layer. We sequentially build up the fundamental concepts and then apply them in order to discuss quantitatively the thermodynamic and fluid dynamic processes that occur in magma chambers.

The simplest situation to analyse, in which a one-component, or pure, melt solidifies, is known as the Stefan problem [1,2]. A semi-infinite layer of liquid, which is initially at a uniform temperature, is suddenly subjected at its base to a maintained temperature that is below the solidification temperature. Because the subsequent temperature in the liquid increases with height above the base it is assumed that no fluid motion occurs. The (one-dimensional) temperature profile in both the liquid and the resulting solid is determined by solving time-dependent diffusion equations in both media, possibly with a different coefficient of thermal diffusion in each layer. Because there is no length or time scale specified by the problem, the solution can be expressed in terms of a similarity variable and the solid phase grows in thickness at a rate proportional to $t^{1/2}$, where t is the time from the initiation of the temperature at the base.

Cooling a layer of pure melt from above, rather than from below, leads to a temperature profile that increases with depth and hence, in most cases, is thermally unstable. This generally results in fluid motions, in the form of convection, whose vigour is determined by the Rayleigh number based on the (finite) depth of the liquid layer remaining at any time. This situation was first analysed by Turner, Huppert and Sparks [3] who explained all the essential features of the motion and evaluated the rate of solid production. However, their analysis was presented in an appendix whose primary aim was to aid the explanation of a series of experiments in which two-component melts were solidified. In addition, the subject and its interpretation has advanced since that time and the results of the calculations have been occasionally misinterpreted. It thus seems sensible to present a new, careful description of the problem in Section 2. We will also explain there what general principles can be learnt from the analysis about the fluid behaviour and the solidification processes in lava lakes.

In Section 3, we commence a discussion of the influence of compositional effects on both the solidification processes and the resulting fluid motions. We consider the cooling from above of a layer of a two-component melt that releases less dense fluid upon solidification. As solidification proceeds, a mushy layer develops, which consists of sparse interconnected solid with interstitial melt. Our initial analysis of this mushy layer, following Kerr *et al.* [4,5], considers the solidification to take place under strict thermodynamic equilibrium. In Section 4 we replace this constraint by one that incorporates kinetic effects into the solidification process [6]. Possibly the most important conclusion of this section is that vigorous convective motions can still result in the fluid even though there is no initial superheat; that is, even though the fluid lies initially on the liquidus and carries no additional thermal energy.

We quantitatively consider the consequences of this conclusion, and the other results that are derived from the model, for the two-component system made up of anorthite and diopside in Section 5. This simple system can be considered to have

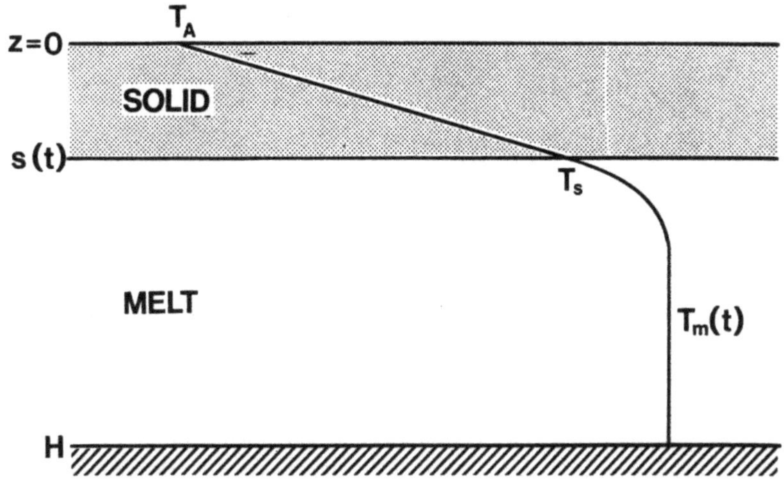

Figure 2: A schematic diagram of a layer of fluid of depth H being cooled and solidified from above. The upper boundary, $z = 0$, is maintained at a constant temperature T_A that is lower than the solidification temperature of the melt T_s, while the lower boundary is insulated. The temperature of the melt $T_m(t)$ is kept uniform by the convective motions of the melt and decreases with time owing to the convective heat transfer to the solidifying roof.

many properties similar to those of a multi-component magma. In Section 6 we extend this analysis to consider further the cooling and solidification of lava lakes. The paper ends with a summarizing set of conclusions.

2. Cooling a pure melt from above. We consider in this section the (one-dimensional) response of a layer of pure melt, at initial temperature T_0, which is subject to a maintained temperature T_A at the top, $z = 0$, and, for simplicity, is insulated at the base, $z = H$, as sketched in figure 2. The solidification temperature is taken to be T_s with $T_A < T_s < T_0$. Because thermal energy is continually being removed from the system, the melt gradually solidifies and at a time, t_s say, after the initiation of the top temperature T_A the system is completely solid.

At the unknown interface, $z = s(t)$, between the solid and liquid, two equations must be satisfied. One is that the temperature at the interface is given by the solidification temperature, which is expressed mathematically by

$$T[z = s(t), t] = T_s. \tag{2.1}$$

The other equation reflects the physical fact that there is conservation of heat across the interface. On the assumption, subject to *a posteriori* verification, that the melt convects vigorously, the conservation relationship becomes [1]

$$k_s \frac{\partial T}{\partial z}\bigg|_{s-} = F_T + [c_m(T_m - T_s) + \mathcal{L}]\frac{ds}{dt}, \tag{2.2}$$

where

$$F_T = \gamma k_m (\alpha g/\kappa_m \nu)^{1/3}(T_m - T_s)^{4/3}, \tag{2.3}$$

k_s and k_m are the thermal conductivities of the solid and the melt respectively, F_T, the thermal flux from the vigorously convecting fluid to the interface, is given by the oft-used four-thirds law (2.3) [7], c_m is the specific heat per unit volume of the melt, T_m is the mean temperature of the melt, $\mathcal{L}$ is the latent heat released as melt transforms to unit volume of solid, γ is an empirical constant with a value of 0.14, α is the coefficient of thermal expansion of the melt, g is the acceleration due to gravity, κ_m is the thermal diffusivity of the melt and ν is the viscosity of the melt, which is assumed to be constant. The assumption has also been made that any change in volume upon solidification can be neglected. The thermal flux F_T leads to a decrease in temperature of the melt layer which is expressed by

$$F_T = -c_m(H - s)\frac{dT_m}{dt}. \tag{2.4}$$

The conductive flux which appears on the left-hand-side of (2.2) needs to be evaluated from the solution of the diffusion equation in the solid. Alternatively, on the assumption that the thickness of the solid $s(t)$, which is governed by the release of latent heat, is very much less than the scale of thermal diffusion in the solid, $(\kappa_m t)^{1/2}$, the temperature profile in the solid is very nearly linear and

$$T_z \mid_{s-} = (T_s - T_A)/s. \tag{2.5}$$

The mathematical problem thus consists of solving (2.2)–(2.5) together with the initial conditions

$$T_m = T_0, \quad s = 0 \qquad (t = 0). \tag{2.6a, b}$$

It is convenient at this stage to define dimensionless variables by

$$u = (T_m - T_s)/(T_0 - T_s), \quad v = s/H \quad \text{and} \quad \tau = \kappa_m t/H^2, \tag{2.7a, b, c}$$

where time has been non-dimensionalised by the conductive time scale, H^2/κ_m. The variable u represents the remaining superheat of the melt, while v represents the fraction of the layer solidified. Substituting (2.7) into (2.2)–(2.6), we obtain

$$\frac{1}{v} = \mathrm{Q}\mathcal{N}u^{4/3} + (\mathrm{Q}u + \tilde{\mathcal{S}})\frac{dv}{d\tau}, \tag{2.8}$$

[1]The term involving c_m on the right-hand side of (2.2) was inadvertently omitted in [3]; however, in most cases it plays a minor role. Further details of the differences between the current presentation and that in [3] are discussed in footnote "2".

(2.9) $$(1-v)\frac{du}{d\tau} = -\mathcal{N}u^{4/3}$$

and

(2.10) $$u = 1 \quad v = 0 \qquad (\tau = 0),$$

in terms of the three dimensionless parameters that describe the system

(2.11a) $$\mathrm{Q} = \frac{k_m}{k_s}\frac{(T_0 - T_s)}{(T_s - T_A)},$$

a modified Stefan number

(2.11b) $$\tilde{\mathcal{S}} = \frac{k_m}{k_s}\frac{\mathcal{L}}{c_m(T_s - T_A)},$$

and the initial Nusselt number for the melt

(2.11c) $$\mathcal{N} = \gamma\left[\alpha g(T_0 - T_s)H^3/\kappa_m\nu\right]^{1/3}$$

(2.11d) $$\equiv \gamma \mathrm{Ra}_{\mathrm{init}}^{1/3},$$

where Ra_{init} is the initial Rayleigh number of the system. The physical interpretation of the governing equation (2.8) is that the conductive heat flux through the solid balances the three terms on the right-hand side of (2.8) which represent respectively: the convective flux that is transferred from the melt; the sensible heat that is lost as the melt is cooled to its solidification point; and the latent heat that is released upon solidification.

In general, solutions to the non-linear coupled set of equations (2.8)–(2.10) can only be obtained numerically. However, useful approximate analytical solutions can be derived in the geologically relevant situation of $\mathcal{N} \gg 1$, with Q and $\tilde{\mathcal{S}}$ fixed. The analysis is based on the principle that, since each term in (2.8) is positive, the term on the left-hand side is always important. Initially, because of the singularity of this term, as indicated by (2.10b), it is balanced by the last two of the three terms that make up the right-hand side of (2.8). As τ increases and the rate of advance of the interface $dv/d\tau$ decreases, the first term of (2.8) becomes important. Later, its influence wanes as the superheat u decays, and the balance of terms returns to the initial one.

In physical terms, this variation is expressed as: an initial balance between a large conductive flux through the thin solid phase and latent heat release; a period during which the convective flux from the interior exceeds the latent heat release and retards growth of the solid; and a final period during which latent-heat effects again overpower the dwindling convective flux. The analysis below demonstrates that the initial phase occurs for $0 < \tau < O(\mathcal{N}^{-2})$, the intermediate stage for $O(\mathcal{N}^{-2}) < \tau < O(\mathcal{N}^{-6/7})$ and the final stage for $O(\mathcal{N}^{-6/7}) < \tau < \tau_s$, where τ_s is the (dimensionless) time, of order unity, for total solidification to occur.

The analysis proceeds in three stages. In the first, at early times, (2.8)–(2.10) can be approximated as

$$\frac{1}{v} = (Qu + \tilde{S})\frac{dv}{d\tau} \tag{2.12}$$

$$\frac{du}{d\tau} = 0 \tag{2.13}$$

$$v = 0, \quad u = 1 \qquad (\tau = 0), \tag{2.14a, b}$$

the solution to which is

$$v = \left(\frac{2}{Q + \tilde{S}}\right)^{1/2} \tau^{1/2} \qquad \text{and} \qquad u = 1. \tag{2.15a, b}$$

This solution is valid until

$$Q\mathcal{N} \sim (Q + \tilde{S})\frac{dv}{d\tau} \tag{2.16a}$$

$$= \left(\frac{Q + \tilde{S}}{2}\right)^{1/2} \tau^{-1/2}, \tag{2.16b}$$

that is, until

$$\tau \sim \frac{Q + \tilde{S}}{2Q^2}\mathcal{N}^{-2}, \tag{2.17}$$

at which time

$$v \sim (Q\mathcal{N})^{-1} \ll 1. \tag{2.18}$$

For a while thereafter, the first term dominates the other terms of the right-hand side of (2.8), and equations (2.8) and (2.9) can then be written as

$$\frac{1}{v} = Q\mathcal{N}u^{4/3} \tag{2.19}$$

$$\frac{du}{d\tau} = -\mathcal{N}u^{4/3} \tag{2.20}$$

with matching to (2.8) giving the initial condition

$$u(0) = 1 \tag{2.21}$$

to leading order. The solution of (2.19)–(2.21) can be readily obtained as

$$u = \left(1 + \tfrac{1}{3}\mathcal{N}\tau\right)^{-3} \quad \text{and} \quad v = (Q\mathcal{N})^{-1}\left(1 + \tfrac{1}{3}\mathcal{N}\tau\right)^{4}. \tag{2.22a, b}$$

This solution remains valid until

$$Q\mathcal{N}u^{4/3} \sim \tilde{S}\frac{dv}{d\tau}, \tag{2.23}$$

with u and v given by (2.22). That is, until

(2.24)
$$\tau \sim \left(\frac{3Q^2}{4\tilde{\mathcal{S}}}\right)^{\frac{1}{7}} \mathcal{N}^{-\frac{6}{7}}.$$

At this time

(2.25a, b)
$$u \sim \left(\frac{3Q^2}{4\tilde{\mathcal{S}}}\right)^{-\frac{3}{7}} \mathcal{N}^{-\frac{3}{7}} \quad \text{and} \quad v \sim \frac{1}{Q}\left(\frac{3Q^2}{4\tilde{\mathcal{S}}}\right)^{\frac{4}{7}} \mathcal{N}^{-\frac{3}{7}}.$$

In the short period of time given by (2.24), convection causes the superheat u to decay to the small value given by (2.25a) while little solid has grown, as indicated by (2.25b).

In the final stage, the thermal balance is similar to that expressed in equation (2.8), except that now $u \sim 0$ to leading order. Thus

(2.26)
$$\frac{1}{v} = \tilde{\mathcal{S}}\frac{dv}{d\tau},$$

while equation (2.9) is most conveniently written in the form

(2.27)
$$\frac{1-v}{v}\frac{du}{dv} = -\mathcal{N}u^{\frac{4}{3}}.$$

The solution to equations (2.26) and (2.27) that matches with the solutions at earlier times is given by

(2.28a, b)
$$v = \left(\frac{2}{\tilde{\mathcal{S}}}\right)^{\frac{1}{2}} \tau^{\frac{1}{2}} \quad \text{and} \quad u = \left[1 - \tfrac{1}{3}\mathcal{N}\tilde{\mathcal{S}}(\ln(1-v)+v)\right]^{-3}.$$

Note that this asymptotic expression for u is uniformly valid for all times [2].

Some results of integrating equations (2.8)–(2.10) numerically are shown in figures 3 and 4. The general behaviour of this system of equations is best seen in figure 3, for which the parameter values used were $\tilde{\mathcal{S}} = 1$, $Q = 1$, and $\mathcal{N} = 100$. This value of Q is much larger than would typically apply to magma chambers but the values chosen here serve to illustrate the structure of the solutions. The curve for v in figure 3a has two points of inflexion, corresponding to the transitions between the three stages of evolution represented by (2.15), (2.22), and (2.28). These transitions can be seen more easily in figure 3b in which $\log v$ is plotted as a function of $\log \tau$. The transition times of $O(\mathcal{N}^{-2})$ and $O(\mathcal{N}^{-6/7})$ are indicated with dashed lines. Note that v increases like $\tau^{1/2}$ in both the first and third stages of evolution. During the second stage of evolution, the large convective heat flux from the melt serves to slow down the rate of growth until the superheat has decayed to small values and the convective

[2] Most of the solutions obtained in the current presentation have their antecedents in the solutions derived in [3]. The two main differences in the presentations are, first, the different non-dimensionalizations of time that are employed, and second, that, in [3], solutions for different ranges of the dimensionless parameters are discussed, whereas, in the current paper, we consider all the parameters to be fixed and determine the temporal evolution of the solutions under the assumption that $\mathcal{N} \gg 1$

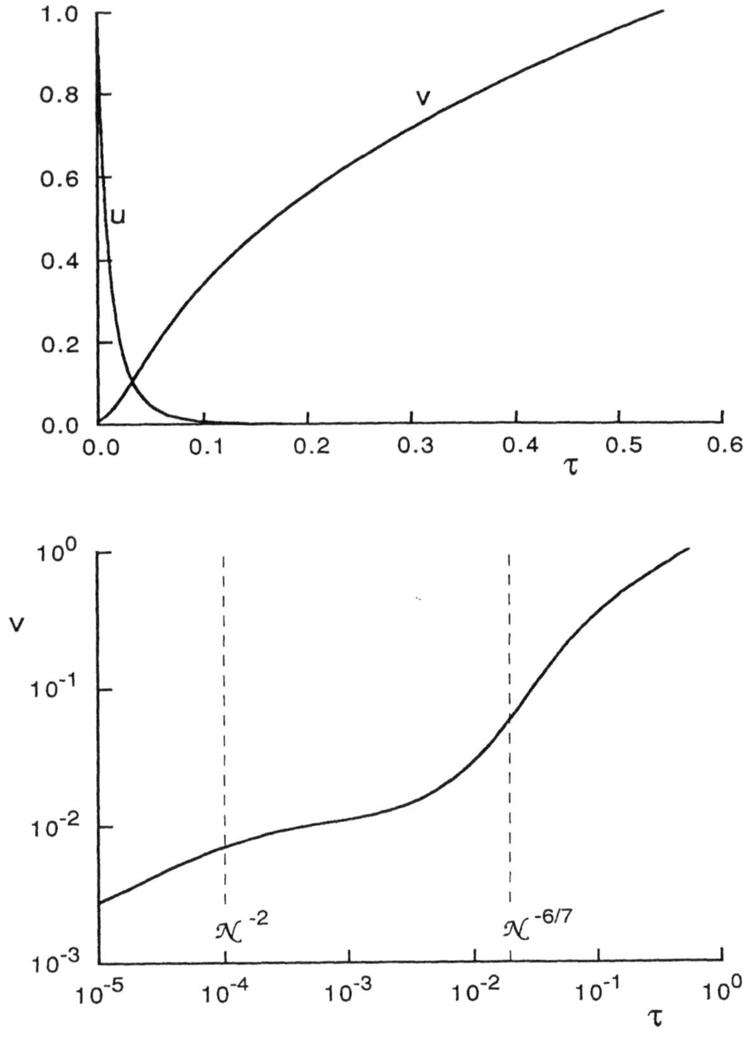

Figure 3: Results of numerical calculations of the solidification of a layer of liquid cooled from above using dimensionless parameter values of $\tilde{\mathcal{S}} = 1$, $\mathcal{N} = 100$, and $Q = 1$. (a) The dimensionless temperature of the melt u and the depth of solid at the roof v as functions of the dimensionless time τ. (b) The depth of the solid layer as a function of time shown on logarithmic axes to illustrate the different stages of evolution. The vertical dashed lines indicate the approximate transition times as determined by the asymptotic analysis in section 2.

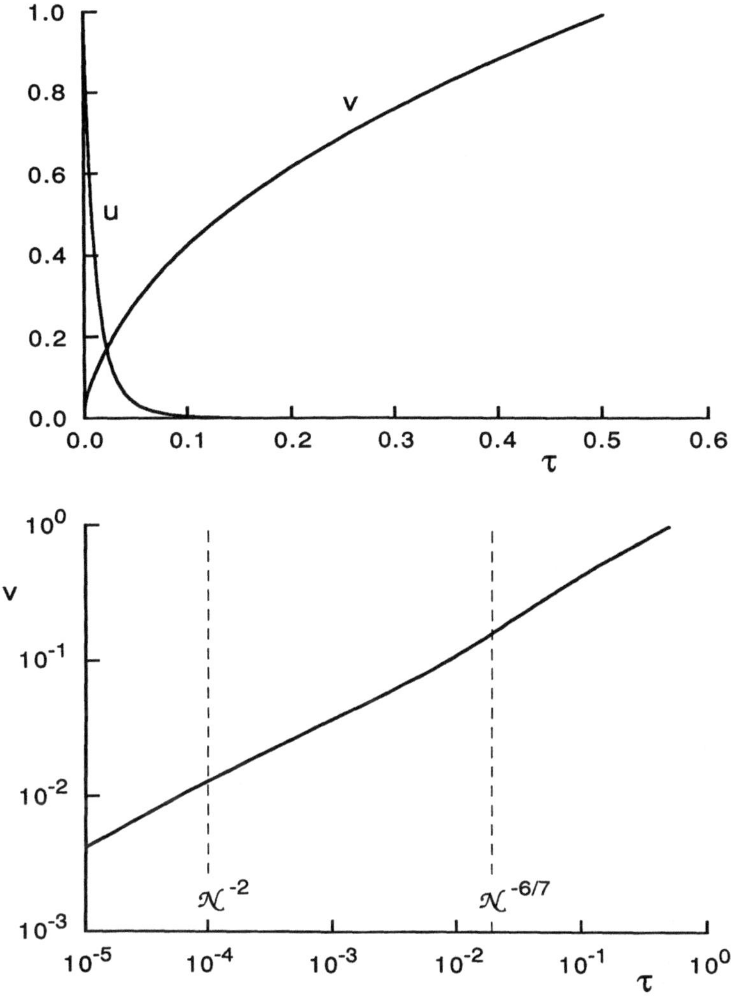

Figure 4: Results of numerical calculations of the solidification of a layer of liquid cooled from above using dimensionless parameter values of $\tilde{\mathcal{S}} = 1$, $\mathcal{N} = 100$, and Q = 0.1. (a) The dimensionless temperature of the melt u and the depth of solid at the roof v as functions of the dimensionless time τ. (b) The depth of the solid layer as a function of time shown on logarithmic axes to illustrate the different stages of evolution. The vertical dashed lines indicate the approximate transition times as determined by the asymptotic analysis in section 2. Note that, in contrast to figure 2b, the growth is approximately proportional to $\tau^{1/2}$ for all time. This is because the dimensionless initial superheat is smaller in this case, which causes the convection to have less influence on the solidification at the roof.

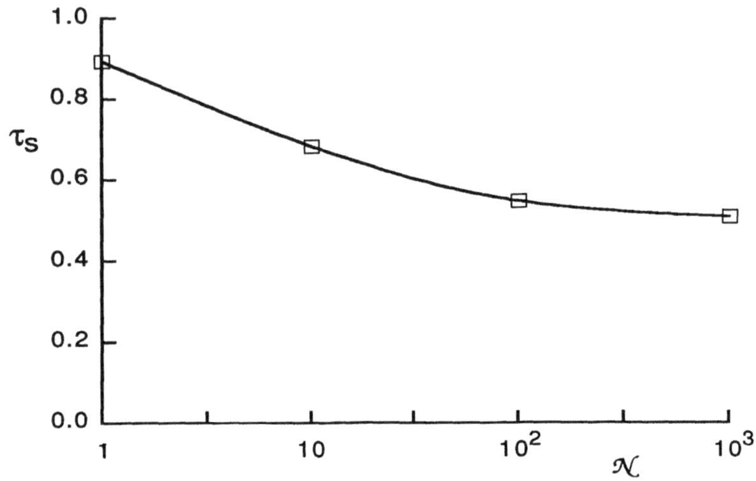

Figure 5: The dimensionless time τ_s for complete solidification as a function of the initial Nusselt number $\mathcal{N}$. Note that, as the convective intensity increases, the solidification time decreases relative to the timescale for thermal diffusion.

heat flux has diminished. The overall effect of convection, however, is to remove the superheat rapidly and thus to allow the system to solidify more quickly. Figure 3a shows the superheat u decaying rapidly to small values by the end of the second stage of evolution. Note, however, that this decay is only algebraic and that the Rayleigh number based upon these small values of the superheat remains large until dimensionless times of order unity.

Figure 4 shows the results obtained with parameter values of a size appropriate to magma chambers. In particular, the dimensionless initial superheat Q was taken to be only 0.1, while the other parameter values were kept the same as used for figure 3. We again see the superheat u decaying rapidly in figure 4a. Now, however, although the Rayleigh number based upon the value of the superheat is still very large, the convective heat flux is always small in comparison with the heat flux conducted through the roof, and the growth of v is almost proportional to $\tau^{1/2}$ for all time, as seen in figure 4b. Convection still plays a role in reducing the time for complete solidification, however, as can be seen in figure 5, which shows τ_s as a function of the initial Nusselt number $\mathcal{N}$.

The analysis above can be considered as an approximate model for the evolution of a lava lake. These form when lava from an eruption flows into a depression in the earth's surface. The lakes may be a few meters to a few tens of meters deep [8,9] and are cooled by contact with the atmosphere at the top surface. There are three major implications of the model just presented when the results are applied to a lava lake. The first is that any superheat is removed rapidly. It then follows that the lake solidifies in a time comparable to the conductive time scale H^2/κ. Finally, during this time, vigorous convective motions will take place in the interior of the lake. All three of these conclusions are in broad agreement with field data obtained from lava lakes, and remain valid even as we add more detail to the model in the following sections.

Further discussion of lava lakes is deferred until section 6.

3. Compositional effects. When the melt that is being cooled and solidified is composed of more than one pure component, various interesting additional phenomena occur. In general, the solid product is of a different composition from the melt. Thus compositional gradients are generated in the melt as it becomes locally depleted of the components forming the solid phase. The compositional gradients often result in density gradients that can drive convection. These same gradients of composition can be the cause of morphological instabilities of the solid–liquid interface that ultimately result in the formation of a mushy zone separating the completely solid region from the melt (see figure 6). Mushy zones can also be anticipated from the equilibrium phase diagram of a multi-component system. Whereas there is a single temperature (the melting temperature) at which a pure solid is in equilibrium with its melt, there is a range of temperatures in which solid can coexist in equilibrium with a multi-component melt. This range is bounded by the solidus temperature, below which the system is completly solid, and the liquidus temperature, above which the system is completely liquid. These temperatures as functions of composition are often plotted to form the equilibrium phase diagram for the multi-component system that shows which phases are in equilibrium in a sample of given uniform temperature and bulk composition. The particular phase diagram for the anorthite–diopside system is shown in figure 7, which is also a typical phase diagram for many simple two-component systems that freeze to form an almost pure solid (i.e. the solidus curves are vertical). Between the solidus and liquidus temperatures, solid and liquid typically coexist in intimate contact, such that the microstructure has a lengthscale that is very small compared with the lengthscale of the mushy layer as a whole. This fact is the basis for our mathematical modelling of the mushy region.

In this paper, we restrict attention to systems that are cooled from above and which leave a buoyant residual as solid is formed from the melt. Since the effect of compositional variations upon the density of the liquid is typically much greater than the effect of the associated temperature variations within mushy regions, the interstitial fluid within the mushy layer that forms near the cooled roof is stably stratified and we assume therefore that it is stagnant. The model presented here was developed in reference [5], where a careful and detailed account of its assumptions and conditions can be found.

The two-component system that we consider first is similar in many ways to the single-component system analysed in section 2 . The solid region of figure 2 is replaced by a mushy region in figure 6. The region of melt is modelled in the same way and its temperature obeys equation (2.4) with the heat flux given by equation (2.3), except that the solidification temperature T_s is replaced by the (unknown) temperature T_i of the interface between the mushy and liquid regions. If the diffusivity of solute D in

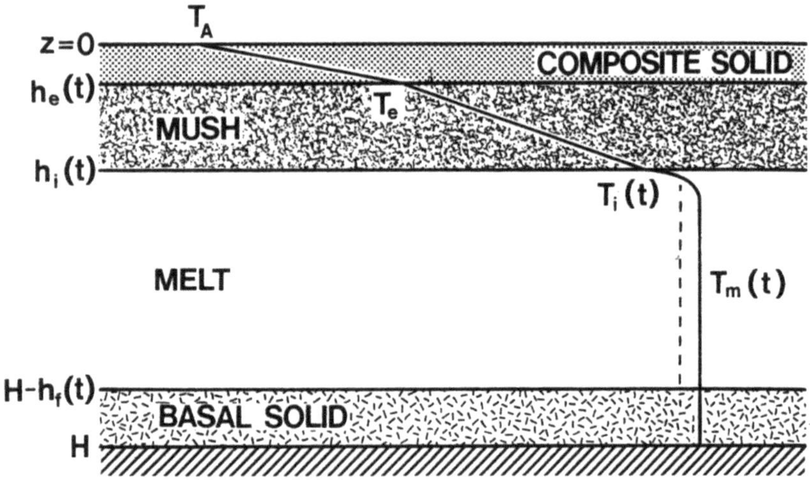

Figure 6: A schematic diagram of a solidifying layer of binary melt that is cooled from above. A mushy layer comprising interconnected crystals and interstitial liquid grows from the melt near the roof. If the temperature of the cooled roof is lower than the eutectic temperature then a completely solid, composite layer forms adjacent to the roof, comprising crystals of both end members of the binary system. A layer of crystals is observed to form near the insulated base in laboratory experiments and is predicted by the theory once the kinetic undercooling at the mush–liquid interface is taken into account.

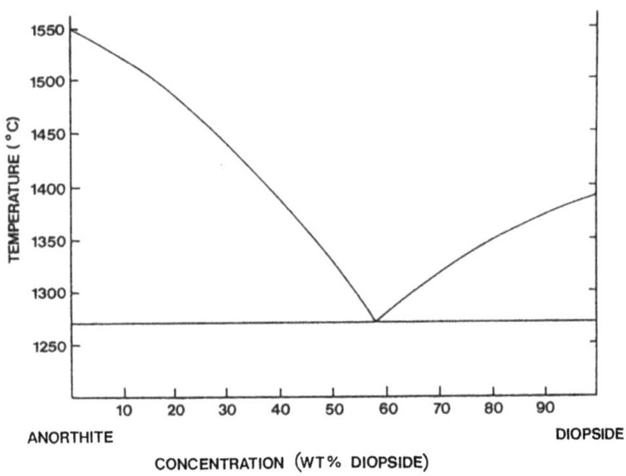

Figure 7: The equilibrium phase diagram for the anorthite–diopside system. The eutectic temperature of 1274°C is attained on the liquidus curve at a concentration of 58 wt% diopside.

the melt is small compared with the diffusivity of heat κ, as is typically the case, then the compositional boundary layer ahead of the mush–liquid interface is negligible and the interfacial temperature is given by

$$T_i = T_L(C_m), \tag{3.1}$$

where $T_L(C)$ is the liquidus temperature of the two-component system and C_m is the composition of the melt. Equation (3.1) is equivalent to equation (2.1) and is appropriate under the assumption of local equilibrium. In section 4 we shall replace equation (3.1) with a non-equilibrium growth law and discover important consequences of disequilibrium. For now we use the simpler condition of equilibrium thermodynamics in order to focus attention on effects related solely to having a multi-component melt, rather than a single-component melt.

The equation expressing conservation of heat at the mush–liquid interface (2.2) is replaced by

$$k_m T_z|_{h_i -} = F_T + (T_m - T_i)\frac{dh_i}{dt}, \tag{3.2}$$

where $z = h_i$ is the position of the mush–liquid interface. Note that there is no release of latent heat at this interface since the solid fraction in the mushy layer there is zero when $D/\kappa \ll 1$ and thermodynamic equilibrium is assumed [5,11]. Instead, the release of latent heat is distributed throughout the mushy layer as expressed by the thermal-diffusion equation, which now takes the form

$$\bar{c}\frac{\partial T}{\partial t} = \frac{\partial}{\partial z}\left(\bar{k}\frac{\partial T}{\partial z}\right) + \mathcal{L}\frac{\partial \phi}{\partial t}. \tag{3.3}$$

The release of latent heat is proportional to the rate of change of the solid fraction ϕ within the mushy layer, while the mean thermodynamic properties of the layer are functions of ϕ, which are taken to be

$$\bar{c} = \phi c_s + (1-\phi)c_m, \tag{3.4a}$$

$$\bar{k} = \phi k_s + (1-\phi)k_m. \tag{3.4b}$$

Note that equation (3.4a) is exact, while equation (3.4b) is an approximation that has led to good results in earlier studies [4,10,11]. An additional equation is required in order to determine how the solid fraction varies within the mushy layer. Since the microstructure of the mushy layer is so fine and the relative surface area of phase boundaries within the layer is consequently so large, it is a very good approximation to assume that

$$T = T_L(C) \tag{3.5}$$

within the layer; that is, the mushy region is considered to be in local thermodynamic equilibrium. The concentration C of the interstitial liquid is found by conservation of solute to satisfy

$$(1-\phi)\frac{\partial C}{\partial t} = (C - C_s)\frac{\partial \phi}{\partial t}, \tag{3.6}$$

where we have ignored the diffusion of solute.

Equations (3.1)–(3.6), together with equations (2.3) and (2.4), can be conveniently combined into the dimensionless system of equations and boundary conditions [5]

$$c\frac{\partial \theta}{\partial t} = \frac{\partial}{\partial z}\left(k\frac{\partial \theta}{\partial z}\right) \qquad (0 \leq z \leq h_i), \tag{3.7a}$$

$$\theta = -1 \qquad (z = 0), \tag{3.7b}$$

$$\theta = 0 \qquad (z = h_i), \tag{3.7c}$$

where

$$k \equiv \phi\frac{k_s}{k_m} + 1 - \phi, \tag{3.8a}$$

$$c \equiv \phi\frac{c_s}{c_m} + 1 - \phi + \frac{S}{\mathcal{C}}(1-\phi)^2, \tag{3.8b}$$

and

$$\phi \equiv \frac{-\theta}{\mathcal{C} - \theta}. \tag{3.8c}$$

with

$$\theta_m \dot{h}_i = \left.\frac{\partial \theta}{\partial z}\right|_{h_i -} - \tilde{\mathcal{N}}\theta_m^{\frac{4}{3}}, \tag{3.9}$$

$$(1 - h_i)\dot{\theta}_m = -\tilde{\mathcal{N}}\theta_m^{\frac{4}{3}}, \tag{3.10}$$

$$h_i = 0, \quad \theta_m = \theta_0 \qquad (t = 0). \tag{3.11a, b}$$

In these dimensionless equations, lengths and time have been scaled as in section 2, while the dependent variable θ is defined by

$$\theta = \frac{T - T_L(C_0)}{T_L(C_0) - T_A}. \tag{3.12}$$

Note that, here, the temperature has been scaled with the temperature difference across the solidifying region, rather than that across the thermal boundary layer in the melt as it was in section 2. Consequently, the parameter in equations (3.9) and (3.10) is $\tilde{\mathcal{N}} = \theta_0^{-1/3}\mathcal{N}$, where $\mathcal{N}$ is the initial value of the Nusselt number, and should not itself be misinterpreted as a Nusselt number. A new parameter is introduced in these equations,

$$\mathcal{C} = \frac{C_s - C_0}{C_0 - C_A}, \tag{3.13}$$

where $T_L(C_A) = T_A$, which represents the difference in composition between the solid and liquid phases relative to the variations in concentration of the liquid phase

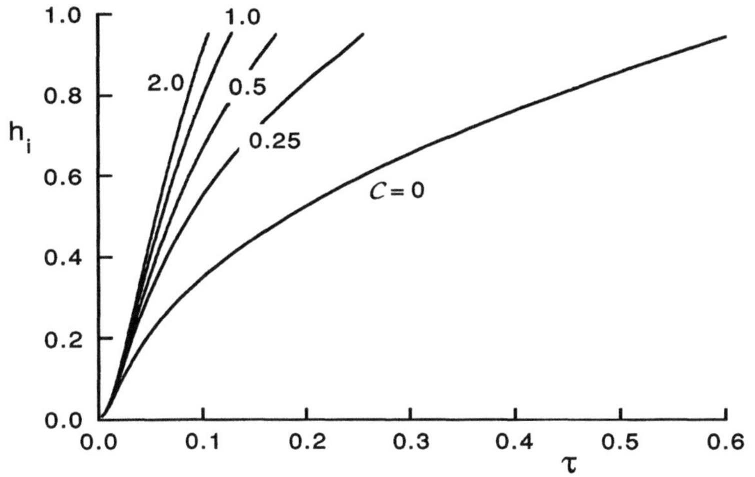

Figure 8: The dimensionless depth h_i of a mushy layer grown at thermodynamic equilibrium as a function of time τ for various values of the concentration ratio $\mathcal{C}$. The other dimensionless parameter values used were $\mathcal{S} = 1$, $\tilde{\mathcal{N}} = 100$, and $\theta_0 = 1$. Note that, as $\mathcal{C}$ increases, the solid fraction in the mushy layer decreases, which allows the depth of the mushy layer to grow more rapidly.

within the mushy layer. The value of $\mathcal{C}$ is zero for a pure melt and is always positive otherwise. In this system of equations, the Stefan number $\mathcal{S} = \mathcal{L}/c_m(T_L(C_0) - T_A)$ only appears divided by $\mathcal{C}$, so we can deduce that positive values of $\mathcal{C}$ serve to reduce the effective value of the Stefan number. This is because, as seen from equation (3.8c), the solid fraction in the mushy layer decreases as $\mathcal{C}$ increases, so less latent heat is required to be removed in order to increase the depth of the mushy layer. This is the principal effect of composition on the solidification of a multi-component melt and it causes the mush–liquid interface to advance more rapidly than the solid-liquid interface in a pure melt. This is illustrated by the results presented in figure 8. Note that increasing $\mathcal{C}$ produces similar results to those obtained by decreasing $\mathcal{S}$ [5].

A consequence of the fact that the release of latent heat is distributed throughout a mushy layer, rather than just at the mush–liquid interface, is that the rate of advance of the interface is controlled to a greater extent by a balance between the heat conducted through the mushy layer and the heat flux from the melt. Indeed, equations (3.7)–(3.11) predict an infinite rate of advance of the mush–liquid interface $\dot{h}_i$ when the superheat θ_0 is zero [11] so that there is no convective flux from the melt. In general, the decaying superheat of the region of melt can cause the interface to advance very much faster than it would if latent heat were released there. However, this runaway behaviour is associated with very small values of the solid fraction in the regions of the mushy layer nearest the melt [12] and one must question whether such regions will remain mechanically robust. In addition, we shall see in section 4 that kinetic effects at the mush–liquid interface restrict this runaway behaviour by causing non-zero values of the solid fraction at the mush–liquid interface.

Numerical solutions of the equations presented in this section have been found to agree very well with the results of laboratory experiments [5] in which mushy layers of solid ice were formed from aqueous solutions of isopropanol. This agreement, which can be seen in figure 9, gives support to the various approximations invoked in the development of the model. However, these same authors found that the region of melt became supersaturated during the course of the experiments (see figure 9b), an observation that is inexplicable by a theory employing the condition of equilibrium thermodynamics expressed by equation (3.1). Just as in section 2, the convection associated with equilibrium solidification at the roof can only reduce the superheat to small values; the melt cannot thereby be cooled below its liquidus temperature.

4. Kinetic effects. Mathematical models of convection driven by solidification that assume equilibrium thermodynamics, such as the models outlined in the previous two sections, show that the temperature of the melt decays towards the equilibrium solidification temperature (liquidus temperature) that is assumed to pertain to the solidification front. Convection ceases at some small value ΔT_c of the superheat, which is defined by the criterion that the Rayleigh number based upon that temperature difference and the full depth of the melt is less than its critical value. Brandeis and Marsh [13] coined the phrase 'convective liquidus' to apply to the temperature that the melt has when this occurs. They found experimentally that the 'convective liquidus' is almost equal to the actual liquidus temperature of the melt. Indeed, a simple scaling analysis shows that ΔT_c is given by

$$\Delta T_c \sim Ra_c \frac{\kappa \nu}{\alpha g H^3}, \tag{4.1}$$

where Ra_c is the critical value of the Rayleigh number ($Ra_c \approx 10^3$). This temperature difference is extremely small in magma chambers owing to the very large typical values of their depth H. For the parameter values in Table 1, $\Delta T_c \approx 10^{-9}\,{}^\circ$C.

Table 1. Parameter values corresponding approximately to those appropriate to a diopside melt in a layer 100m deep.

Quantity	Value	Units
κ	8×10^{-3}	$\mathrm{cm^2\,s^{-1}}$
μ	26	poise
ν	10	$\mathrm{cm^2\,s^{-1}}$
α	10^{-4}	${}^\circ\mathrm{C}^{-1}$
g	10^3	$\mathrm{cm\,s^{-2}}$
H	10^4	cm

However, the assumption of equilibrium thermodynamics is only an approximation; the actual temperature of an advancing phase boundary must always be slightly below the equilibrium liquidus temperature of the melt in order for the crystal to

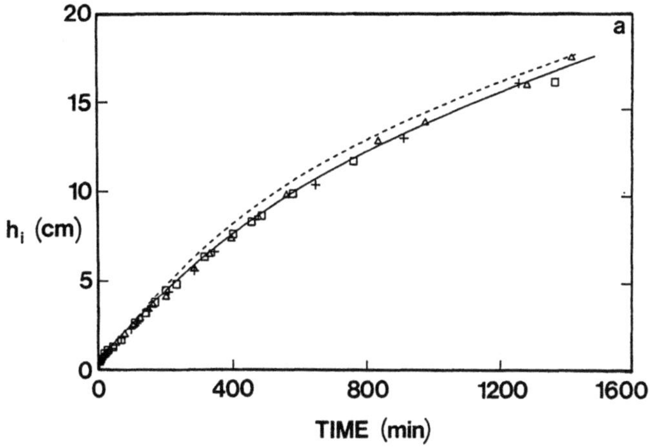

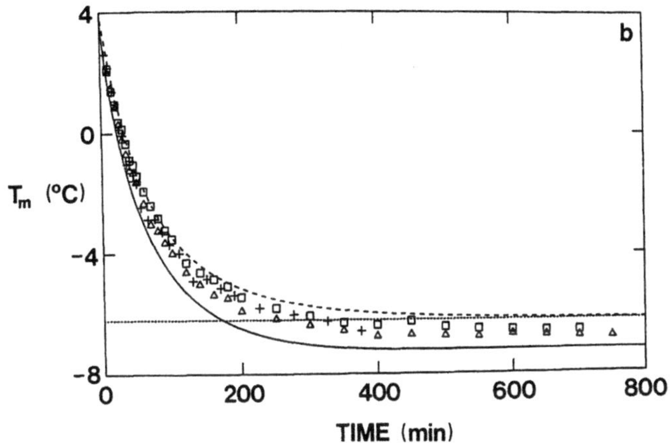

Figure 9: Comparisons of theoretical predictions with results of experiments on the solidification of ice from a mixture of water and isopropanol [5]. The dashed curves are the predictions of the model that assumes equilibrium thermodynamics, while the solid curves are predictions of the model that includes kinetic undercooling at the mush–liquid interface. Excellent agreement between theory and experiment is found for the depth of the mushy layer (figure a), and the theory successfully accounts for the observed supersaturation of the melt (figure b). Note that heat transfer from the laboratory may have caused the experimental data of temperature to be slightly higher than they would have been had the system been perfectly insulated.

grow [14]. This undercooling, ΔT_k, is a function of the rate of advance of the interface. So far as describing convection in the melt is concerned, the approximation of equilibrium thermodynamics is only good while the superheat is much larger than the kinetic undercooling ΔT_k. However, we shall see in section 5 that ΔT_k is typically much larger than ΔT_c, so that the concept of the 'convective liquidus' rarely applies, and convection in magma chambers continues until the depth of remaining melt is very small, even in cases in which there is no initial superheat.

The relationship between the kinetic undercooling and the rate of advance of the solidification front varies with the material that is being solidified and the molecular nature of the solidification process. For a solid that grows by the method of continuous growth [15], the relationship is linear of the form

$$\dot{h}_i = \mathcal{G}(T_L - T_i), \tag{4.2}$$

where $\mathcal{G}$ is a constant, T_L is the liquidus temperature of the melt and T_i is the temperature at the solidification front.

Kerr *et al.* [6] conducted experiments to verify equation (4.2) for the case of ice growing from a mixture of water and isopropanol. They found good agreement with a linear law for undercoolings up to about 3°C and determined the constant $\mathcal{G}$ to be approximately $2.2 \times 10^{-4}\mathrm{cm\,s^{-1}\,{}^\circ C^{-1}}$ for this system. They proceeded to incorporate the kinetic growth law (4.2) into the general model for the growth of mushy layers as follows.

A mathematical model for the solidification of a melt cooled from above that takes account of the kinetics of crystallization follows that presented in section 3 very closely, with equation (3.1) replaced by equation (4.2). The dimensionless system of governing equations and boundary conditions is given by (3.7)–(3.11) with the boundary condition (3.7c) replaced by

$$\theta = \theta_i \qquad (z = h_i), \tag{4.3}$$

and equations (3.9) and (3.10) replaced by

$$(\mathcal{S}\phi_i + \theta_m - \theta_i)\dot{h}_i = \left.\frac{\partial\theta}{\partial z}\right|_{h_i-} - \tilde{\mathcal{N}}(\theta_m - \theta_i)^{\frac{4}{3}}, \tag{4.4}$$

$$(1 - h_i)\dot{\theta}_m = -\tilde{\mathcal{N}}(\theta_m - \theta_i)^{\frac{4}{3}}, \tag{4.5}$$

together with the dimensionless kinetic growth law

$$\dot{h}_i = -\gamma\theta_i. \tag{4.6}$$

Note that, although kinetic undercooling is included at the mush–liquid interface (equation (4.6)) it is still neglected in the interior of the mushy layer since the very large specific surface area of solid–liquid interfaces in the interior allows for very rapid relaxation to equilibrium there.

These equations are solved by treating equation (4.4), with $\dot{h}_i$ replaced by the right-hand side of (4.6), as an algebraic equation for the unknown interface temperature θ_i and treating equation (4.6) as an evolution equation for h_i. Note that, since the interfacial temperature T_i is no longer zero, the solid fraction at the interface is also non zero, and there is consequently a contribution to the thermal balance at the interface (4.4) due to latent heat.

The incorporation of the kinetic growth law into the theoretical model makes only a small correction to the prediction of the growth of solid from the cooled roof, which is in excellent agreement with the experimental results of Kerr *et al.* [6], as can be seen in figure 9a. Much more importantly, the model now predicts the observed supersaturation of the melt that occurs after a finite time in experiments in which the melt is initially undersaturated (figure 9b).

Apart from maintaining convection, this phenomenon has important consequences for the formation of solid from the melt. Once the melt is supersaturated (supercooled), any nucleation sites within the body of the melt provide locations for further solidification of the melt, regardless of their position relative to the cooled boundary. This was seen dramatically in experiments in which sodium sulphate was crystallized from its aqueous solution [6,16]; crystal growth took place at the floor of the experimental tank, as well as near the roof of the tank, even though the system was cooled only at the roof.

As crystals grow in the interior of the melt, latent heat and residual liquid are released. The latent heat raises the temperature of the supercooled melt, while the residual liquid, which is of a lower concentration, depresses the liquidus temperature of the melt. These effects combine to cause the level of supercooling to be reduced. The crystals at the floor in the laboratory experiments were observed to have a very large surface area exposed to the melt, and this surface area would be even larger if, for example, the crystals were to grow in suspension. In addition, since the residual fluid is buoyant and rises convectively away from the crystals, the internal growth of crystals is very efficient both at reducing the level of supersaturation and in keeping the melt well mixed. A good approximation to the rate of internal crystal growth can therefore be made by assuming that it is sufficiently rapid to keep the melt close to equilibrium. We therefore assume that the uniform temperature T_m and composition C_m of the melt are related by equation (3.1), while conservation of solute demands that the composition of the melt is given by

$$(C_s - C_m)\dot{h}_f = (H - h_i - h_f)\dot{C}_m, \tag{4.7}$$

where h_f is the depth that the crystals would occupy if they were to form a solid layer at the floor of the chamber. Note, however, that the mathematical description takes no explicit account of where the internal crystal growth takes place and therefore equation (4.7) is strictly a prediction for the *volume* of crystals grown, with the interpretation that h_f is the volume per unit horizontal area.

As the composition of the melt changes due to the internal growth of crystals, so the composition of the liquid incorporated into the mushy layer also changes. The

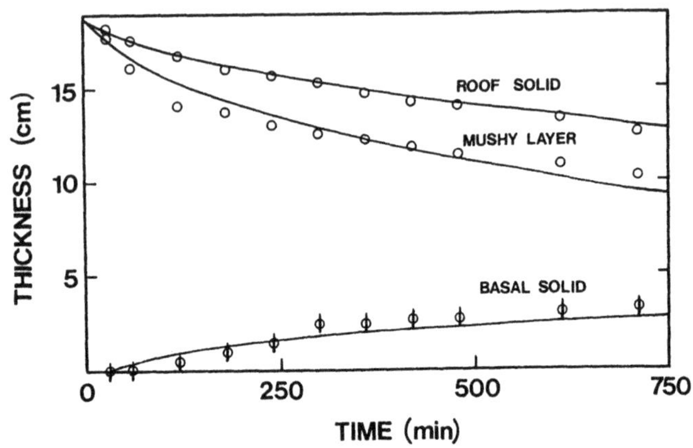

Figure 10: Comparisons of theoretical predictions with experimental results for the depths of various layers that formed during the complete solidification of an aqueous solution of sodium sulphate. The kinetic-growth parameter $\mathcal{G}$ was adjusted in the theoretical model in order to achieve satisfactory agreement with the experimental results.

bulk composition C_H of the mushy layer at depth z is given by

$$C_H(z) = C_m(t_i) \quad \text{where} \quad h_i(t_i) = z. \tag{4.8}$$

Thus there is macrosegregation of the solid caused by the internal growth of crystals.

Kerr *et al.* [6] used this mathematical model to predict successfully the rates of growth of crystals at the roof and at the floor (figure 10). They could also predict the compositional stratification of the solid [16] in experiments with aqueous systems that were solidified completely by maintaining the temperature of the roof below the eutectic temperature of the solution. A photograph of one of their experiments is shown in figure 11 (see Plates section). The kinetic coefficient for crystals of sodium sulphate decahydrate growing from an aqueous solution of sodium sulphate was found [6] to be approximately $1.5 \times 10^{-4}\mathrm{cm\,s^{-1}\,{}^{\circ}C^{-1}}$. In the following sections, we shall describe some calculations made of the solidification of magma chambers using this same system of equations.

5. Magma chambers: the anorthite-diopside system. The previous sections have used theoretical arguments confirmed by careful laboratory experiments to present the main concepts involved when a binary solution is cooled from above with the release of less dense fluid. A composite layer forms at the roof (as long as the temperature there is less than the eutectic temperature) with a stagnant mushy layer beneath it. At the interface between the mushy layer and the underlying convecting fluid, the temperature is slightly below the liquidus temperature, as quantified by the kinetic growth law, (4.2) for example, which is necessary for the solidification to continue. After the temperature of the convecting fluid has decreased from its initial (superheated) value to the liquidus temperature, solidification can also take place at the floor, with the accompanying latent heat that is released being transported by convection across the fluid layer and into the mushy layer. With these fundamental concepts in mind, we can make some quantitative calculations appropriate to the cooling and solidification that occurs in magma chambers. Our desire to carry out such calculations acted as one of the major motivations behind the research.

Magma chambers come in a variety of shapes and sizes (see, for example, [17,18,19,20,21]. Many, however, are believed to have a horizontal scale which is considerably in excess of the vertical scale. This suggests that for these chambers the two- (and possibly three-) dimensional effects which occur at the ends of the chamber will be much less pronounced than those which occur at the floor and roof. We shall consider these to be horizontal so that a one-dimensional problem results. However, there may well be further details involved due to the irregular shape of the roof and floor of all magma chambers, though this is beyond our present analysis. We shall consider the base of the chamber to be insulated, though this too is an approximation, and note simply that any heat lost through the floor will only enhance the effects discussed in this section. Some investigations of the purely thermal aspects of this problem have been considered in [22,23]. At the top of the chamber, heat will be transferred by pure conduction into the (supposedly infinite) country rocks above. This necessitates coupling the partial differential equation of thermal diffusion to the heat flux through the composite layer at the roof, as is described quantitatively below.

Magma is a very complicated, multi-component crystallizing silicate melt. The composition of magma varies significantly with location within the earth and few concise measurements of physical and chemical properties exist. A simplified system, which has been studied considerably by geologists in the past, consists of a melt of anorthite and diopside for which there are well-documented values for thermochemical properties [24] and kinetic growth rates [25] and for which the phase diagram is reproduced in figure 7. We shall confine our analysis to this particular system. We concentrate on anorthite-diopside systems with more diopside than the eutectic value, 58%, which, above the eutectic temperature, cool to form solid diopside and release a less dense (diopside-depleted) melt.

We shall assume that an appropriate anorthite-diopside melt, at its liquidus temperature, is instantaneously emplaced in the chamber. The fact that the melt is at its liquidus temperature, or in other words that there is initially zero superheat in the system, is in accord with the standard geologists' view of silicate melts within the earth. That the melt be emplaced instantaneously is an oft-used theoretical simplification. To our knowledge there has been no theoretical analysis of the effects due to a continual input of magma, and indeed we know of only one experimental investigation [26] of such a situation.

With these assumptions on the physical aspects of the system to be studied in mind, we are ready to consider the quantitative analysis of the problem. This will demonstrate that even with no initial superheat whatsoever, the kinetic undercooling is sufficient to drive vigorous convective motions in the melt. Further, this leads to

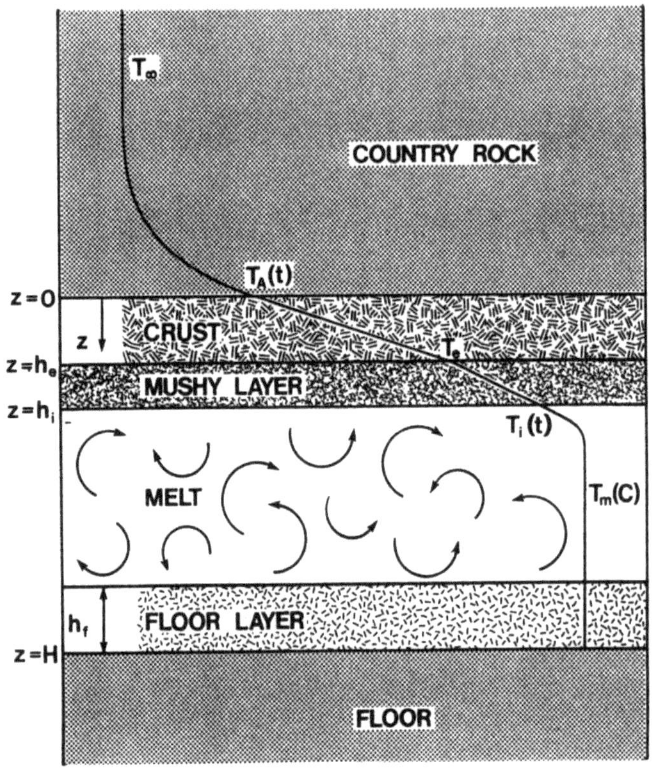

Figure 12: A schematic diagram of the solidification of a magma chamber that is losing heat predominantly by conduction through the country rock at its roof. The temperature $T_A(t)$ at the upper contact of the magma chamber with the country rock varies with time as the conductive temperature profile in the country rock decays. It is assumed in the present study that no heat is lost through the floor of the chamber.

solidification at the floor, even though there is cooling only at the roof, in agreement with geological observations which we shall discuss briefly at the end of this section. A sketch of the solidifying system and the resulting temperature profile is presented in figure 12.

The kinetic growth law for crystals of diopside growing into a supersaturated melt was determined by Kirkpatrick et al. [25] to be quadratic of the form

(5.1) $$\dot{h}_i = \mathcal{G}\mu^{-1}(T_L - T_i)^2,$$

where μ is the dynamic viscosity of the melt. Such a law is appropriate for crystals growing by screw dislocations [15]. Thus, in a magma chamber of depth H, in which crystals are growing by screw dislocations, the kinetic undercooling has a magnitude of order

(5.2) $$\Delta T_k \sim \left(\frac{\kappa\mu}{H\mathcal{G}}\right)^{\frac{1}{2}}.$$

Plates 1 and 2. See "Turbulent thermal convection" by S. Balachandar.

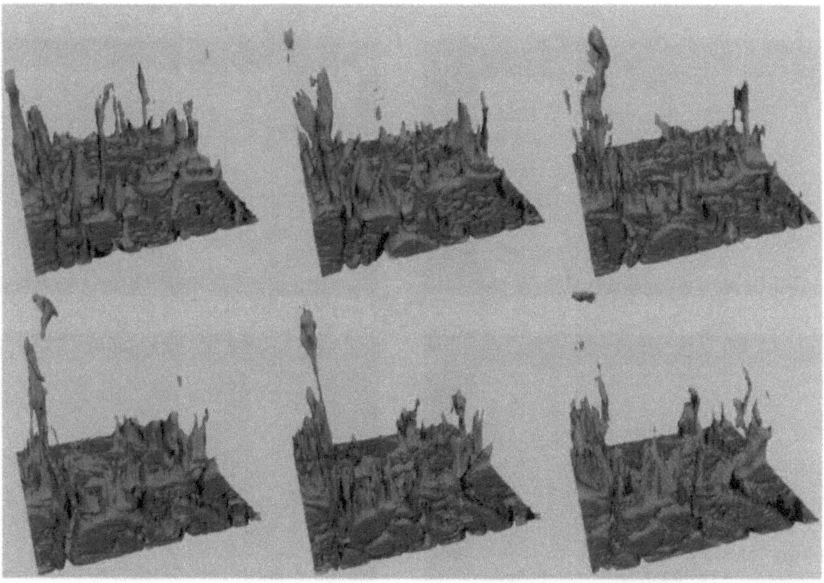

Plate 1. Surface of constant total temperature equal to 0.25 evaluated from the complete flow field at six different time instants. The total temperature is normalized to be 0.5 at the bottom and −0.5 at the top. Each frame is separated by 200 time steps and the frames are arranged left to right, top row followed by the bottom.

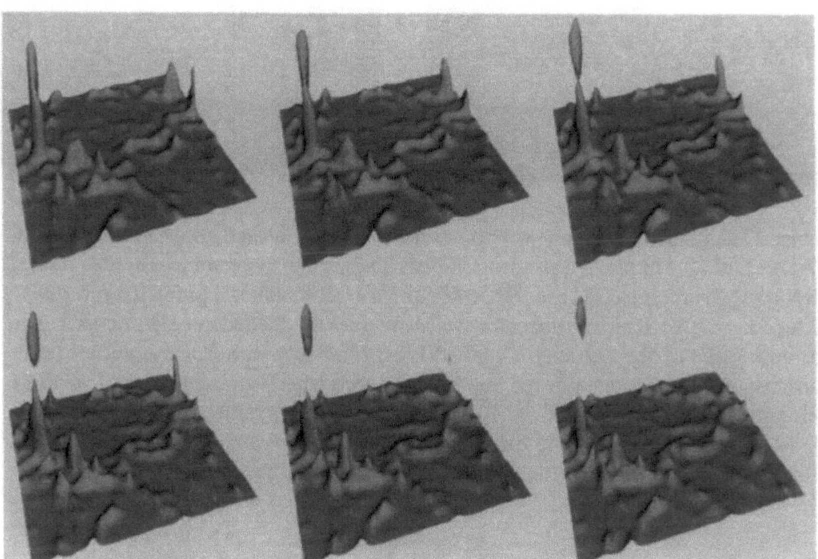

Plate 2. Surface of constant total temperature equal to 0.25 evaluated from the reconstructed flow field based on 1231 eigen modes. The total temperature is normalized to be 0.5 at the bottom and −0.5 at the top. Each frame is separated by 4 time steps and the frames are arranged left to right, top row followed by the bottom.

Figures 1, 6, and 7 (this and following page). See "Strongly chaotic Newtonian and non-Newtonian mantle convection" by David A. Yuen and Andrei V. Malevsky.

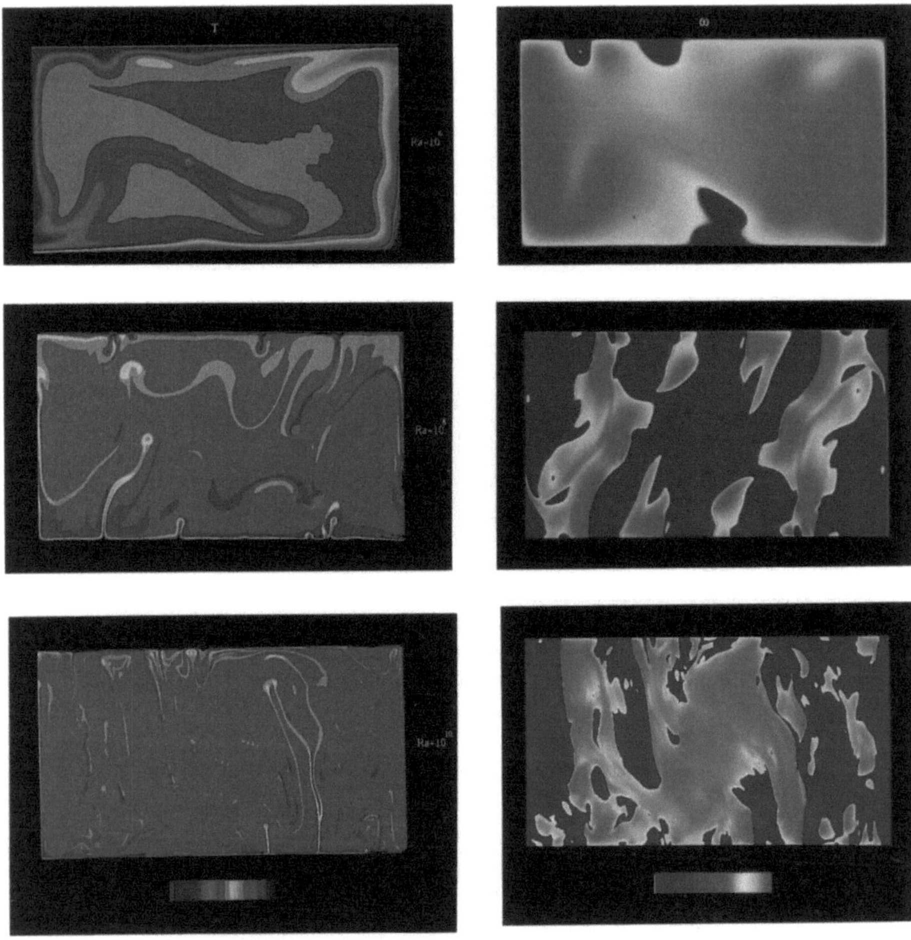

Figure 1. Temperature T and vorticity ω fields of base-heated convection for Ra = 10^6, 10^8 and 10^{10} in an aspect-ratio 1.8 box. These snapshots are taken after several overturns. Gridpoints used are 160×48 for Ra = 10^6, 400×140 for 10^8 and 800×70 for Ra = 10^{10}. Temperature color scales are linearly divided into 25 intervals. The absolute value of the vorticity is plotted logarithmically with positive and negative vorticity fields represented by red and blue, respectively. The magnitudes of vorticity fields increase with Ra. For Ra = 10^{10} maximum vorticity values are $0(10^7)$, while for Ra = 10^6 vorticity values are around $0(10^4)$, and Ra = 10^8 ω's are $0(10^5)$. Small values of ω are indicated by the middle level of the color bar.

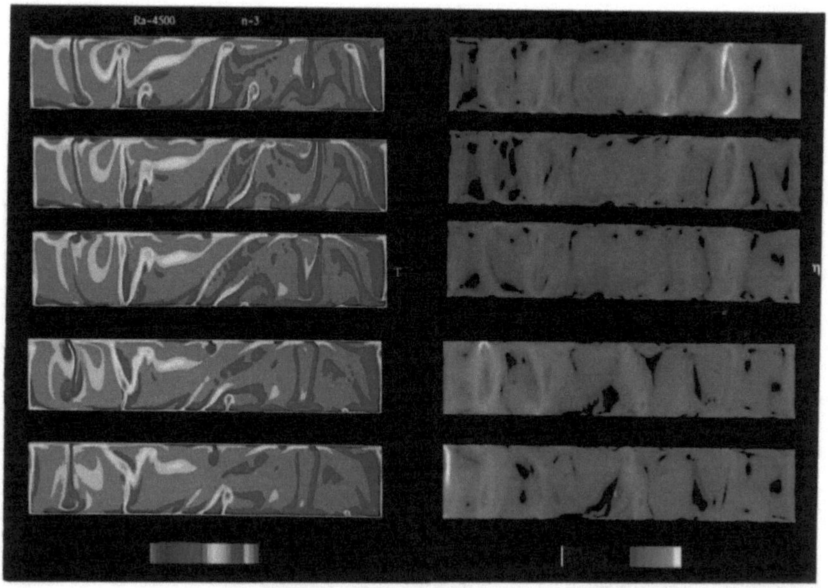

Figure 6. Evolution of the temperature and viscosity fields for Ra′ = 4500. Times are for t = 0.00238, 0.00241, 0.00254, 0.00267 and 0.00273. Color bars for temperature are the same as in Fig. 1. The viscosity fields are given logarithmically. The color bars are scaled linearly in the logarithmic scale, with white and black indicating low and high viscosities, respectively.

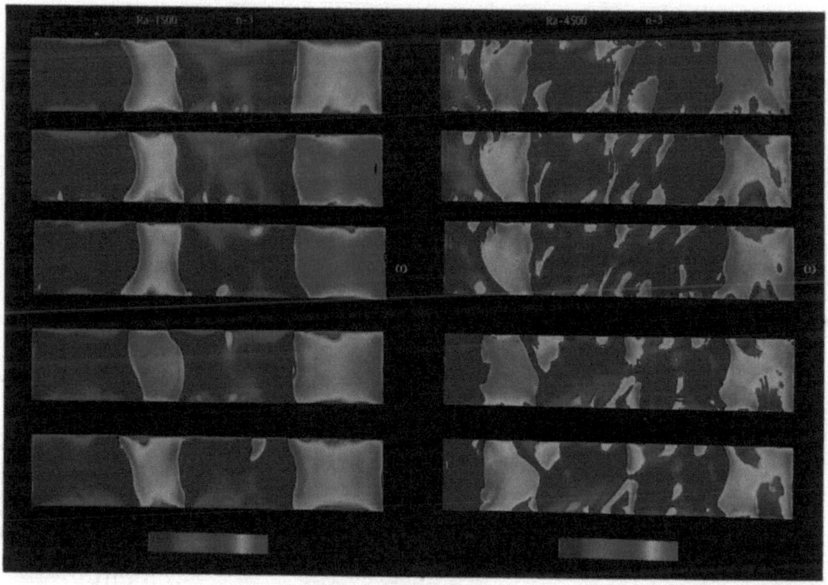

Figure 7. Temporal evolution of vorticity fields for non-Newtonian convection. The two Ra's are 1500 and 4500, yielding time-averaged Nu's of 13 and 23, respectively. Logarithms of the vorticity are displayed with the same convention for the color bar as in Fig. 1. The maximum magnitudes of ω for Ra′ = 1500 and 4500 are, respectively, $O(10^4)$ and $O(10^5)$. Times for Ra′ = 1500 are 0.00363, 0.00476, 0.00554, 0.00615, 0.00754. Times for Ra′ = 4500 are given in Fig. 5.

Figures 1 (A and B), 11, and 15 (this and following two pages). See "Mantle phase transitions, layered chaotic convection and the viscosity of the deep mantle" by W.R. Peltier and L.P. Solheim.

1A

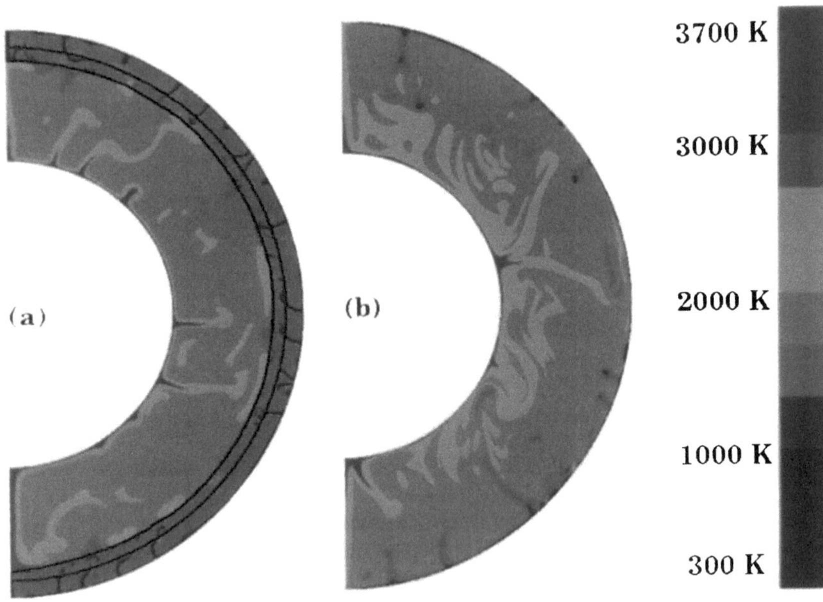

1B

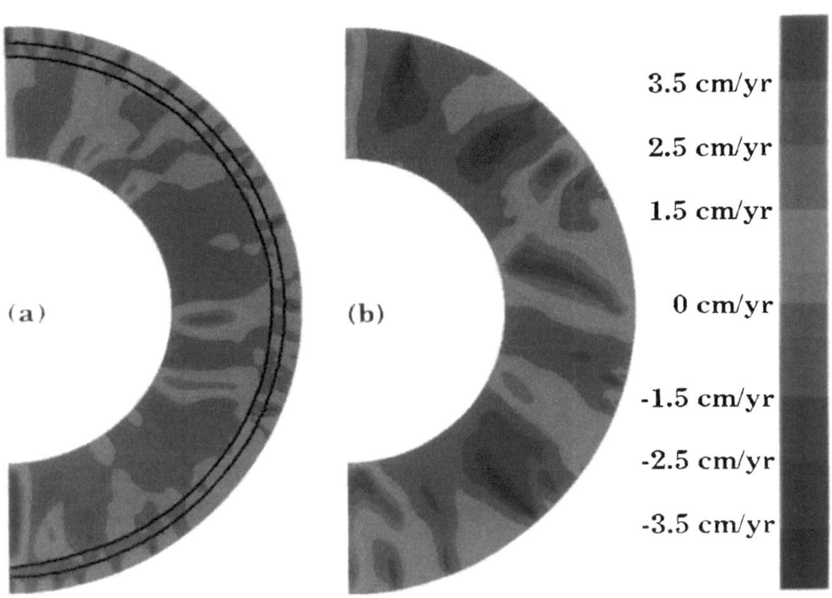

Figures for Peltier and Solheim (*continued*).

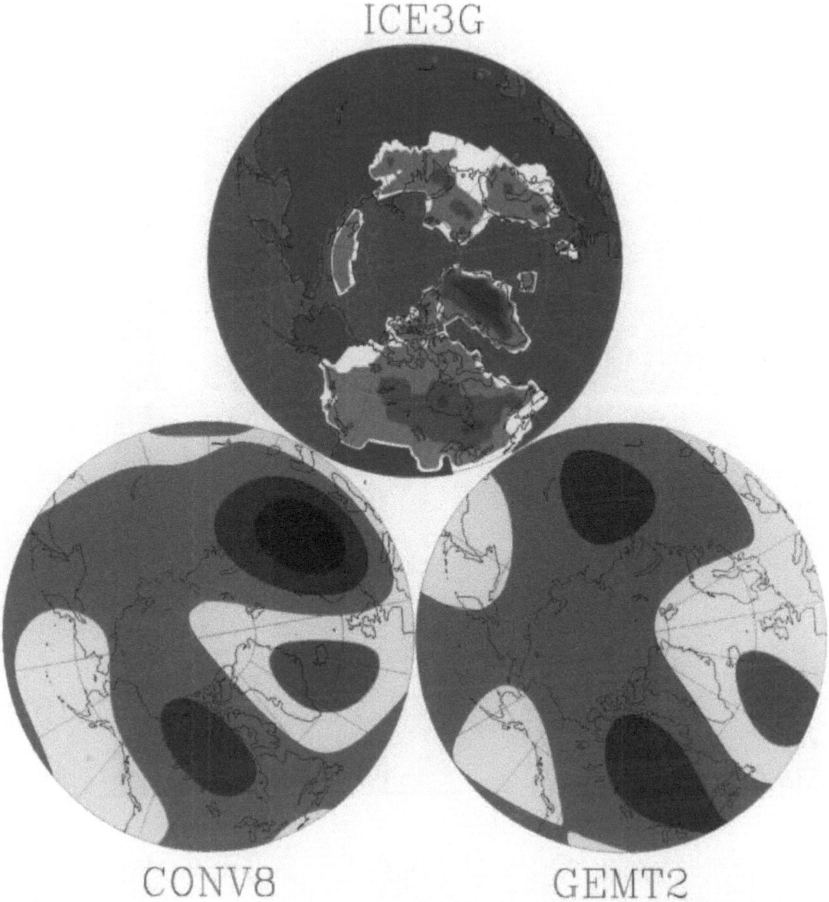

Figure 11. The thickness of Northern Hemisphere ice at the last glacial maximum (apex) according to the recently presented ICE-3G reconstruction. The colour bar employed to represent the field changes discontinuously at 1 km intervals. The greatest thickness of ice occurs over Hudson Bay, where it exceeds 3 km. The lower hemispheres present the satellite observed (GEMT2) and convection predicted (CONV8) free air gravity anomaly fields for the Northern Hemisphere.

Figure 1(a). Instantaneous temperature fields from two axi-symmetric simulations of the mantle convection process at a Rayleigh number of 10^7. Both flows are heated entirely from below. One contains no phase transitions whereas the second contains both a model Olivine-Spinel transition and a model Spinel-post Spinel transitional. The thicknesses of the phase loops are both set at 50 km. Note that neither hot ascending plumes nor cold descending plumes appear to penetrate the 670 km discontinuity in the latter case. Note also the "puddles" of high temperature that form on the underside of the deeper phase transition.

Figure 1(b). Instantaneous radial velocity fields corresponding to the instantaneous temperature fields shown on Figure 1(a).

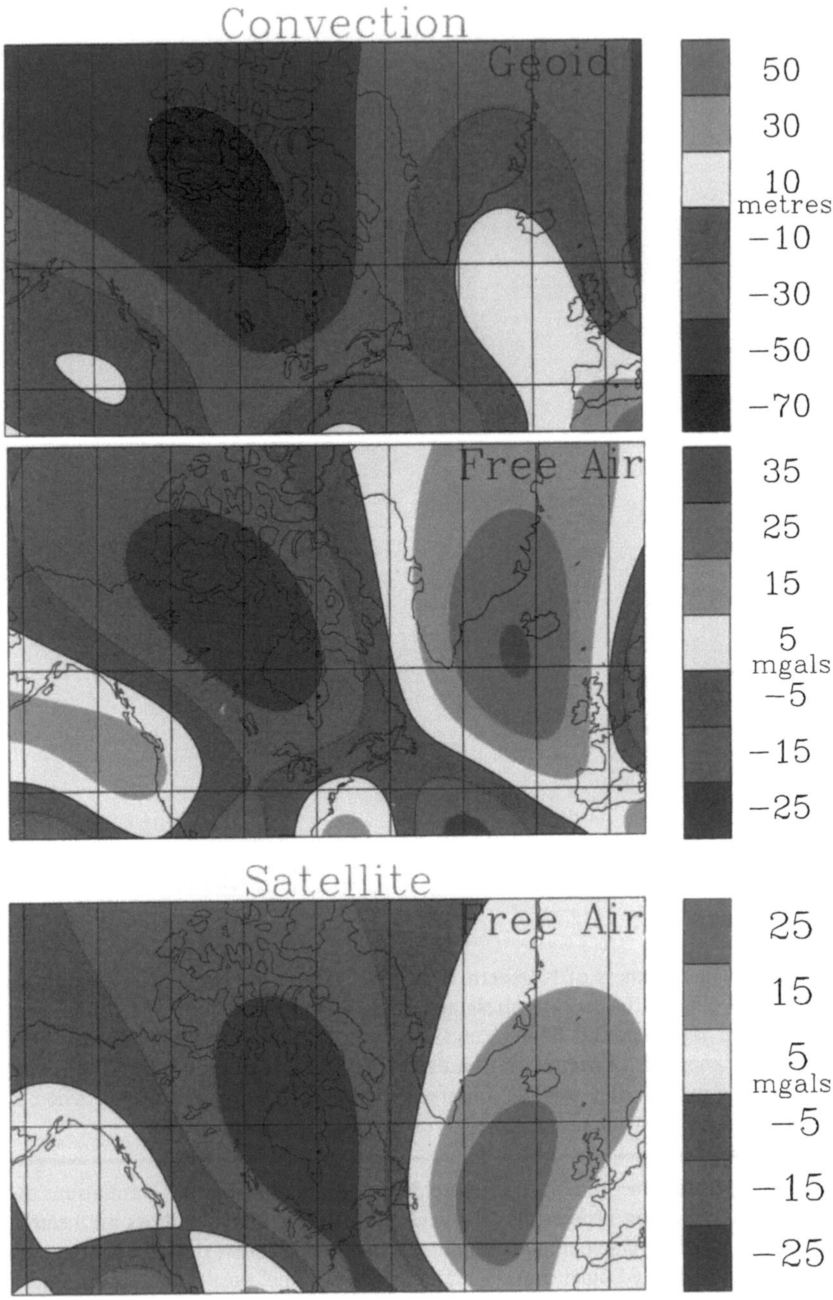

Figure 15. Geoid and free air gravity anomaly maps for the North American continent predicted using the two layer mantle convection model described in the text. Both maps are based upon spherical harmonic expansions truncated at degree and order 8. The model of the internal density heterogeneity employed to make the predictions is model SH425.2 described in Su and Dziewonski (1991). Plate (c) shows the GEMT2 satellite field over the same geographic region and at the same resolution.

Figures 1 and 11. See "Vigorous motions in magma chambers and lava lakes" by Herbert E. Huppert and M. Grae Worster.

Figure 1. Lava fountaining during the eruption of Pu'u O'o, Hawaii, on 2 June 1986. The scale of the fountaining can be ascertained from the helicopter which appears as a mere speck two-thirds of the way up the fountain. Photograph by J.D. Griggs, United States Geological Survey.

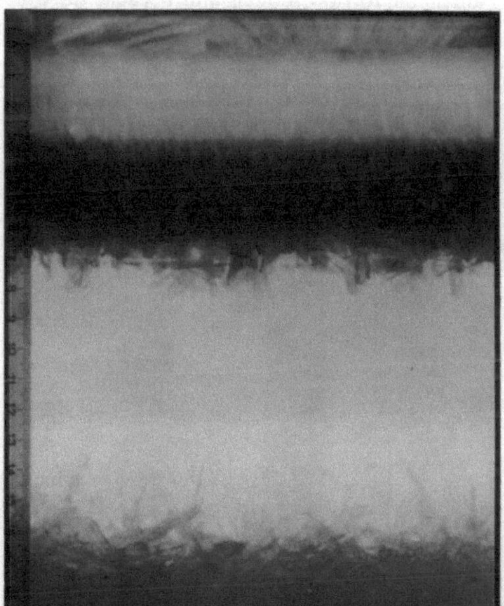

Figure 11. A photograph taken during an experiment in which an aqueous solution of Na_2SO_4 was cooled from above to form a composite layer next to the roof of the tank (the white region), a mushy layer below this and a layer of faceted Na_2SO_4 $.10H_2O$ crystals at the base. The experimental conditions were C_0 = 16wt% Na_2SO_4, T_0 = 30.5°C, T_B = −17°C, and H = 18.8 cm.

Plate 1. See "The eruption and spreading of lava" by Claude Jaupart.

Plate 1. A photograph of Mount St. Helens after its destructive eruption of May 1980, taken by Colin Wilson. The smoking grey mound in the middle of the crater is not a pile of rubble, but an active flow of dacite lava. This kind of eruption is very useful because it is amenable to flow models and because the time-evolution of the eruption rate and flow thickness leads to constraints on conditions in the magma reservoir located at depth.

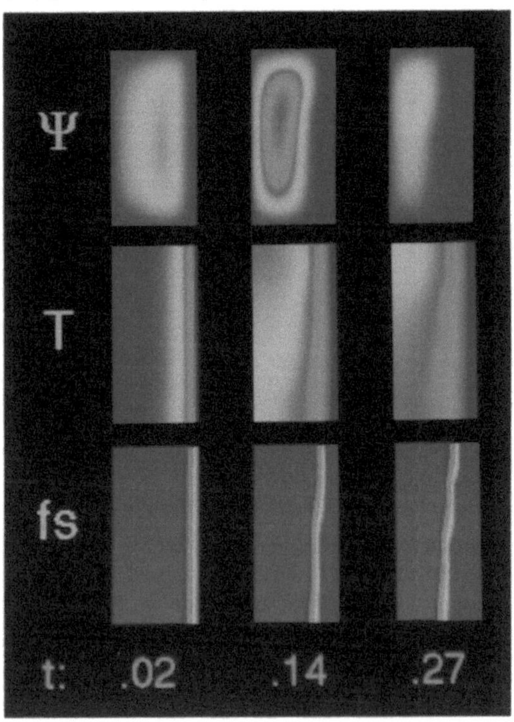

Figures 3 and 5. See "Convection in porous media with thermal and chemical buoyancy: A comparison of two models for solute dispersion" by N.D. Rosenberg and F.J. Spera.

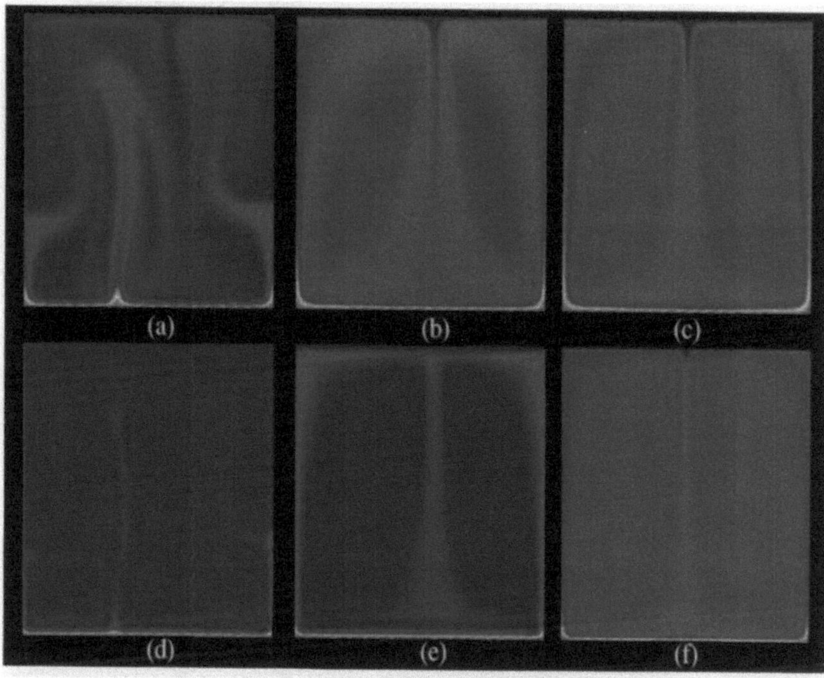

Figure 3. Solute fields for $R\rho = 0$ for simple dispersion model at (a) $\hat{t} = 0.1$, (b) $\hat{t} = 0.25$ and (c) at steady-state and for complex dispersion model at (d) $\hat{t} = 0.1$, (e) $\hat{t} = 0.25$ and (f) at steady-state.

← Color Plate 1. See "Modeling transport processes in nonlinear systems: The example of solidification and convection" by Curtis M. Oldenburg and Frank J. Spera.

Color Plate 1. Stream function, temperature, and fraction solid fields for the sidewall solidification and convection of pure tin at $t = .02, .14, .27$. The right-hand sidewall is held at a temperature of 265 K. Convection is clockwise in an ever-decreasing region as the solid grows inward from the right-hand sidewall. Note that in this pure tin system, the fraction solid is either 1 or 0; the narrow mush region visible in the color plate is due to the space discretization of the numerical method.

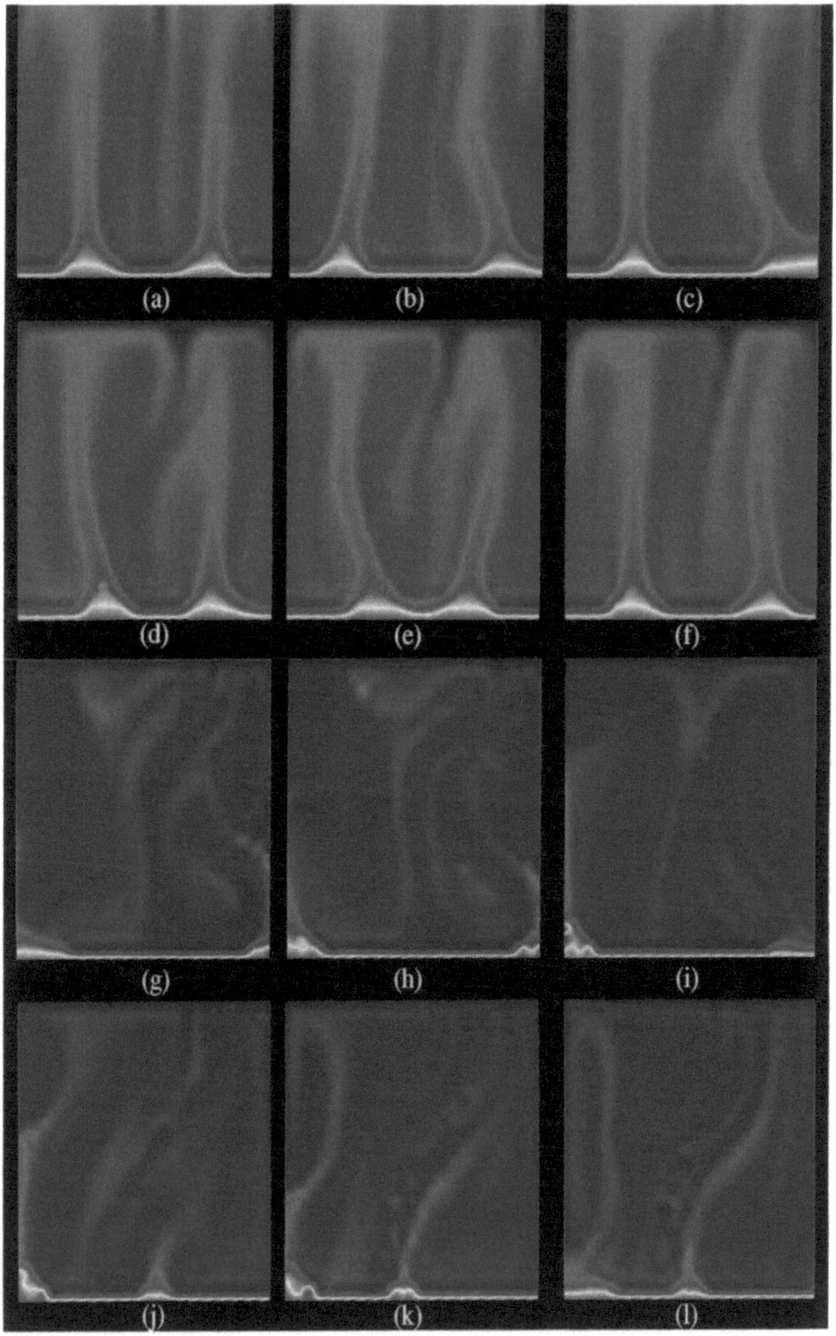

Figure 5. Solute fields for $R\rho = 0.5$ for simple dispersion model at $\hat{t} = 0.360–0.385$ (a–f) and for complex dispersion model at $\hat{t} = 0.360–0.385$ (g–l).

Figures 3, 4, and 6 (a,b,c). See "The spatial structure of isotropic turbulence and the related three-dimensional graphics problems" by A. Vincent, M. Meneguzzi and J.D. Villasenor.

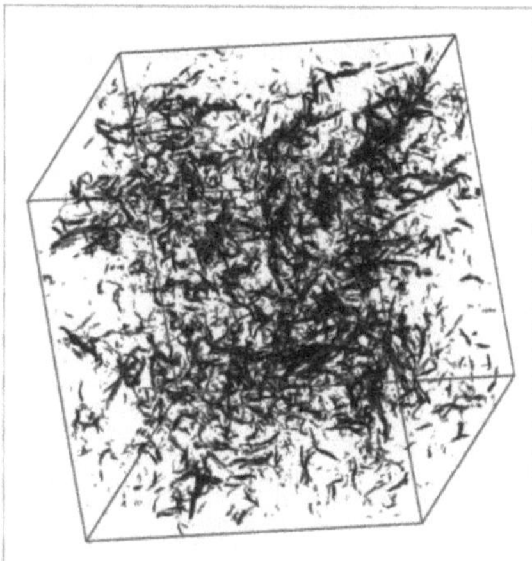

Figure 3. View of the vorticity field, represented by a vector of length proportional to the vorticity amplitude at each grid point. Only vectors larger than a given threshold value are shown. Color scale goes from blue (lowest intensities) to red (highest intensities).

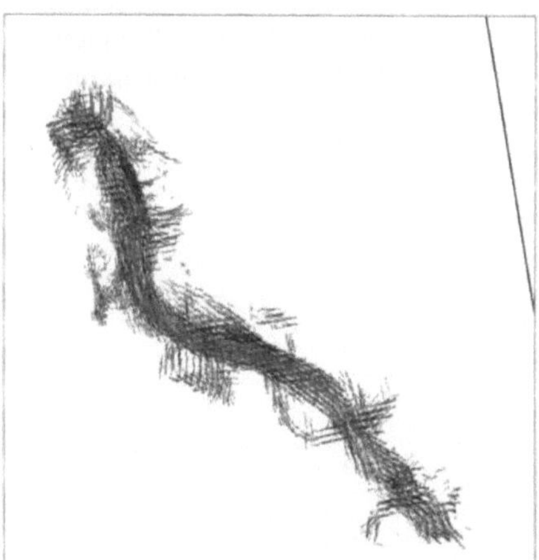

Figure 4. A view of a single vorticity tube after it has been extracted by a pattern recognition technique.

Figures for Vincent, Meneguzzi and Villasenor (*continued*).

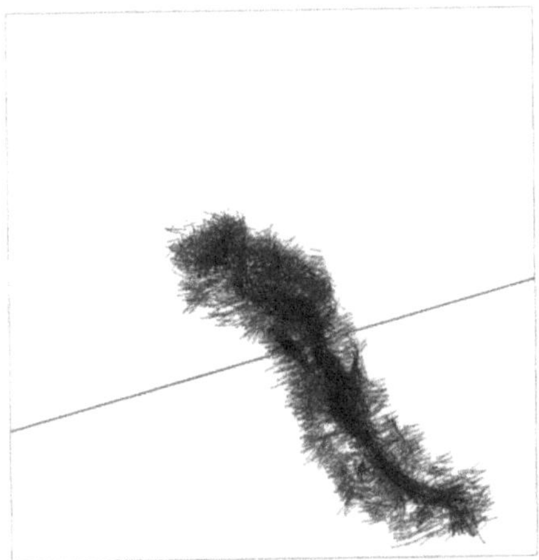

(a)

Figure 6a. Vorticity field, in blue, and eigenvector $\mathbf{e}_3$ associated with the largest (positive) eigenvalue, in green.

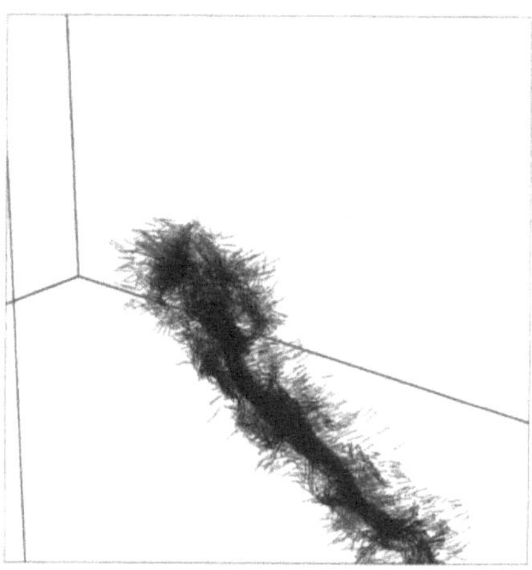

(b)

Figure 6b. Vorticity field, in blue, and eigenvector $\mathbf{e}_1$ associated with the smallest (negative) eigenvalue, in magenta.

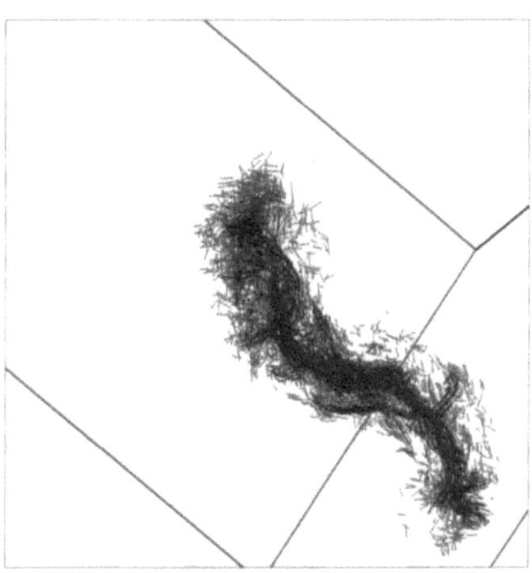

(c)

Figure 6c. Vorticity field, in blue, and eigenvector $\mathbf{e}_2$ associated with the intermediate eigenvalue, in red.

The value of $\mathcal{G}$ is approximately 5.6×10^{-5}poise cm s^{-1} °C^{-2} for diopside [25], so a typical value of ΔT_k is about 0.6°C (using the values in Table 1). Convection will continue until ΔT_k is about equal to the critical undercooling ΔT_c given by equation (4.1), that is until the depth of the remaining melt h is about that given by

$$\frac{h}{H} \sim \left(Ra_c^2 \frac{\kappa \nu^2 \mathcal{G}}{\alpha^2 g^2 \mu H^5} \right)^{\frac{1}{6}}. \tag{5.3}$$

This ratio is typically very small given the large dimensions of most magma chambers. It is therefore appropriate to model the evolution of a magma chamber using the assumption of vigorous convection of the melt driven solely by the small kinetic undercooling in cases when there is no initial superheat.

The governing equations in the composite layer, mushy layer, convecting melt and floor layer are given by (3.1)–(4.10) with the kinetic growth law (4.2) replaced by (5.1). The dynamic viscosity μ of the molten diopside, which varies with the temperature of the melt, can be represented by

$$\mu = \exp[(12 - 52.5x + 62.5x^2)\ln 10], \tag{5.4}$$

where $x = 1 - 1000T^{-1}$ and T is the absolute temperature of the melt (in Kelvin), a relationship obtained [27] by curve fitting the graphical data presented by Kirkpatrick [28].

Since heat is conducted into the overlying country rock, the temperature at the upper boundary of the magma chamber T_A is no longer constant. It can be found in terms of a Green's function solution [29] of the thermal diffusion equation in the country rock, and is given by

$$T_A(t) - T_\infty = \frac{1}{k_r}\sqrt{\frac{\kappa_r}{\pi}} \int_0^t \frac{f(\tau)}{\sqrt{t-\tau}}\, d\tau \tag{5.5a}$$

$$= \frac{1}{k_r}\sqrt{\frac{\kappa_r}{\pi}} \left\{ 2\sqrt{t} f(t) + \sqrt{t} \int_0^1 \frac{f(tu) - f(t)}{\sqrt{1-u}}\, du \right\}, \tag{5.5b}$$

where T_∞ is the far-field temperature in the country rock, subscript 'r' denotes properties of the country rock and

$$f(t) = \left[k \frac{\partial T}{\partial z} \right]_{z=0+} \tag{5.6}$$

is the heat flux through the boundary $z = 0$ at time t, which is evaluated in terms of the temperature field and thermal properties of the solidifying roof below $z = 0$. The second expression (5.5b) for T_A, in which the singularity in the Green's function has been removed, was used in the numerical scheme.

Some of the results of our numerical integrations of the governing differential equations described above are presented in figures 13–16[3]. The curves in figure 13

[3] Since publication of [27], we have found an error in the computer program used to calculate the results presented in that paper. The error produced results equivalent to giving a falsely low value to the conductivity of the country rock, and therefore cooling and solidification times were over predicted throughout, by about 40%. None of the general conclusions of that paper are affected by the correction of this error. Figures 13–16 correspond to figures 2,3b, 4 and 6 of [27], and show calculations made using the corrected program.

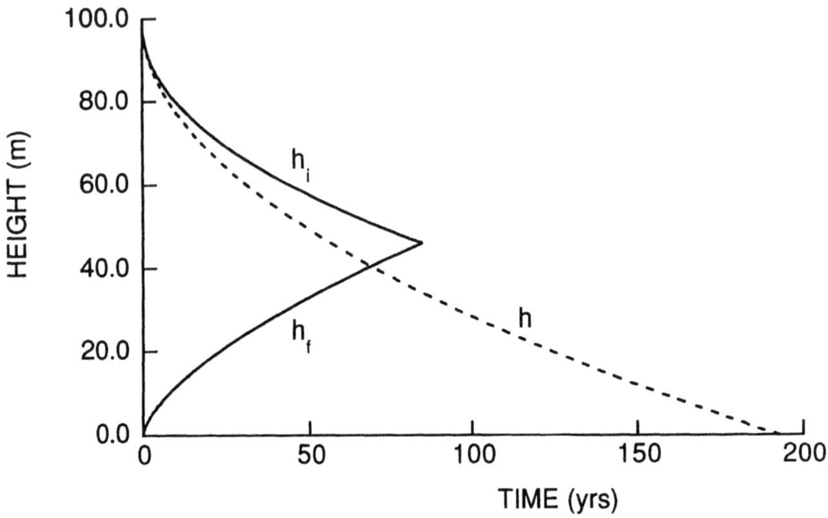

Figure 13: The predicted solidification of a layer of pure diopside cooled at its roof. The solid curves show the results of the model that includes convection and kinetic effects. The dashed curve shows, for the sake of comparison, the results of an unrealistic calculation in which convection is ignored.

depict as functions of time, the depth of solid grown at the roof and the floor from a layer of pure diopside initially 100m thick emplaced into country rock at 0°C. Also included for comparison is the resultant thickness of the roof layer if convective (and kinetic) effects are totally neglected, which means that no solid is grown at the floor. We see that the presence of convection reduces the time for complete solidification by about 56 % over the physically (and geologically) inappropriate case when convection is neglected (c.f. Section 2). In addition convection allows approximately 45% of the chamber to solidify at the floor. The reason that convection decreases the time for complete solidification in this case is that it transports the latent heat of solidifcation very efficiently from the growth at the floor to the cooled roof. In contrast, when all the solidification takes place at the roof, all the latent heat must be conducted back through the much-thicker, growing crust.

The result of the calculations for an initial 100m thick layer with a 20% anorthite composition is shown in figure 14. Very rapidly after the initiation of the convective motions, a mushy layer develops beneath the composite solid layer near the roof, with a solid layer evolving from the floor. As the temperature in the melt decreases to the eutectic temperature (1274°C), the mushy layer disappears, the layer growing from the roof becomes completely solid, and a composite layer of diopside and anorthite crystals begins to grow on the floor. With time a growing layer of eutectic composition (42% anorthite) extends into the remaining melt until total solidification has occurred.

The dimensionless time $\kappa t_s/H^2$ for complete solidification of a layer of initial

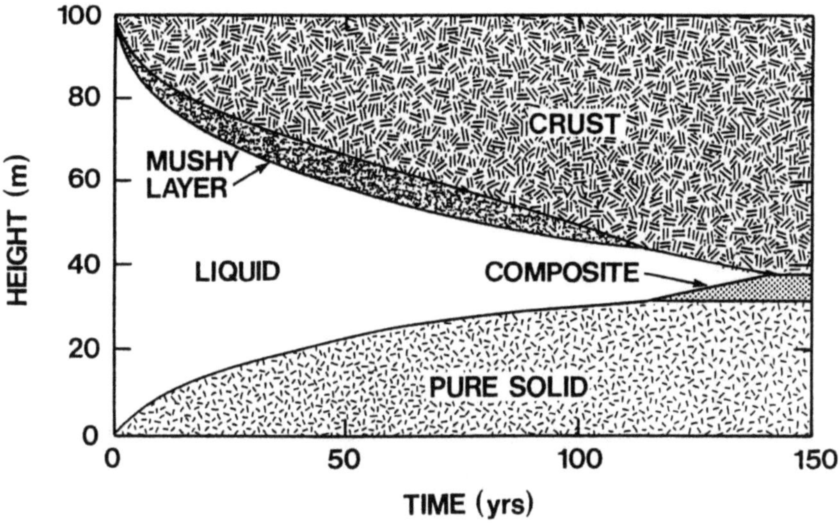

Figure 14: The predicted solidification of a layer of melt of composition $An_{20}Di_{80}$, one hundred meters thick emplaced at its liquidus temperature 1352°C into country rock at 0°C.

depth H is graphed in figure 15a The increasing importance of convection as the layer thickness increases is seen from the decreasing time for solidification to take place compared with the purely conductive timescale H^2/κ. Increasing the height H, and thus the intensity of the convective motions, also increases the fraction of solid deposited at the floor, as seen in figure 15b.

The gradual decrease in the temperature of the melt is presented in figure 16a. Because the melt is constrained to lie on the liquidus, the decreasing melt temperature is related to the decreasing diopside concentration. After about 180 years, the temperature of the melt has decreased to the eutectic temperature and it then remains at that value until solidification is complete. The temperature of the interface between the mushy layer and the melt, which is always less than the temperature of the melt itself, is also graphed in figure 16a with their difference presented in figure 16b. Note that, with typical geological parameter values, even though the depicted temperature differences are of order of only a few tenths of a degree centigrade, this leads to values of the Rayleigh number, shown in figure 16c, which are sufficiently large for vigorous convective motions to occur until the layer has almost completely solidified.

6. Lava lakes. The foregoing analysis can also be used, with only minor alteration, to interpret the solidification history of lava lakes. The principal mathematical difference between the models for lava lakes and magma chambers is in the boundary condition that is applied at the upper boundary. Whereas for a magma chamber the upper boundary condition reflects the conduction into the country rock (equation (5.5)), the appropriate condition for lava lakes is that of constant temperature, since they lose heat very efficiently into the well-mixed air or water above them, which is at

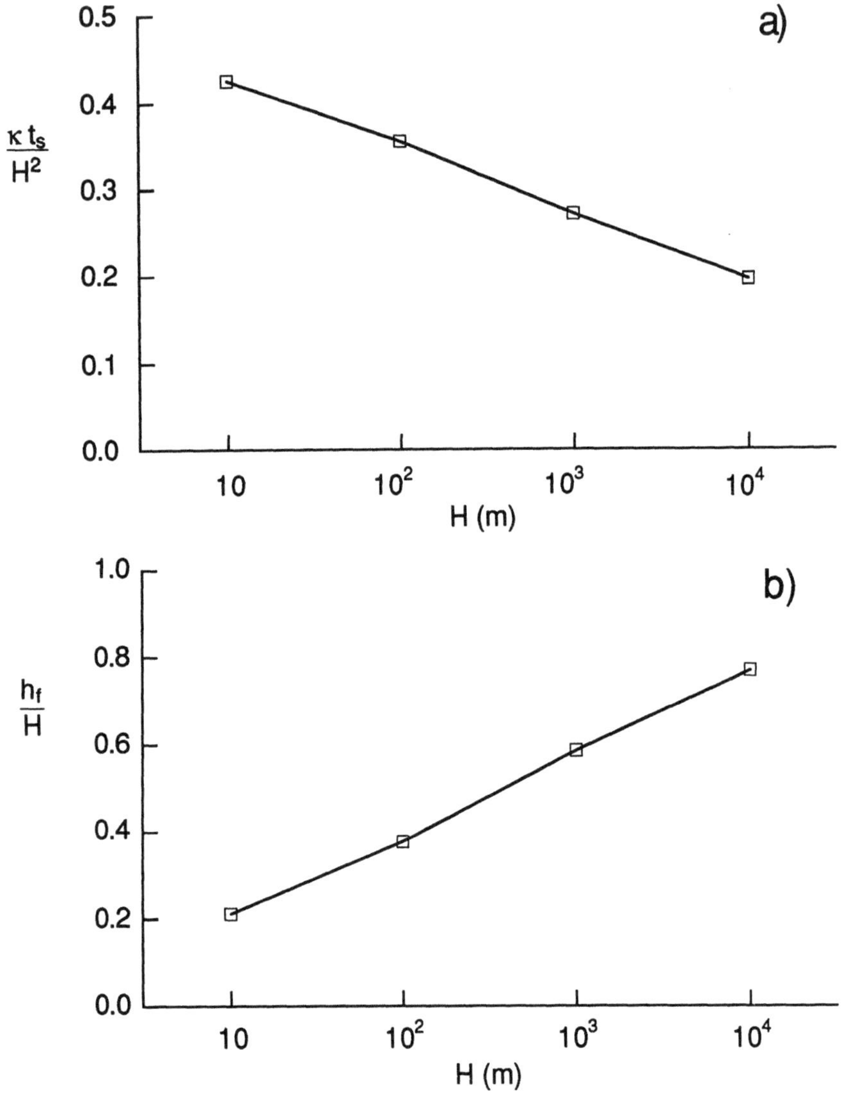

Figure 15: (a) The ratio of the time for complete solidification t_s to the conductive timescale H^2/κ as a function of the initial depth H of a layer of melt of composition $An_{20}Di_{80}$. As the depth increases, the vigour of convection increases and causes the solidification time to decrease relative to the conduction time. The stronger convection also causes the relative depth of crystals grown at the floor to increase, as shown in figure (b).

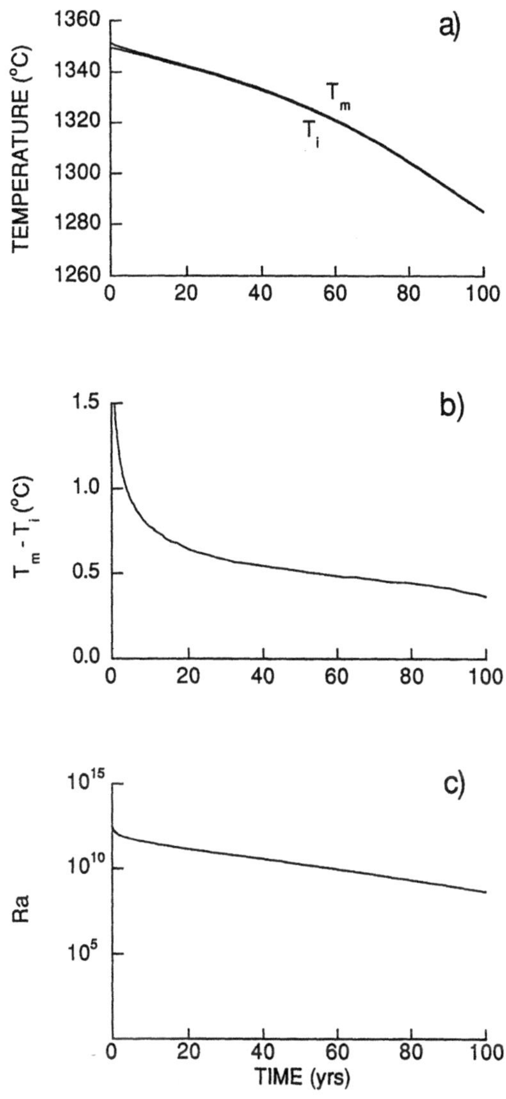

Figure 16: (a) The temperature of the melt T_m and the temperature of the mush-liquid interface T_i as functions of time for a layer of melt of composition $An_{20}Di_{80}$ emplaced at its liquidus temperature 1352°C into country rock at 0°C. Their difference is emphasized in (b). Although this temperature difference is only a fraction of a degree Centigrade, it is sufficient to give the very large Rayleigh numbers shown in (c).

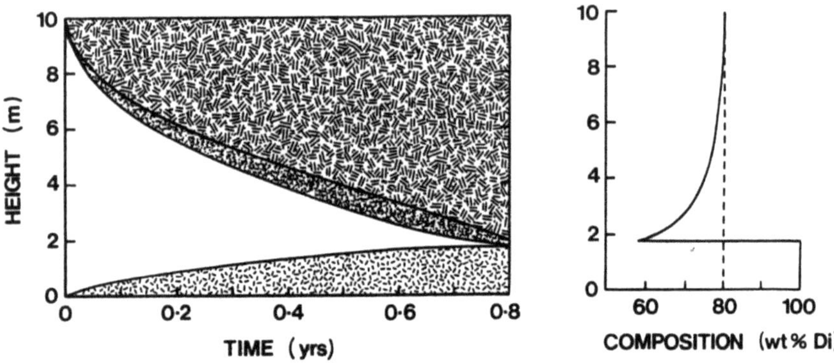

Figure 17: The predicted solidification of a lava lake of initially uniform composition $An_{20}Di_{80}$, ten meters thick, whose upper surface is maintained at a constant temperature of 0°C. Also shown is the final compositional profile of the solidified lake, as predicted by the theoretical model. Such macrosegregation of the components of the melt is a consequence of the coupled effects of convection and kinetic undercooling.

approximately 0°C. This change in boundary condition affects the quantitative predictions of the model somewhat, but the general evolution is similar to that described for magma chambers and the orders of magnitude of the results are not much altered.

More important differences may arise once the additional cooling through the floor is taken into account and in response to different physical properties, such as the viscosity and kinetic coefficient, between various magmas and lavas. We intend to publish such results in the near future. For now, we will present results based upon the current model, with the parameter values of the Di–An system and a top boundary condition of constant temperature. In addition, we shall briefly counter some misrepresentations regarding lava lakes that have appeared in the recent literature.

The depth of the solidified crust, the depth of the mushy layer and the depth of crystals grown internally are shown as functions of time in figure 17. The calculations were made for a lava lake of initial depth 10m containing a melt of uniform composition $Di_{80}An_{20}$ and initial temperature equal to its liquidus temperature of 1352°C. Also shown in the figure is the final compositional profile, which results due to the differentiation that occurs during internal growth of crystals. This differentiation is a consequence of the coupling of kinetic undercooling at the mush–liquid interface with the convective stirring of the melt. It would not be predicted if either of these physical phenomena were omitted from the mathematical model.

In figure 18a, we reproduce figure 17 of a paper by Marsh [30] in which some field data on the cooling of lava lakes and a number of theoretical curves are presented. Marsh does not make explicit what calculations or parameter values have been used to construct these curves, and it is our belief that the initial and boundary conditions used in each case are different, which invalidates the common scaling used to compare them, as we explain below.

The field data are for cases in which the lavas contained phenocrysts initially and hence were not superheated. These data are compared with a theoretical curve (labelled 'stagnant; latent heat; no overburden') derived from a purely-conductive model, which has the property that the interior temperature remains almost constant. Marsh contrasts this with the measurements of Jaupart et al. [22] (labelled 'JBA (1984)') and what is claimed to be a theoretical curve from Turner et al. [3] (labelled 'THS (1986)') which both show the temperature decreasing rapidly. However, both these studies correspond to systems that are initially superheated, and the results should correctly be represented on figure 17 of [30], which corresponds to our figure 18a, by curves starting above $T_c' = 1.0$ and decaying rapidly towards $T_c' = 1.0$. Such rapid decay is shown in figures 3a and 4a of section 2. Correct application of the model of Turner et al. [3] (presented in 2) to the case of a lava that is initially at its freezing temperature would predict that the interior temperature would remain constant at the freezing temperature $T_c' = 1.0$, and that there would be no convection.

In contrast, the model of Kerr et al. [6,16] presented in sections 3 and 4, which includes kinetic effects, can be applied to the case of a lava lake that is not initially superheated and, in that case, predicts vigorous convection driven by kinetic undercooling and also predicts the dimensionless temperatures shown in figure 18b. These results, for lava lakes of different initial depths containing melts of composition $Di_{80}An_{20}$, show a slow decay of the interior temperature corresponding to the slow evolution of the composition of the magma as crystals of diopside form in the interior or at the base of the lake. Note that the calculations were stopped once the temperature reached the eutectic temperature of the Di–An system.

These results, which are presented in terms of the same dimensionless variables used in [30], are in approximate agreement with the field data reproduced in figure 18a. This is also true of the purely conductive model. But, whereas Marsh [30] uses this agreement to argue against the existence of convection in magma chambers and lava lakes we have shown that these particular field data are equally consistent with a model that includes vigorous convection of the melt. Thus, given only the presented field data of temperature, it is not possible to decide between a model involving only conduction or one incorporating convective motions. However, if one also incorporates the well-established, fluid-mechanical principle that large Rayleigh numbers imply vigorous convection, then a distinction can be drawn. In addition, we believe that the field evidence for the convective model will be found in terms of the type of compositional segregation predicted and presented in figure 17, which cannot be explained simply in terms of conduction and phase equilibria.

7. Conclusion. This paper has discussed a series of one-dimensional mathematical problems that describe the cooling and resultant solidification of a one- or two-component melt. (A partial introduction to the associated two- and three-dimensional problems is presented in two reviews by Huppert [31,32]). The two most

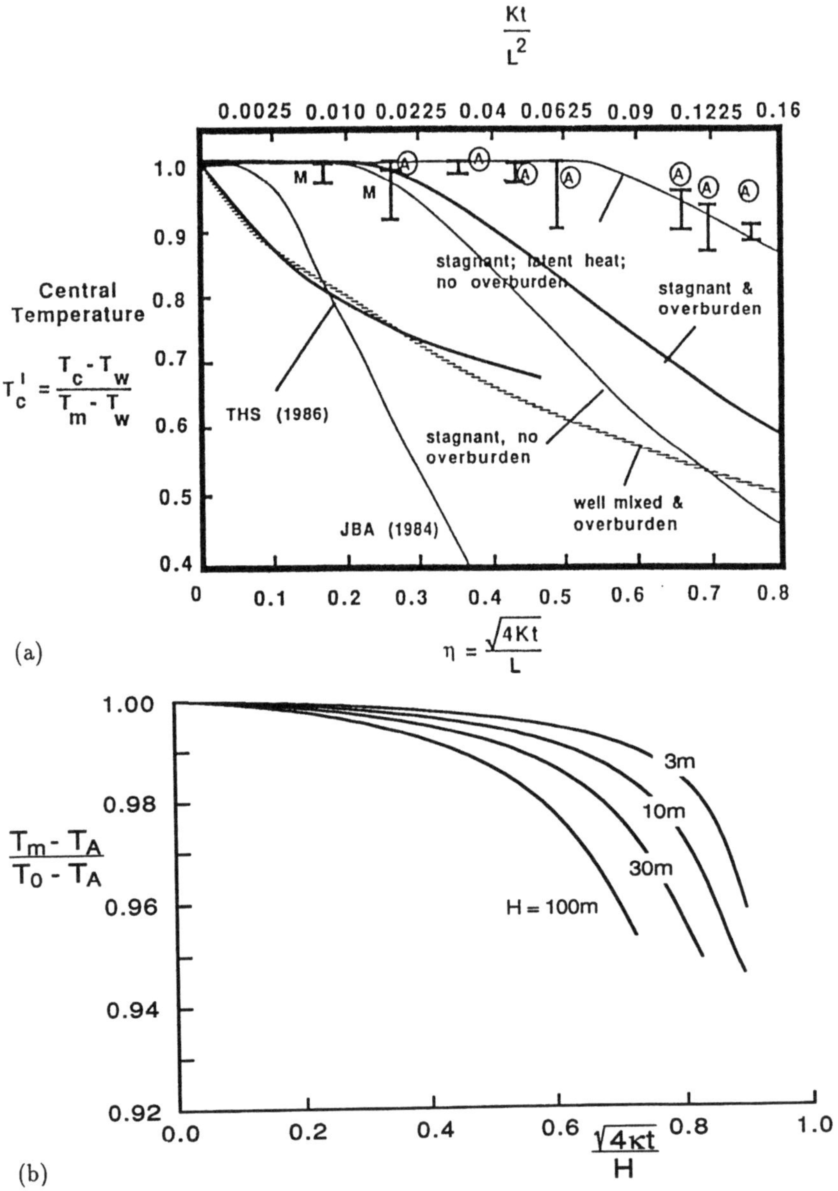

Figure 18: (a) A reproduction of figure 17 of [30], which includes some field data from lava lakes and a number of curves that are explained in the text. (b) The predicted interior temperature of lava lakes of different initial depths according to the model presented in the present paper. The scaling used in this figure is the same as that used in (a) above. The calculations used parameter values appropriate to a melt of composition $An_{20}Di_{80}$ so no direct quantitative comparison with the field data should be made.

important concepts that run through the paper are that the timescale for solidification in the described systems is given by the conductive time scale, while, at the same time, convective motions in the solidifying melt play an essential role. For example, the removal of superheat from the melt takes place on the much-more-rapid convective timescale, and the presence of convection driven by kinetic undercooling at the solidifying interface of a two-component melt allows cooling at the roof of a system to lead to solidification in the interior of the melt and also at the floor. From a quantitative point of view, the convective heat transfer decreases the actual time taken for complete solidification, as indicated for particular examples in figures 5, 13 and 15. In addition, the combination of convection and kinetic undercooling results in the vertical differentiation of the solidified magma or lava, an effect that is predicted by the current model.

Many of the specific results of our mathematical models are in excellent agreement with data obtained from laboratory experiments. In addition, the general results are in qualitative agreement with the somewhat limited data available from geological observations. Calculations that are based on the two-component anorthite-diopside system predict a compositional gradient in the final solidified product which is similar to that observed in sills and old magma chambers throughout the world. The extension of the calculations made in order to describe the evolution of a lava lake indicates that the fluid part of the lava is in vigorous motion, while the thickness of the solidifying crust gradually increases. The observed rate of increase of the thickness of this crust agrees well with the results predicted by the theoretical model.

Certainly, more research is still needed to explain many of the details of the evolution of magma chambers and lava lakes. We look forward to the introduction of new physical concepts, more powerful mathematical and numerical techniques and fresh geological data to unravel the many chaotic features of solidification problems in particular and geophysical problems in general.

Acknowledgments. We are grateful to our colleagues M.A. Hallworth, R.C. Kerr, R.S.J. Sparks, J.S. Turner and A.W. Woods for contributing to the research reported in this paper, and would like, in addition, to thank them, along with J.R. Lister, for their helpful comments on an earlier version of the manuscript. The research work of H.E. Huppert is supported by a grant from Venture Research International, ably directed by Don Braben. M.G. Worster is grateful for partial support from the Thermal Systems Program of the National Science Foundation.

REFERENCES

[1] J. Crank, *Free- and Moving-Boundary Problems*, Clarendon Press, Oxford, 1984.

[2] J. M. Hill, *One-dimensional Stefan Problems: An Introduction*, Longman, Harlow, 1987.

[3] J. S. Turner, H. E. Huppert and R. S. J. Sparks, *Komatiites II: Experimental and theoretical investigations of post-emplacement cooling and crystallization*, J. Petrol., 27 (1986), pp. 397–437.

[4] R. C. KERR, A. W. WOODS, M. G. WORSTER AND H. E. HUPPERT, *Disequilibrium and macrosegregation during solidification of a binary melt*, Nature, 340 (1989), pp. 357–362.

[5] R. C. KERR, A. W. WOODS, M. G. WORSTER AND H. E. HUPPERT, *Soldification of an alloy cooled from above. Part I Equilibrium growth*, J. Fluid Mech., 216 (1990), pp.323–342.

[6] R.C. KERR, A.W. WOODS, M.G. WORSTER AND H.E. HUPPERT, *Solidification of an alloy cooled from above. Part II Non-equilibrium interfacial kinetics*, J. Fluid Mech., 217 (1990), pp.331–348.

[7] J. S. TURNER, *Buoyancy Effects in Fluids*, Cambridge University Press, Cambridge, 1979.

[8] T. L. WRIGHT AND R.T. OKAMURA, *Cooling and crystallization of tholeiitic basalt, 1965 Makaopuhi lava lake, Hawaii*, U.S. Geol. Survey Prof. paper, 1004 (1977), 78pp.

[9] T. L. WRIGHT, D. L. PECK AND H. R. SHAW, *Kilauea lava lakes: natural laboratories of study of cooling, crystallization and differentiation of basaltic magma* in The Geophysics of the Pacific Ocean Basin and its Margin (G. H. Sutton, M. H. Manghnani and R. Moberly, Eds) American Geophysical Union, Washington D.C. (1976), pp. 375–390

[10] H.E. HUPPERT AND M.G.WORSTER, *Dynamic solidification of a binary melt*, Nature, 314 (1985), pp. 703–707.

[11] M.G. WORSTER, *Solidification of an alloy from a cooled boundary*, J. Fluid Mech., 167 (1986), pp. 481–501.

[12] M.G. WORSTER, *Natural convection in a mushy layer*, J. Fluid Mech., 224 (1991), pp. 335–359.

[13] G. BRANDEIS AND B.D. MARSH *The convective liquidus in a solidifying magma chamber: a fluid dynamic investigation*, Nature, 339 (1989), pp.613–616.

[14] W. KURZ AND D.J. FISHER *Fundamentals of Solidification*, Trans. Tech. Publications, Aedermannsdorf (1986)

[15] R.J. KIRKPATRICK, *Crystal growth from the melt: a review*, American Mineralogist, 60 (1975), pp.798–814.

[16] R.C. KERR, A.W. WOODS, M.G. WORSTER AND H.E. HUPPERT, *Solidification of an alloy cooled from above. Part III Stratification of the solid*, J. Fluid Mech., 218 (1990), pp.337–354.

[17] M.P. RYAN, R.Y. KOYANAGI AND R.S. FISKE, *Modelling the three-dimensional structure of macroscopic magma transport systems: Application to Kilauea Volcano, Hawaii*, J. Geophys. Res., 86 (1981), pp.7111–7129.

[18] R.S. DETRICK, P. BUHL, E.ERA, J MUTTER, J. ORCUTT, J. MADSEN AND T. BROCHER, *Multi-channel seismic imaging of a crustal magma chamber along the East Pacific Rise*, Nature, 326 (1987), pp.34–41.

[19] K.C. MACDONALD, *Anatomy of the magma reservoir*, Nature, 339 (1989), pp.178–179.

[20] M.S. BURNETT, D.W. CARESS AND J.A. ORCUTT, *Tomographic image of the magma chamber at 12° 50′ N on the East Pacific Rise*, Nature, 339 (1989), pp.206–208.

[21] G.M. KENT, A.J. HARDING AND J.A. ORCUTT, *Evidence for a smaller magma chamber beneath the East Pacific Rise at 9° 30′ N* Nature, 344 (1990), pp.650–653.

[22] C. JAUPART, G. BRANDEIS AND C.J. ALLEGRE, *Stagnant layers at the bottom of convecting magma chambers*, Nature, 308 (1984), pp.535– 538.

[23] C. JAUPART AND G. BRANDEIS, *The stagnant bottom layer of convecting magma chambers*, Earth and Planet. Sci. Lett., 80 (1986) pp.183–199.

[24] D.F. NEILL, R. HON AND A. NAVROTSKY, *The igneous system* $CaMgSi_2O_6 - CaAl_2Si_2O_3 - NaAlSi_3O_8$*: variations on a classic theme by Bowen*, in Physics of Magmatic Processes (R.B. Hargraves, Ed), Princeton University Press, New Jersey, 1980.

[25] R.J. Kirkpatrick, G.R. Robinson and J.F. Hays, *Kinetics of crystal growth from silicate melts: anorthite and diopside*, J. Geophys. Res., (1976), pp.5715–5720.

[26] H.E. Huppert, J.S. Turner and R.S.J. Sparks, *Replenished magma chambers: effects of compositional zonation and input rates*, Earth and Planet. Sci. Lett., 57 (1982), pp.345–357.

[27] M.G. Worster, H.E. Huppert and R.S.J. Sparks, *Convection and crystallization in magma cooled from above*, Earth and Planet. Sci. Lett. , 101 (1990), pp. 78–89.

[28] R.J. Kirkpatrick, *Kinetics of crystal growth in the system* $CaMgSi_2O_6 - CaAl_2SiO_6$, American Journal of Science, 274 (1974), pp.215–242.

[29] H.S. Carslaw and J.C. Jaeger, *Conduction of Heat in Solids*, Cambridge University Press, Cambridge, 1959.

[30] B.D.Marsh, *On convective style and vigor in sheet-like magma chambers*, Journal of Petrology, 30 (1989), pp.479–530.

[31] H.E. Huppert *The intrusion of fluid mechanics into geology*, J. Fluid Mech., 173 (1986), pp.557-594.

[32] H.E. Huppert *The fluid mechanics of solidification*, J. Fluid Mech., 212 (1990) pp.209-240.

THE ERUPTION AND SPREADING OF LAVA

CLAUDE JAUPART*

Abstract. Lava eruptions are probably the least dangerous volcanic manifestations and have been monitored in detail, providing long time-series useful for the development and testing of flow models. Lavas contain gas bubbles and hence have two specific properties: compressibility and dilatational viscosity. The effects of dilatational viscosity are only significant for eruptions with high mass flux. On the contrary, compressibility is always important. For ascent in a volcanic conduit, the compressible nature of vesicular magma implies that (1) the mass flux is not related simply to pressure conditions at the conduit ends, (2) flow pressure and void fraction vary with height in a complex manner. In general, pressure is higher in the flow than in the surroundings, which may lead to failure of the conduit walls and horizontal injection of magma. Quantitative analysis of lava eruptions allows constraints on the evolution of the magma reservoir which feeds them. During flow away from a volcanic vent at the Earth's surface, compressibility affects the dynamics of spreading, leading to smaller variations of flow thickness and hence to slower rates of advance.

Lava eruptions present a puzzle because they require very small volatile contents for which there is no direct chemical evidence. A simple model is proposed in which the rising lava loses a fraction of its gas phase through fractured country rock. This leads to a set of coupled differential equations for pressure and void fraction which have a variety of solutions. The general result is that the void fraction decreases together with the eruption mass flux, in relation to a decrease of the driving overpressure in the volcano chamber. This explains why lava eruptions usually occur at the end of eruptive phases, as volcanic activity wanes. In this model, the void fraction is highly sensitive to driving pressure, leading to bifurcations for the evolution of gas content with height in the conduit.

1. Introduction

Volcanic eruptions offer a wealth of fascinating phenomena ranging from explosions and associated shock propagation, gas jets and plumes carrying large mass fractions of particles and finally lava flows. Sometimes, the same volcano may produce all these phenomena during the course of a single eruption. Research in volcanology is concerned with relating observed phenomena to variables such as the original volatile content of the melt in the chamber which feeds the eruption and to the behaviour of the host rocks as the eruption proceeds. Ultimate goals are to (1) understand how and when magma is sampled by an eruption, (2) assess hazards in geologically active parts of the world, and (3) use ancient volcanic deposits to obtain constraints on the rate of melt production in early geological periods and on the dynamics of planetary interiors.

Physical models of volcanic phenomena began to multiply fifteen years ago, when phenomenology was well established and large amounts of field data started to accumulate. Applying the concepts and methods of fluid dynamics to volcanology is therefore relatively new and there is no general introduction to the subject. Early studies [e.g. refs. 1–4] dealt essentially with volcanic blasts, columns and plumes in the atmosphere and on the formation of ash and pumice deposits from them. Since

---

*Department of Geology, University of Bristol, Wills Memorial Building, Bristol BS8 1RJ, United Kingdom. This research was supported in part by EEC grant ST–2000484 and in part by the Institute for Mathematics and its Applications with funds provided by the National Science Foundation.

then some progress has been made on gas/particle flows, with both theoretical studies and scaled laboratory experiments [e.g. refs. 5–7]. In comparison, less effort has been spent on lava eruptions which are the subject of this paper. These eruptions can be observed safely and have been monitored in more detail than their explosive counterparts. They provide continuous data series over long times which are useful to test flow models and to derive constraints on the evolution of the feeding system.

Lavas have high viscosity and have been studied using low Reynolds number hydrodynamics theory. Past investigations have dealt with the effects of density decrease due to pressure release [3], with conduit size variations due chilling or melting [8] and with the spreading of lava away from volcanic vents [9,10]. Lavas contain gas bubbles, and hence are viscous compressible liquids. The flow of such liquids seems to be a neglected subject in fluid dynamics, notable exceptions being Illingsworth [11] and White [12]. Another property of lavas which is due to the presence of gas bubbles is dilatational viscosity [13,14], such that expansion due to pressure release is retarded. In this paper, we analyze these two properties and derive a few simple solutions to the compressible flow equations of relevance to lava eruptions. In section 2, a brief review of the physical properties of lavas is given, with emphasis on the gas volume fraction and its effect on density and viscosity. In section 3, we investigate the ascent of lava from a chamber located at depth in the Earth. We first demonstrate that the effects of dilatational viscosity can be safely ignored in lava eruptions, but *not* in explosive eruptions with high mass flux. We then calculate the mass flux of lava eruptions and compare it to observations. Finally, we use data on lava eruptions to calculate how pressure evolves in the volcano chamber. In section 4, we study the flow of lava at the surface away from a volcanic vent, and show how compressibility affects the dynamics of spreading. In the final section, we discuss how lava eruptions may be due to gas loss through fractured host rocks, and give a simple mathematical model to emphasize the range of possible eruption regimes and ascent paths for lava in a volcanic conduit. The conclusion suggests some topics worthy of further investigations. Table 1 lists all the symbols used in the paper.

2. The Nature and Physical Properties of Lava

Lavas[1] are silicate melts carrying gas bubbles and crystals, and hence are three-phase mixtures. Their rheological properties have seldom been investigated [15,16] and are not well known. However, the properties of silicate melts are well understood [17] and it may be assumed that lavas behave as suspensions. In the following, we neglect the influence of crystals and focus on the influence of void fraction on physical properties.

2.1. The density of lava. Silicate melts contain several volatile species in solution. The major species are water and carbon dioxide, which have concentrations

[1] Strictly speaking, the term lava only applies to an erupted liquid, derived from magma which has vesiculated during ascent towards the Earth's surface. For the sake of simplicity, we shall also refer to vesicular magma as lava.

of a few percent and a few tens of a percent respectively [18,19]. Their solubility can be described by Henry's law. For our present purposes, it is sufficient to consider one volatile species with only pressure dependent solubility. At a pressure P which is lower than the saturation pressure P_s, assuming that nucleation of gas bubbles is instantaneous and neglecting volatile concentration gradients in the melt, lava density ρ is given by:

$$\frac{1}{\rho} = \frac{x(P_s) - x(P)}{\rho_g} + \frac{1 - [x(P_s) - x(P)]}{\rho_l} \tag{2.1}$$

where ρ_g and ρ_l are the gas and melt densities respectively, and x denotes solubility in weight fraction. At the high temperatures obtaining in volcanic systems and pressures smaller than 10^8 Pa , corresponding to depths of less than about 4 km, water and carbon dioxide can be treated as perfect gases with little error, such that:

$$\rho_g = \frac{P}{\frac{R}{M}\,T} \tag{2.2}$$

where R is the perfect gas constant, M molar mass and T temperature. Throughout the following, we consider eruptions of silicic lavas derived from melts with high silica contents, like andesites, dacites and rhyolites, and erupted by volcanoes such as Mount at St Helens on the western coast of the United States of America and Montagne Pelee in the French Antilles islands. Further, we focus on water because it is the most abundant volatile species, and keep temperature fixed at a value of 1200 K. The solubility law may be given by an empirical equation of the form:

$$x(P) = s\,P^n \tag{2.3a}$$

with coefficients taking the following values when pressure is expressed in Pascals:

$$s = 4.11 \times 10^{-6} \tag{2.3b}$$

$$n = 0.5 \tag{2.3c}$$

Figure 1 shows how lava density and gas volume fraction vary as a function of pressure for different water contents. The results emphasize that lava density is a strong function of pressure. An important point is that lava eruptions and flows have less than 75% vesicularity at atmospheric pressure and hence contain very small amounts of volatiles: for the solubility law chosen here, water mass fractions less than 2×10^{-3} are required, corresponding to saturation pressures smaller than 2 bars. Thus, for the largest part of their ascent, these magmas do not vesiculate and hence are bubble-free.

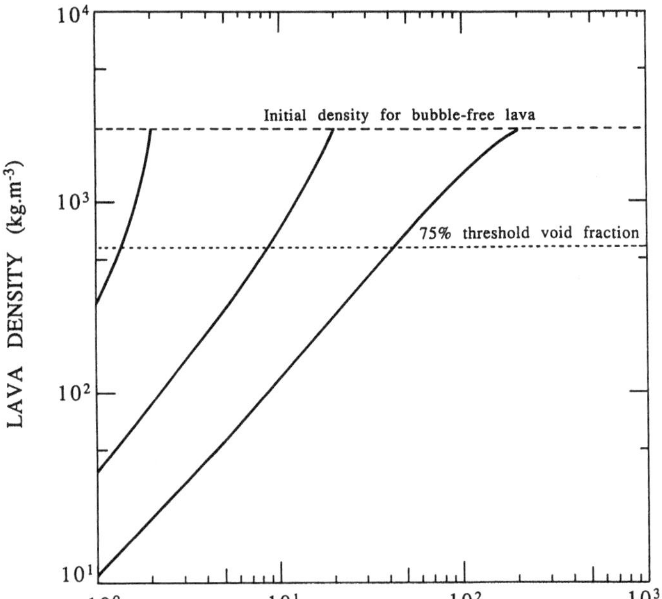

(a)

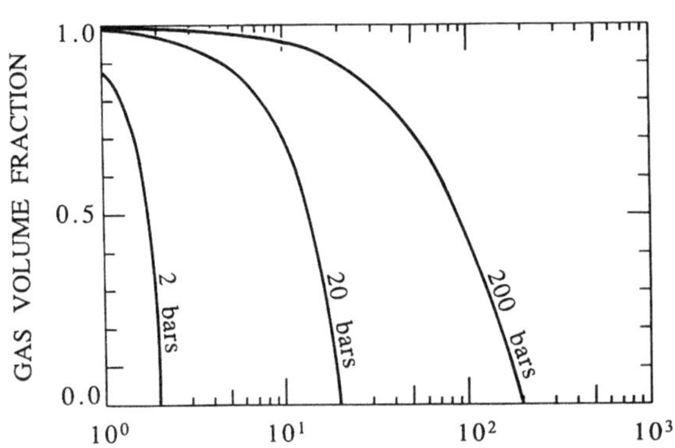

(b)

Figure 1. (a) Lava density as a function of confining pressure. Three curves are shown corresponding to silicic lavas saturated with water at pressures of 2, 20, 200 bars. Parameters used in the calculations are given in the text. The upper horizontal dashed line corresponds to bubble-free melt, and the lower dashed line corresponds to 75% void fraction, which is supposed to be the maximum value for stable bubbly liquid. (b) Void fraction as a function of confining pressure for the same three cases.

Many eruptions involve silicic melts with water contents larger than 2% [20], for which the gas volume fraction at the vent, i.e., at atmospheric pressure, may exceed 0.99. In these, the melt becomes gradually more vesicular as it rises towards the surface. At some stage, the bubbles burst and turn the ascending mixture into a gas jet carrying liquid droplets. This phenomenon, called fragmentation, is usually assumed to occur when the gas volume fraction reaches the value of 0.75. Justifications for this specific threshold are that it is equal to the average vesicularity of erupted pumice samples [21], and that it is near the close packing limit for undeformed and touching spheres [22]. However, bubbles achieve more efficient packing arrangements when they have a variety of sizes and when they deform [22]. Furthermore, some pumice samples have very high vesicularities, notably so-called 'reticulite' ejecta erupted during fire fountaining at Hawaii in which the void fraction may reach 98%. The physics of the fragmentation process have not been studied and remain a matter of conjecture and hypothesis. For this reason, it is difficult to build reliable flow models of volatile-rich eruptions.

2.2. The viscosity of lava. We shall consider a shear stress-strain rate relationship of the form:

$$\overline{\overline{\tau}} = P\overline{\overline{i}} - 2\mu\overline{\overline{e}} + \lambda(\nabla \cdot v)\overline{\overline{i}} \tag{2.4}$$

where P is thermodynamic pressure. $\overline{\overline{e}}$ and $\overline{\overline{i}}$ are the strain rate and identity tensors respectively. μ is the first coefficient of viscosity, also called extensional viscosity. The second viscosity coefficient λ is equal to:

$$\lambda = \frac{2}{3}\,\mu - K \tag{2.5}$$

Coefficient K has been called dilatational viscosity. Both μ and λ vary as a function of void fraction [14,23].

In bubbly liquids with small gas volume fraction α, the shear stress-strain rate relationship is linear with viscosity given by [24]:

$$\frac{\mu}{\mu_1} = 1 + m\,\alpha \tag{2.6}$$

where μ_1 is the liquid viscosity, and m is a coefficient which depends on both the liquid and gas viscosities and on the boundary conditions at the gas-liquid interface. For gas bubbles with negligible viscosity, m is usually taken to be 1, but the behaviour of the interface may lead to larger values [23]. The limiting case is when bubbles behave as rigid spheres, for which $m = 5/2$.

For gas volume fractions between values of 0.5 and 0.7, an old set of measurements indicates that the shear stress-shear strain relationship is also linear [25]. The available data can be interpreted with a critical law of the form:

$$\frac{\mu}{\mu_1} = \left(\frac{1}{1 - \alpha/\alpha_c}\right)^m \tag{2.7}$$

α_c is the critical volume fraction above which the foam breaks down. Foams with α values in excess of 0.98 have been prepared [22] and, as stated above, observed in Hawaiian eruptions. Thus $\alpha_c \approx 1$. The critical exponent m is determined by noting that equation (2.7) reduces to (2.6) in the dilute limit and by using the experimental results of Sibree [25]. As shown in Figure 2, this leads to $m = 5/2$.

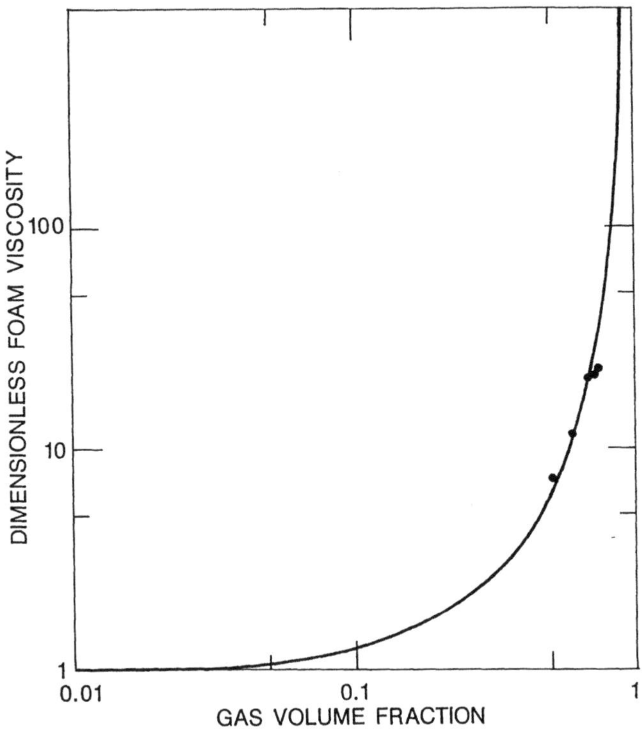

Figure 2. Viscosity increase due to the presence of bubbles in a viscous liquid, as a function of void fraction. The dots are experimental determinations from Sibree [25] and the curve is from equation (2.7) with parameters α_c and m equal to 1 and 5/2 respectively.

At large gas volume fractions in excess of close packing, bubbles are deformed and separated by very thin liquid films. This leads to shear-thinning behaviour with an effective viscosity which exceeds that of the liquid by large amounts at low shear rate [26]. In these conditions, it is observed that foams exhibit a yield stress below which there is no flow, and the phenomenon of slip at the wall, where flow is lubricated by a liquid film which wets the conduit [23, 27].

An additional effect of importance for silicic lavas is that the melt viscosity increases strongly as the amount of dissolved volatiles decreases [17]. This, together with the influence of gas bubbles, implies that lava viscosity increases by a few orders of magnitude as it rises towards the surface and undergoes pressure release.

A recent theoretical study of the dilatational viscosity of a Newtonian liquid containing gas bubbles yields the following equation [14]:

$$\frac{K}{\mu_1} = \frac{4}{3}\frac{1-\alpha}{\alpha} \tag{2.8}$$

This equation differs only slightly from the earlier one by Taylor [13] and shows that dilatational viscosity is high at low bubble contents. It remains finite if bubbles are present in the liquid, as emphasized by Prud'homme and Bird [14]. Without bubbles, of course, the fluid has no dilatational viscosity.

This discussion shows that theory and experiments on the rheological properties of bubbly liquids are only available in the dilute limit and in the "dry" limit where the volume of liquid is negligible. Thus, there is no reliable framework over the large void fraction range which is necessary for the analysis of volcanic eruptions. Furthermore, the fragmentation process which is responsible for the formation of explosive discharges of gas and tephra is not understood. For this reason, most of our attention will be focussed on compressible effects in constant viscosity liquids, with application to lava-forming eruptions.

3. The Ascent of Vesicular Magma in a Volcanic Conduit

Consider a cylindrical conduit with constant radius a connecting a reservoir located at depth H to the Earth's surface. For the sake of simplicity, we consider that the rising magma does not lose heat to the surroundings, and further that the work of pressure forces is negligible. Thus, temperature stays constant during ascent, and all state variables are functions of pressure only. The only non-zero velocity component is along the vertical (directed upwards). In steady-state conditions, which is a reasonable approximation for most long-lived types of volcanic activity [3], the equations of motion reduce to:

$$0 = \frac{\partial}{\partial z}(\rho w) \tag{3.1a}$$

$$0 = -\frac{\partial P}{\partial r} - \frac{\partial}{\partial r}\left(\lambda\frac{\partial w}{\partial z}\right) + \frac{\partial}{\partial z}\left(\mu\frac{\partial w}{\partial r}\right) \tag{3.1b}$$

(3.1c)

$$\rho w\frac{\partial w}{\partial z} = -\frac{\partial P}{\partial z} + \frac{1}{r}\frac{\partial}{\partial r}\left(r\mu\frac{\partial w}{\partial r}\right) + \frac{\partial}{\partial z}\left([2\mu-\lambda]\frac{\partial w}{\partial z}\right) - \rho g$$

3.1. Dilatational viscosity effects. To proceed, we assume that the viscosity coefficients are constant, which is valid as long as the gas volume fraction does not change by large amounts. Equation (3.1b) becomes:

$$0 = -\frac{\partial P}{\partial r} - (\mu-\lambda)\frac{\partial}{\partial r}\left(\frac{\partial w}{\partial z}\right) \tag{3.2}$$

This can be integrated with respect to r, using the boundary condition that $w(a,z) = 0$:

$$P(r,z) - P(a,z) = (\mu-\lambda)\left(\frac{\partial w}{\partial z}\right) \tag{3.3a}$$

or, using equation (2.5) for λ:

$$P(r,z) - P(a,z) = \left(\frac{\mu}{3} + K\right)\left(\frac{\partial w}{\partial z}\right) \tag{3.3b}$$

This equation shows that pressure is in general not uniform across the conduit width and maximum at the conduit axis.

Substituting for this equation in (3.1c) yields:

$$\rho w \frac{\partial w}{\partial z} = -\frac{dP_a}{dz} + \mu\left\{\frac{1}{r}\frac{\partial}{\partial r}\left(r\frac{\partial w}{\partial r}\right) + \frac{\partial^2 w}{\partial z^2}\right\} - \rho g \tag{3.4}$$

where P_a stands for $P(a,z)$. This equation has the same form as the vertical momentum balance for an incompressible liquid, save for the body force term which is a function of r.

To proceed further, we first estimate the typical value of the rate of expansion, given by $\frac{\partial w}{\partial z}$. Continuity equation (3.1a) is rewritten as:

$$\rho w = G \tag{3.5}$$

where G is a function of r only. Thus:

$$\frac{\partial w}{\partial z} = G\frac{d}{dP}\left(\frac{1}{\rho}\right)\frac{\partial P}{\partial z} \tag{3.6}$$

In this problem, the mixture rises in a conduit through country rock of density $\rho_r \approx \rho_l$ and a reasonable estimate for the vertical pressure gradient is $(\rho_l g)$. Using further equation (2.1) and Figure 1 to estimate the derivative of density with respect to pressure, we find that the expansion rate takes a representative value of:

$$\frac{\partial w}{\partial z} \approx G10^{-7} \tag{3.7}$$

This shows that the magnitude of the expansion rate is a function of the eruption rate. In silicic lava eruptions, velocities have a typical value of 10^{-3} m/s [9, 28]. For a density of $10^3 kg.m^{-3}$, this yields a mass flux per unit area of $1 kg.m^{-2}.s^{-1}$. Viscosity estimates for silicic melts are in the range of $10^4 - 10^7$ Pas, depending on chemical composition and crystal content. For small values of the gas volume fraction of 10^{-3}, representative for a lava which leaves the chamber during eruption [29], dilatational viscosity takes values in the range of $10^7 - 10^{10}$ Pas. Substituting for these numerical estimates, it is found that the horizontal variation of pressure is negligibly small.

The same conclusion does not hold for more powerful eruptions, such as Plinian ones. As already stated before, these are not well understood and may far from steady-state flow conditions. A rough estimate of the mass flux in the volcanic conduit may be obtained as follows. The exit velocity has a typical value of 300 m/s [3]. For a water mass fraction of 4% and an atmospheric exit pressure, the density of the erupted material is $4kg.m^{-3}$. Thus, at the vent, the mass flux per unit area takes a typical value of $10^3 kg.m^{-2}.s^{-1}$. Assuming steady-state and using the same viscosity estimates as before, equation (3.7) implies that the horizontal variation of pressure may reach $10^6 Pa$, which is important for predicting accurately the distribution of void fraction (Fig. 1) and for understanding the process of fragmentation. Further, this leads to horizontal variations of density implying different velocity profiles. This feature deserves more attention and may have important volcanological implications.

3.2. The mass flux of lava eruptions. In the case of lava eruptions with small mass flux, the effects of dilatational viscosity can be neglected and the solution is simple. Pressure depends only on z and hence so are the void fraction, density and viscosity for a given volatile content. Noting that the length of a volcanic conduit is considerably larger than its width, these flows can be studied in the lubrication approximation. Further, the Reynolds number stays small, and hence inertial terms can be neglected. The vertical momentum balance reduces to:

$$0 = -\frac{dP_a}{dz} + \frac{1}{r}\frac{\partial}{\partial r}\left(r\mu\frac{\partial w}{\partial r}\right) - \rho_a g \tag{3.8}$$

where ρ_a stands for density at pressure P_a. Using boundary conditions $\frac{\partial w}{\partial r}(0,z) = 0$ and $w(a,z) = 0$, this equation yields:

$$w = \frac{1}{4\mu}(r^2 - a^2)\left\{\frac{dP_a}{dz} + \rho_a g\right\} \tag{3.9}$$

Using continuity equation (3.1a) and integrating across the conduit width:

$$8\mu Q = -\pi a^4 \rho_a \left\{\frac{dP_a}{dz} + \rho_a g\right\} \tag{3.10}$$

where Q is the mass flux of the eruption. This equation is integrated as follows:

$$\int_{P_i}^{P_f} \frac{dP_a}{\dfrac{8\mu Q}{\pi a^4 \rho_a} + \rho_a g} = -H \tag{3.11}$$

where P_i and P_f stand for the initial pressure at the conduit bottom ($z = 0$) and at the surface vent ($z = H$). In this equation, the unknown is mass flux Q.

Given the complex form of the pressure dependence of density, equation (3.11) has no straightforward analytical solution. However, as noted in section (2.1), lava eruptions are such that no gas phase is present during the largest part of the ascent and density and viscosity changes can be neglected when evaluating the integral on the left-hand side of the equation. The solution is then simply:

$$Q = \frac{\pi a^4 \rho_a}{8\mu}\left\{\frac{(P_i - P_f) - \rho_a g H}{H}\right\} \tag{3.12a}$$

Assume first that the vent pressure P_f is atmospheric. The initial pressure P_i is the sum of the equilibrium lithostatic pressure at depth H plus an unknown overpressure ΔP:

$$P_i = P_{atm} + \rho_r g H + \Delta P \tag{3.12b}$$

where ρ_r is the density of country rock. Substituting into (3.12), the mass flux is:

$$Q = \frac{\pi a^4 \rho_a}{8\mu} \left\{ \frac{\Delta P + (\rho_r - \rho_a) g H}{H} \right\} \tag{3.12c}$$

This shows that an eruption is driven by an overpressure in the volcano chamber and by the buoyancy of lava. The mass flux depends critically on the conduit size and on the viscosity of lava, which varies by several orders of magnitude depending on chemical composition. Although the other variables involved in this equation may vary from one eruption to the next as well as from one volcano to the other, they are found to lie within quite restricted ranges. The chamber overpressure is necessarily smaller than the strength of country rock, and the largest part of the ascent is done in the volcanic edifice itself, such that $\rho_r \approx \rho_a$. For eruptions through conduits of given size, the mass flux is predicted to be a decreasing function of melt viscosity. Indeed, observed values are $5 \times 10^4 kg.s^{-1}$ for basalts at Hawaii [30], $5 \times 10^3 kg.s^{-1}$ for basaltic andesite at La Soufriere de Saint Vincent [9] and $5 \times 10^2 kg.s^{-1}$ for dacite at Mount St Helens [28]. In these three cases, the conduits had dimensions between 15 and 30 meters.

Once Q is determined, equation (3.11) can be rewritten for any height z in the conduit:

$$\int_{P_i}^{P} \frac{dP_a}{\dfrac{8\mu Q}{\pi a^4 \rho_a} + \rho_a g} = -z \tag{3.13}$$

which gives the variation of pressure as a function of z. It is found that pressure varies non-linearly with height and differs significantly from the far-field pressure in surrounding country rock. In general, the solution is such that there is a pressure excess, which may lead to failure of the conduit walls and the horizontal injection of lava.

3.3. Pressure variations in magma chambers. Understanding volcanic phenomena is difficult in part because they involve complex materials and in part because the conditions in the volcano chamber are unknown. Some researchers assume that eruptions result directly from an input of magma from a deep source and use the erupted mass to calculate the rate of melt production at large depth in the Earth (e.g. ref. 31). Others consider that some eruptions empty the volcano chamber [29]. We now discuss how lava eruptions allow an insight into this problem.

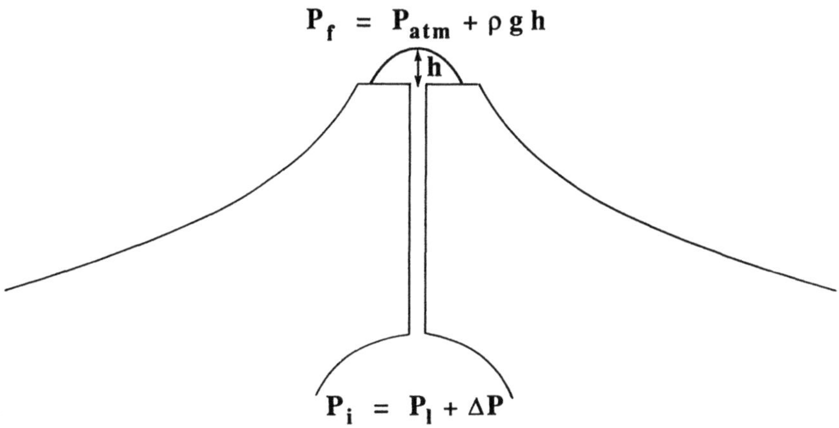

Figure 3. Schematic diagram showing the various parts of a volcanic system involved in an eruption. Lava rises from a reservoir through a conduit of small dimensions and spreads at the surface as a dome or flow of height h. The reservoir is at higher pressure than the surroundings.

In the case of silicic magmas with high viscosity, it is observed that a lava dome builds up whose height grows with time. This acts to change the vent pressure at $z = H$ (Figure 3). This may also act to change the chamber pressure (at $z = 0$) if country rock deforms in response to the applied load. For small lava domes, however, such deformation is negligible and the chamber pressure is unaffected. Neglecting end pressure effects due to lava spreading out of the vent of small aperture, the exit pressure P_f is given by:

$$P_f = P_{atm} + \rho_a g h \tag{3.14}$$

where h is the height of lava above the vent. In this case, equation (3.12) shows that the mass flux varies as a function of h. Both variables can be measured in the field, which allows constraints on the other variables, such as the chamber pressure.

Huppert et al. [9] studied the lava dome built by the Soufriere de Saint Vincent volcano in 1979. They found a linear relationship between the volume flux and the dome height, which according to equations (3.12) and (3.14) implies that the chamber pressure did not vary. Because the chamber is embedded in elastic country rock, this also implies that its volume did not vary. Thus, the analysis indicates that the chamber was being replenished with new magma as the eruption proceeded. However, the total erupted volume in this particular case was about $5 \times 10^7 m^3$, which is probably a very small fraction of the chamber volume. There are unfortunately no direct constraints on the Soufriere chamber, but in all volcanoes where chambers have been located and imaged, their volumes were found to be several $10^{10} m^3$ [29]. Thus, it may be that the pressure changes associated with this lava dome were too small to have measurable effects on the flow rate.

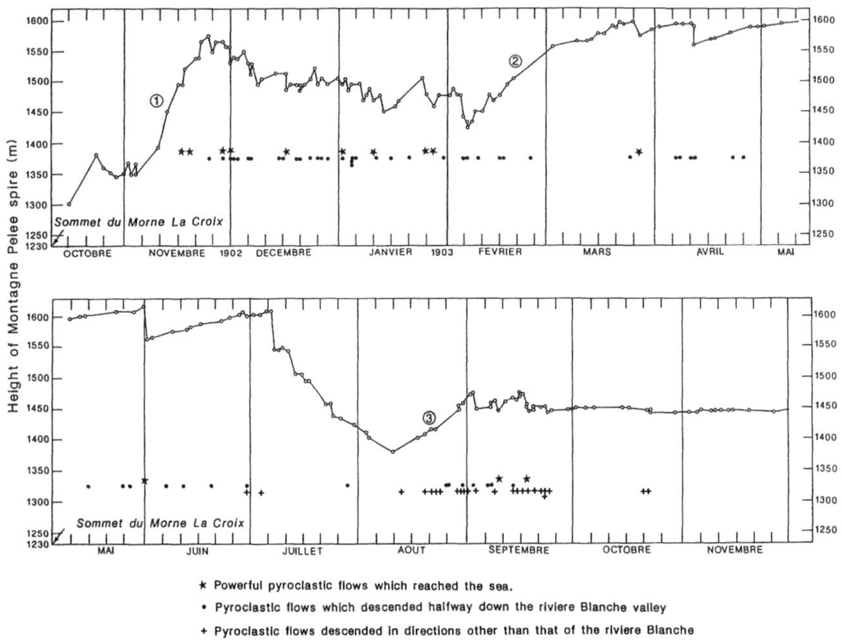

Figure 4. The height of the Montagne Pelee spine in 1902–1903 as a function of time, from Lacroix [32]. Note the three successive episodes of growth, labelled from 1 to 3, which span similar height ranges.

A particularly useful phenomenon occurred at Montagne Pelee (Martinique island, French Antilles) after its famous 1902 climactic eruption. A spine rose out of the vent and reached a height of three hundred meters (Figure 4). Its rate of growth steadily decreased with time and it was gradually destroyed by erosion and explosions. A new spine then emerged, which followed the same cycle. Finally, a third spine was formed and the eruption ground to a halt. The height of these spines was measured systematically with telemeters and the data are reproduced in Figure 4. The interesting feature is that the volcano reproduced the same experiment three times with the same conditions. Thus, variations of eruption rate could only be due to changes of conditions in the plumbing system. Equations (3.12) and (3.14) are combined to yield:

$$Q = \frac{\pi a^4 \rho_a}{8\mu} \left\{ \frac{\Delta P + (\rho_r - \rho_a) g H - \rho_a g h}{H} \right\} \tag{3.15}$$

Q is also given by the observed rate of growth:

$$Q = \pi a^2 \rho_a \frac{dh}{dt} = \pi a^2 \rho_a v \tag{3.16}$$

where v is vertical velocity outside the vent. At the reference height of 1430 meters, one finds values of $1.8 \times 10^{-4}, 9.7 \times 10^{-5}$ and $3.7 \times 10^{-5} m.s^{-1}$ for the rate of growth of the three successive phases. Assuming that the properties of the spine and the conduit dimensions did not change, which is supported by the field observations [32], these results demonstrate that the chamber overpressure decreased from phase 1 to 3. To get a quantitative estimate, we note that the implied pressure changes are small. Indeed, equations (3.15) and (3.16) imply that:

$$\frac{\Delta P_2 - \Delta P_3}{\Delta P_1 - \Delta P_2} = \frac{v_2 - v_3}{v_1 - v_2} = 0.7 \tag{3.17}$$

where v_i and ΔP_i denote the vertical velocity and chamber overpressure for phase i at the reference height. Thus, it is sufficient to consider that the overpressure stays constant during each phase. Equations (3.15) and (3.16) yield a differential equation for $h(t)$ which is easily integrated:

$$h(t) - h(0) = v_i \tau [1 - \exp(-t/\tau)] = \left[\frac{\Delta P + (\rho_r - \rho_a) g H}{\rho_a g} - h(0) \right\} \{1 - \exp(-t/\tau)\} \tag{3.18}$$

where initial time $t = 0$ is such that the spine height is $h(0) = 1430m$ and τ is a characteristic time defined as follows:

$$\tau = \frac{8 \mu H}{\rho_a a^2 g} \tag{3.19}$$

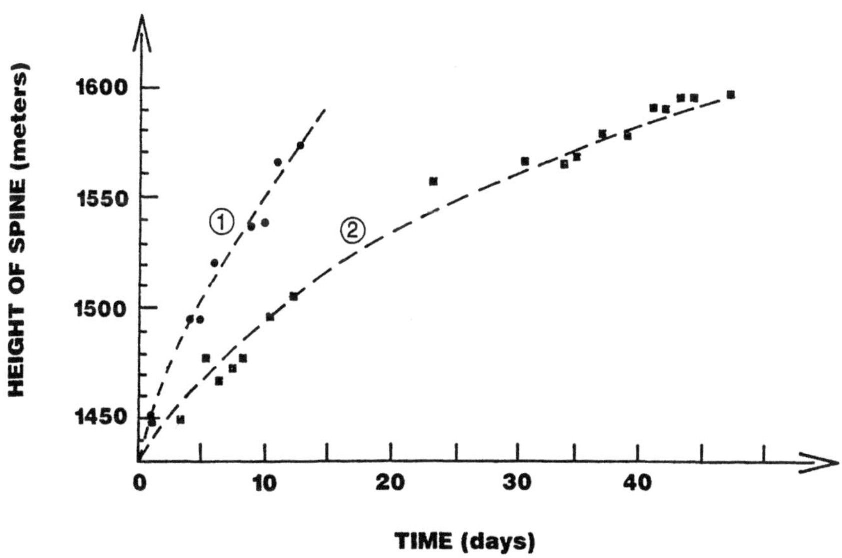

Figure 5. Height of the Montagne Pelee spine as a function of time for episodes 1 and 2 (see Figure 4). The dashed lines are from equation (3.18) with characteristic time τ set to 20 days, and initial velocity values $v_1 = 1.8 \times 10^{-4} m/s$ and $v_2 = 9.7 \times 10^{-5} m/s$.

The spine height can be calculated from equation (3.18) using only the initial velocity and a value for characteristic time τ. A value of 20 days for τ allows a good fit to the data (Figure 5). This allows the calculation of the actual variation of chamber pressure. A mean density value of $2000 kg.m^{-3}$ yields:

$$\Delta P_1 - \Delta P_2 = 2.6MPA$$

$$\Delta P_2 - \Delta P_3 = 1.8MPa$$

The equation for τ can be used to obtain a viscosity estimate. For $a \approx 100m$ [32] and $H \approx 5km$, which cannot be off by more than a factor of two, $\nu = \mu/\rho$ is $4 \times 10^5\ m^2.s^{-1}$. Using again a density of $2000kg.m^{-3}$, this yields a viscosity of about 10^9 Pas, which is reasonable for such lavas.

Finally, an assumption on the country rock-magma density contrast allows an estimate of the actual values of the chamber overpressure. In a volcanic edifice, country rock is made of older solidified and penetratively fractured lava, and hence the density contrast is small. For a reasonable estimate of $400kg.m^{-3}$, the initial overpressure is $\Delta P_1 = 14MPa$.

4. Compressible Viscous Currents

The thickness of the lava flow affects the eruption process through its effect on the conduit pressures. Thus, a complete analysis of lava eruptions must deal with spreading away from the vent. This problem has been tackled for an incompressible fluid by Huppert [33], using both laboratory experiments and theory, and applied to the Soufriere de Saint Vincent 1979 eruption [9]. The field measurements agreed with the theoretical predictions, but implied a viscosity value which was considerably higher than direct estimates. Many effects have been invoked to explain this discrepancy, including increased viscosity due to cooling or to the formation of a solid carapace and the piling up of rubble at the flow front [9,10]. Part of the solution to this problem, however, is the compressible nature of lava, as shown below.

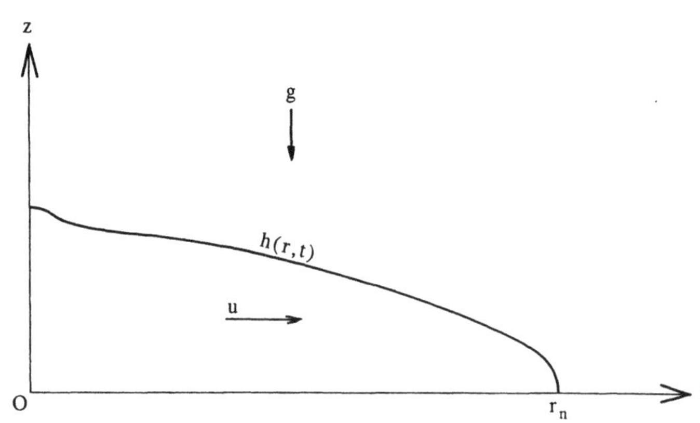

Figure 6. Diagram showing the coordinate system to analyze the spreading of a compressible liquid.

4.1. Lubrication equations for compressible viscous flow. The problem is illustrated in Figure 6. A compressible liquid of constant viscosity μ spreads out of a vent of small dimensions. Its density is given by the following equation:

$$\rho = \rho_0\{1 + \beta(P - P_0)\} \tag{4.1}$$

where β is compressibility and ρ_0 density at pressure P_0 (taken to be the atmospheric pressure). We assume that β is constant, which simplifies the calculations considerably. In the lubrication approximation, which is applicable once the flow is thin compared to its length, the equations of motion reduce to:

$$0 = -\frac{\partial P}{\partial z} - \rho g \tag{4.2a}$$

$$0 = -\frac{\partial P}{\partial r} + \mu \frac{\partial^2 u}{dz^2} \tag{4.2b}$$

The boundary conditions are the following:

$$\tau_{zz}(r, h) = P_{atm} = P_0 \tag{4.3a}$$

$$\tau_{rz}(r, h) = 0 \tag{4.3b}$$

$$u(r, 0) = 0 \tag{4.3c}$$

Finally, the mass flux of the eruption must be specified. In the present case, this is done by specifying how the total mass of the flow varies:

$$\int_0^{r_n}\int_0^h 2\pi r \rho \; dr \; dz = M(t) \tag{4.4}$$

where $r_n(t)$ is the front position.

Substituting for density in equation (4.2a) gives the density and pressure distribution in a flow of height h:

$$\rho = \rho_0\{1 + \beta(P - P_0)\} = \rho_0 e^{-\rho_0 \beta g(z-h)} \tag{4.5}$$

which yields:

$$\frac{\partial P}{\partial r} = \rho_0 g \frac{\partial h}{\partial r} e^{-\rho_0 \beta g(z-h)} \tag{4.6}$$

Substituting for this equation into (4.2b), and using boundary conditions (4.3), the velocity field is given by:

$$u(r, z) = \frac{\rho_0 g}{\mu} \frac{\partial h}{\partial r} \frac{1}{(\rho_0 \beta g)^2} \left\{ e^{-\rho_0 \beta g(z-h)} - e^{\rho_0 \beta g h} + \rho_0 \beta g z \right\} \tag{4.7}$$

The continuity equation, consistent with the lubrication approximation, is written as follows:

$$\frac{d}{dt}\left(2\pi r \int_0^h \rho \; dz\right) = -\frac{\partial \varphi}{\partial r} \tag{4.8a}$$

where $\varphi(r)$ is the horizontal mass flux at radius r:

$$\varphi(r) = 2\pi r \int_0^h \rho u dz \tag{4.8b}$$

Using equation (4.5) for lava density, and substituting for the velocity field into the continuity equation, one obtains an equation for the flow height:

$$0 = e^{\rho_0 \beta g h} \frac{\partial h}{\partial t} + \frac{1}{r}\frac{\partial}{\partial r}\left\{\frac{\rho_0 g}{\mu(\rho_0 \beta g)^3} r \frac{\partial h}{\partial r}\left[-\frac{3}{2} - \rho_0 \beta g h + 2e^{\rho_0 \beta g h} - \frac{1}{2}e^{2\rho_0 \beta g h}\right]\right\} \tag{4.9}$$

Finally, the total mass of the flow is written as:

$$\int_0^{r_n} 2\pi r \frac{1}{\beta g}[e^{\rho_0 \beta g h} - 1] dr = M(t) \tag{4.10}$$

4.2. Similarity solution for constant mass flux. The system of equations (4.9) and (4.10) can be solved for any mass flux. However, a simple and revealing solution can be obtained in the particular case of a constant mass flux Q, i.e.:

$$M(t) = Qt \tag{4.11}$$

In this case, it can be shown that the system has a similarity solution of the form:

$$h(r,t) = H(\eta) \tag{4.12a}$$

where the similarity variable η is such that:

$$\eta = rt^{-1/2} \tag{4.12b}$$

The interesting feature of this system of equations is that there are two length-scales. One can be deduced from the density distribution equation (4.5):

$$h_c = (\rho_0 \beta g)^{-1} \tag{4.13a}$$

The other is the viscous length-scale from (4.9):

$$h_v = \left(\frac{\mu Q}{\rho_0{}^2 g}\right)^{1/4} \tag{4.13b}$$

Equation (4.9) for the flow thickness is best written using the following scales:

$$H = H^* \left(\frac{\mu Q}{\rho_0{}^2\, g}\right)^{1/4} = H^* h_v \tag{4.14a}$$

$$\eta = \eta^* \left(\frac{\rho_0 g Q^3}{\mu \rho_0{}^3}\right)^{-1/8} \tag{4.14b}$$

Substituting into equations (4.9) and (4.10), and dropping the stars, one obtains:

$$-\frac{1}{2}\eta^2 e^{CH}\frac{dH}{d\eta} + C^{-3}\frac{d}{d\eta}\left\{\eta\frac{dH}{d\eta}\left[-\frac{3}{2} - CH + 2e^{CH} - \frac{1}{2}e^{2CH}\right]\right\} = 0 \tag{4.15a}$$

$$2\pi C^{-1}\int_0^{\eta_n}(e^{CH} - 1)\eta\, d\eta = 1 \tag{4.15b}$$

where C is a dimensionless number equal to:

$$C = \rho_0\beta g\left(\frac{\mu Q}{\rho_0{}^2 g}\right)^{1/4} = \frac{h_v}{h_c} \tag{4.16}$$

The final step is to re-scale these equations with η_n, with new variables:

$$\eta = \eta_n \cdot \eta^{**} \tag{4.17a}$$

$$H = \eta_n^{-2/3} H^{**} \tag{4.17b}$$

and a new dimensionless number S:

$$S = C\eta_n^{2/3} \tag{4.17c}$$

The equations are again rewritten dropping the stars. This leaves equation (4.15a) formally unaffected:

$$-\frac{1}{2}\eta^2 e^{SH}\frac{dH}{d\eta} + S^{-3}\frac{d}{d\eta}\left\{\eta\frac{dH}{d\eta}\left[-\frac{3}{2} - SH + 2e^{SH} - \frac{1}{2}e^{2SH}\right]\right\} = 0 \tag{4.18}$$

Finally, equation (4.15b) is recast as an equation for η_n:

$$\eta_n = \left\{2\pi S^{-1}\int_0^1 (e^{SH} - 1)\eta\, d\eta\right\}^{-3/8} \tag{4.19}$$

The governing equation (4.18) was integrated numerically starting at $\eta = 1$ with the boundary condition $H(1) = 0$, and using the leading order expression for H in the vicinity of that point.

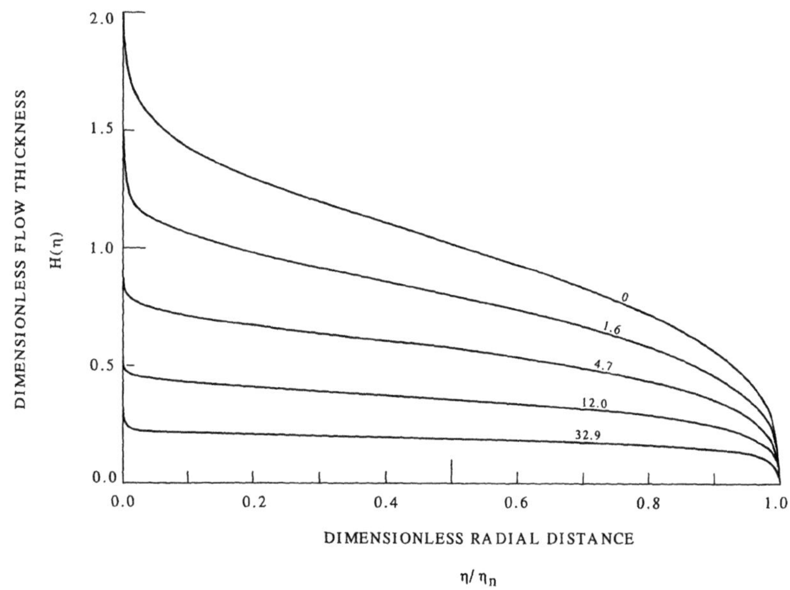

Figure 7. Dimensionless height $H(\eta)$ (as defined by equation 4.14a) as a function of scaled similarity variable η (as defined by equation 4.17a), for several values of dimensionless number C. As C increases, compressibility becomes increasingly more important and the flow develops less topography.

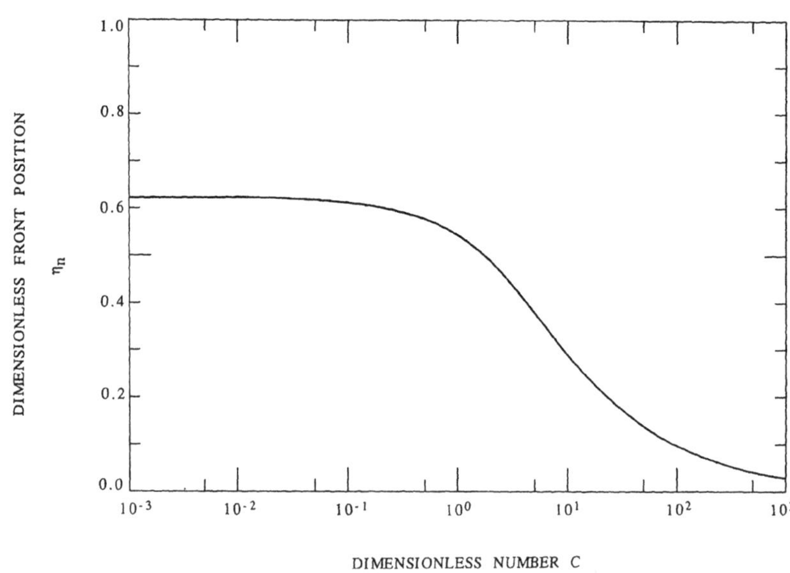

Figure 8. Dimensionless front position η_n (as defined in equations 4.14b and 4.15b) as a function of dimensionless number C. As C increases, the flow spreads less rapidly because thickness variations are reduced. η_n tends to the value 0.623 as C goes to zero, which is the incompressible value.

Figure 7 shows the flow shape in non-dimensional coordinates for various values of dimensionless parameter C. At low values of C, one recovers the incompressible shape. As C increases, the flow becomes thinner, because of the liquid is being compressed more and more efficiently. Also, thickness variations are smaller because the flow expands as it spreads. Figure 8 shows how the dimensionless front position η_n varies as a function of C. As C goes to zero, η_n tends towards the incompressible value of 0.623 (which corresponds to the value of 0.715 given by Huppert [9,33] with a slightly different scaling procedure). As C increases, η_n decreases, reflecting the fact that the compressible liquid flows less rapidly because it develops less topography.

The development of the flow can be monitored in the field, providing data on lava volume and front position as functions of time. The front position measurements can be used to estimate viscosity. The calculation shows that η_n is overestimated if lava is assumed to be incompressible, which implies a higher viscosity value than in reality. Scaling (4.14b) ensures that the viscosity estimate is highly sensitive to the η_n value. In dimensional variables, the font position $r_n(t)$ is equal to:

$$r_n(t) = \eta_n \left(\frac{\rho_0 g Q^3}{\mu \rho_0{}^3} \right)^{1/8} t^{1/2} = A t^{1/2} \tag{4.20a}$$

where A is a numerical constant which is determined from the field observations. Rewriting this equation, we obtain:

$$\mu = \frac{1}{A^8 g} \frac{\rho_0{}^2}{Q^3} \eta_n{}^8 \tag{4.20b}$$

The true value of η_n depends on C, which is the ratio between the viscous and compressible length-scales. The compressible length-scale h_c can be calculated assuming either that thermodynamic equilibrium is maintained between gas and melt with bubbles resorbing as pressure increases, or that there is no resorption and that bubbles only undergo a volume decrease. The first assumption implies that the void fraction drops to negligibly small values at flow depths in excess of about 5 m, which is not consistent with measured values [34]. The second assumption leads to estimates of 30–60 m depending on total gas content. The viscous length-scale h_v depends on the flow rate and is best estimated by the actual mean flow thickness which is close to it (Fig. 8). Lava domes are commonly between 100 and 200 m high. Thus, C takes values between 1.5 and 6, leading to values of η_n between 0.54 and 0.35, instead of the incompressible value of 0.62. Hence, using the incompressible solution, the effective viscosity value for the flow may be overestimated by as much as two orders of magnitude. Further, if one assumes that the flow density is constant and equal to that of surficial samples, i.e. ρ_0, the measured volume yields a mass flux which is smaller than in reality, which also contributes an artificial viscosity increase.

5. Transitions of Eruption Regime

So far, lava eruptions have been studied by assuming that they are due to the ascent of gas-poor magma, such that gas exsolution and expansion occur only at very shallow levels. Such eruptions are usually preceded by explosive phases involving larger quantities of gas. This suggests that the chamber is zoned in volatile content. This interpretation has been challenged recently [34], because samples from fragmented and unfragmented eruptions (Plinian ash falls and lava domes respectively) show no appreciable differences of chemical composition. This suggests that lava may lose gas in the conduit, presumably when its vesicularity becomes high enough for it to be permeable to gas flow. Indeed, volcanic eruptions are often accompanied by fumarolic activity, such that ground cracks release large amounts of volcanic gases [35, 36]. Where observed, the walls of volcanic conduits are found to be penetratively fractured [37]. These fractures, as well as the common occurrence of radial dikes emanating from eruption conduits provide evidence for a pressure contrast between rising lava and country rock. As discussed in section 3.2, the flow equations do indeed predict the occurrence of such a pressure contrast. Thus, there are pathways and a driving pressure difference for gas to flow out of rising lava. The dynamical consequences of this phenomenon are now investigated using a simple model.

5.1. The evolution of gas content. Field evidence suggests that only the gas phase is lost from the rising material, but the mechanism is not obvious. Conduit walls are rough surfaces [37], and hence cannot be perfectly wetted by very viscous liquid, which is known to promote the separation of the vapor phase [38].

We develop an approximate analysis in which quantities are horizontal averages across the conduit and conservation equations are satisfied in integral form. Both liquid and gas have the same average vertical velocity w. At depth z, the rising mixture is taken to be at an average pressure P which is higher than the pressure P_L in undisturbed country rock far from the conduit and to carry on average a volume fraction α of gas bubbles. Mass conservation for the liquid phase gives:

$$Q_l = \pi a^2 \rho_l (1-\alpha) w \tag{5.1}$$

where Q_l is the mass flux of liquid. Rising lava loses gas with mass flux per unit area q, and mass conservation for the gas phase reads:

$$\pi a^2 \left\{ \frac{d}{dz}(\rho_g w \alpha) + \frac{d}{dz}[x \rho_l (1-\alpha) w] \right\} = -2\pi a q \tag{5.2}$$

where x is the mass fraction of gas in solution. Substituting for (5.1) in (5.2) and rearranging, one obtains an equation for the rate of change of the gas volume fraction:

$$\frac{d\alpha}{dz} = (1-\alpha)^2 \frac{\rho_l}{\rho_g} \left[-\frac{dP}{dz} \left(\frac{\alpha}{\rho_l (1-\alpha)} \frac{d\rho_g}{dP} + \frac{dx}{dP} \right) - \frac{2\pi a q}{Q_l} \right] \tag{5.3}$$

This shows that gas loss counteracts the effects of pressure release. The fraction of gas lost is inversely proportional to the eruption rate Q_l, because the flow of gas occurs at a given rate through country rock whilst magma rises. This equation therefore predicts that, for given conditions, a decrease of mass flux leads to smaller values of the void fraction. This is indeed observed: as already discussed in section 3.1, estimates for the mass flux are of the order of $10^3 kg.m^{-2}.s^{-1}$ in fragmented phases with high gas content and less than $1 kg.m^{-2}.s^{-1}$ for lava flows and domes. Further, the analysis of Montagne Pelee developed in section 3.3 shows that this eruption slowed down with time, probably because the chamber pressure decreased. Thus, variations of eruptions regime and gas content may not be due to changes of melt composition.

5.2. Mathematical model. To proceed with the analysis, two additional equations are needed. q, the mass flux of gas through the conduit walls, is equal to:

$$q = -\rho_g \alpha k (P - P_L) \tag{5.4}$$

where k is related to permeability and will be taken as constant. $(P - P_L)$ is the pressure difference between conduit and far field at depth z. q is proportional to the void fraction α, in order to account for the fact that only a fraction of the conduit wall is in contact with the gas phase. Momentum conservation is expressed as:

$$0 = -\frac{dP}{dz} - \rho g - 8\mu \frac{w}{a^2} \tag{5.5}$$

where the horizontal profile of velocity is assumed to be parabolic, as in the absence of gas loss. In effect, gas loss amounts to suction, which in principle should lead to a profile change [39]. However, in the present case, the gas phase represents a very small mass fraction and hence carries little momentum.

Equations (5.3) and (5.5) give a coupling between pressure, mass flux and gas content. No straightforward analytical solution has been found and they have been integrated numerically. Initial conditions for magma as it leaves the chamber are that it is saturated with volatiles, has zero void fraction and that its pressure is larger than the lithostatic value at that depth by an amount ΔP, as in section 3. The problem has a large number of parameters, and it is impractical to discuss all of them independently. We take the following scales for the variables:

pressure scale for the rate of gas loss : ΔP

vertical pressure gradient scale : $\rho_l g$

mass flux : $\rho_l \dfrac{a^4}{\mu} \dfrac{\Delta P - P_f}{H}$

Using these, equations (5.2) and (5.4) yield a non-dimensional number D characterizing the balance between gas loss and pressure release:

$$D = \frac{k\mu H^2}{a^3} \frac{\Delta P}{\Delta P - P_f} \tag{5.6}$$

This summarizes the interplay of the different variables. The effect of pressure is subtle: for small values of the exit pressure P_f, D depends weakly on ΔP, because increasing ΔP acts to increase both the eruption rate and the driving pressure contrast for gas loss. This suggests that the effects of gas loss become important when the chamber overpressure is small, i.e. not large compared to the exit pressure. A larger conduit radius implies a higher eruption rate, because friction is reduced, but also a larger contact area for gas loss. The former effect is dominant, and hence a larger conduit size implies a proportionally smaller amount of gas lost. The effects of the other variables are obvious.

In this model, a gas phase is present throughout the ascent and hence it is no longer possible to realistically assume that density and viscosity remain constant, as was done in section 3. In particular, lava density decreases, which contributes a buoyancy term. The equations can no longer be integrated directly and require the use of numerical techniques, and hence it is worthwhile to also investigate the effect of viscosity changes. Two effects may be taken into account. One is the presence of bubbles according to (2.7) and the other that of dissolved volatiles on the melt phase [17], for which an equation of the following form is chosen:

$$\mu_1 = \mu_i \exp(-[x - x_i]\gamma) \tag{5.7}$$

where x_i and x are the mass fractions of dissolved water at pressures P_i and P respectively. γ is a coefficient which can be determined from available data [17]. The measure of this effect is given by ratio $\delta = \mu_1(P_f)/\mu_1(P_i)$, i.e. the viscosity ratio for saturated liquid at the final and initial pressures. Such density and viscosity changes lead to a different relationship between mass flux and pressure.

For the far field pressure P_L, we assume the simplest conditions of lithostatic equilibrium in country rock whose density is equal to that of pure lava, i.e. ρ_l (taken to be $2400 kg/m^3$). The conduit height and radius are set to 1000 m and 12 m respectively. The starting liquid viscosity is 10^8 Pas. Coefficient δ for viscosity is taken to be either 1 or 10^2. These choices are somewhat arbitrary but allow a summary of an extensive set of calculations. The important point here is that, regardless of the assumptions made, the flow rate decreases as a function of ΔP, and hence there are always solutions with small mass flux such that gas loss is important. The exact form of the viscosity variation does not affect this basic result.

We first discuss cases in which the pressure contrast is set to zero at the conduit bottom, and viscosity depends on void fraction only according to equation (2.7). Figure 9 shows how velocity varies in the conduit for several values of the initial velocity. In general, velocity tends to increase due to the exsolution and expansion of volatiles. However, below the initial velocity of $3 \times 10^{-4} m.s^{-1}$, there is a bifurcation to a regime where gas loss becomes important: velocity starts to decrease because the gas volume fraction decreases. At shallower depths, the gas volume fraction starts increasing again, due to the large amounts of exsolution at low pressures. At even lower initial velocity values, all the gas which is available is lost to the surroundings and velocity stays constant. These eruption paths show a range of behaviours, from all-out expansion and acceleration, to alternating phases of acceleration and deceleration.

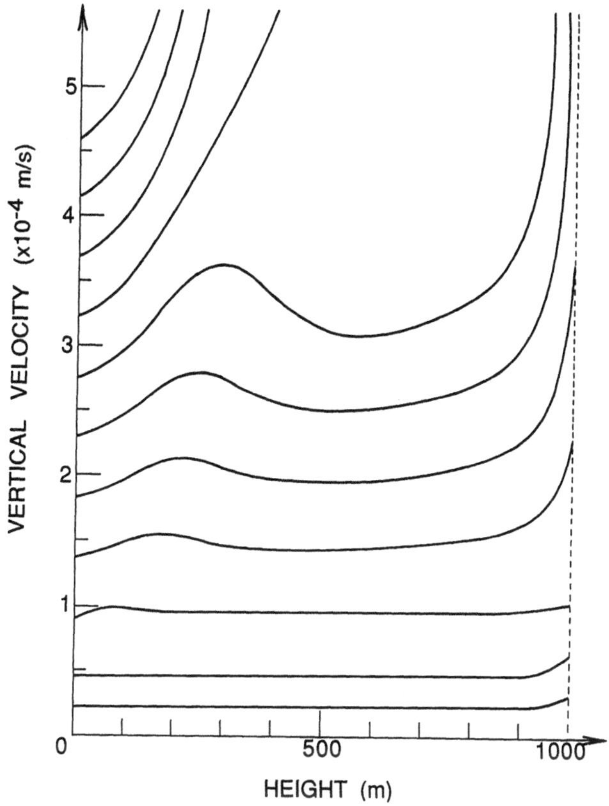

Figure 9. Variation of vertical velocity with height in a volcanic conduit, for several values of the initial velocity with permeability coefficient k set to 10^{-10}. The calculations are for an initial viscosity of 10^8 Pas and take into account the effect of bubbles on lava viscosity (equation 2.7). For low initial velocity, and hence low eruption rate, the rising lava loses gas to the fractured country rock. At high initial velocity, little gas is lost and the flow accelerates due to gas exsolution and expansion. The transition between the two regimes is sharp and reflects the non-linear character of the governing equations.

It is instructive to compare the eruption mass flux Q to that which is calculated with the constant property model in the same pressure conditions with the same initial viscosity value, called Q_0. Figure 10 shows the dimensionless mass flux Q/Q_0 as a function of overpressure ΔP for an exit pressure P_f equal to one atmosphere. In the calculation, lava viscosity depends on both void fraction and the amount of dissolved volatiles. The dimensionless mass flux is less than 1 because of the viscosity increase which accompanies degassing (equation 5.7). It is not constant, which shows that eruption conditions are more sensitive to pressure than in the constant viscosity case. However, it does not vary by large amounts, which may account in part for the success of the constant property calculations of section 3.

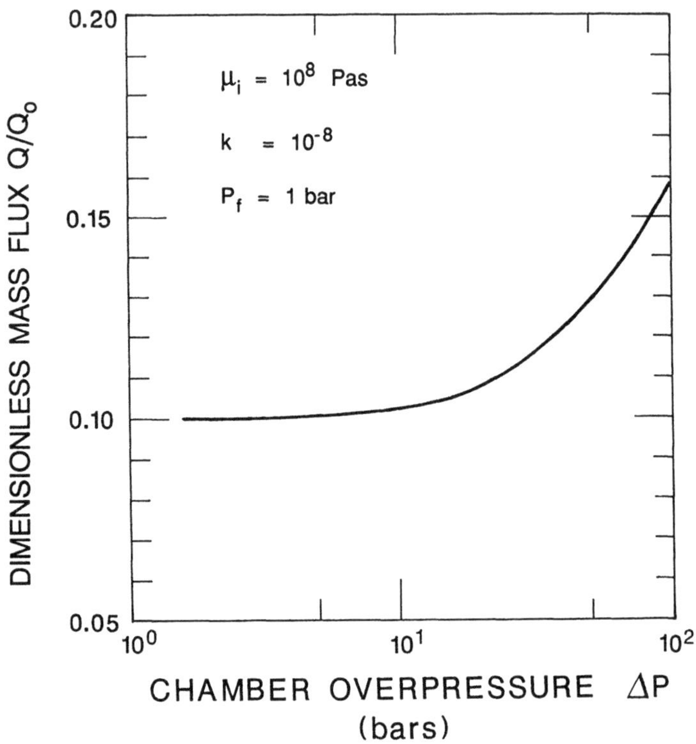

Figure 10. Relationship between mass flux and chamber overpressure for $k = 10^{-8}$. The mass flux is made dimensionless with the constant property mass flux (equation 3.15). The initial lava viscosity in the chamber is 10^8 Pas. The influence of bubbles and dissolved water content on viscosity are both taken into account.

In the initial states of an eruption, the chamber overpressure ΔP is highest and leads to an explosive phase with high gas content involving fragmentation. As the overpressure decreases, the mass flux decreases, implying a decrease of gas volume fraction (Figure 11). Eventually, the gas volume fraction drops below the fragmentation threshold, leading to lava eruption and dome formation. Tiny pressure fluctuations lead to large variations of mass flux and gas content (Figure 11). For example, an overpressure variation of 0.1 MPa (one bar) around the value of 1.5 MPa implies vesicularity variations from 0.74 to 0.64. This suggests that regime transitions are unstable and may explain why some eruptions alternate between fragmented and unfragmented phases before settling to a final phase of continuous lava extrusion. As the overpressure decreases further, generally lower values of void fraction ensure and pressure fluctuations occur without fragmentation. For example, at Mount St Helens in 1981 and 1982, the growing dome was sometimes pierced by 'spines' of frothy lava protruding out to heights of several tens of meters. These extrusion events are evidence for coupled fluctuations in eruption rate and lava physical properties.

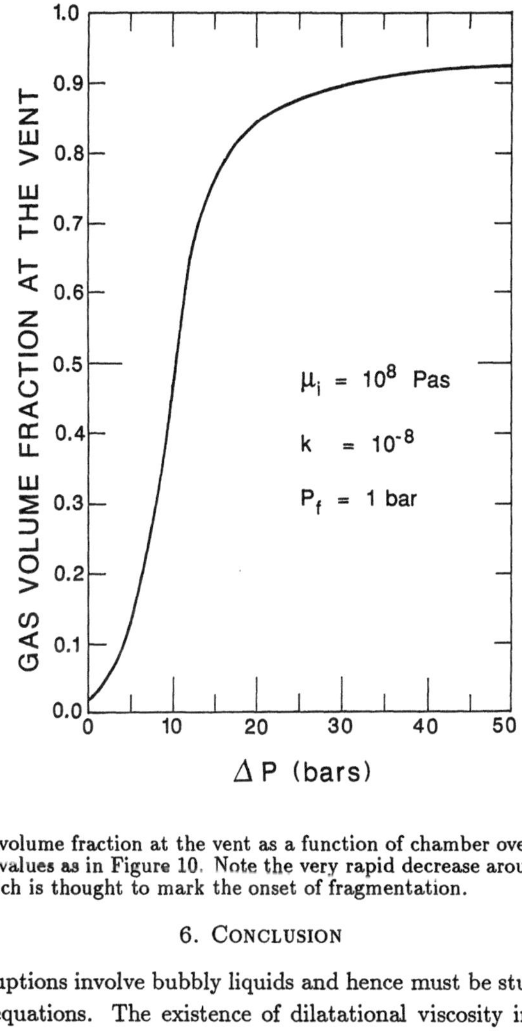

Figure 11. Gas volume fraction at the vent as a function of chamber overpressure for the same parameter values as in Figure 10. Note the very rapid decrease around the threshold of $\alpha \approx 0.75$, which is thought to mark the onset of fragmentation.

6. Conclusion

Volcanic eruptions involve bubbly liquids and hence must be studied with compressible flow equations. The existence of dilatational viscosity implies that the pressure of lava rising in a volcanic conduit is not uniform in a horizontal plane, which leads to horizontal variations of void fraction and lava physical properties. For lava eruptions with small mass flux, the equations of motion can be simplified greatly and allow reliable predictions of mass flux and of pressure evolution in the volcano chamber. The compressible nature of lavas slows down their spreading away from volcanic vents.

Many features of eruptions have not been explained satisfactorily and reflect conditions which are specific to volcanic systems. One example, analyzed briefly in section 5, is the possibility that lava eruptions are due to fumaroles acting to reduce the gas content of the mixture which rises in the conduit. Another example is the collapse of host rocks around magma reservoirs, which leads to eruptions of enormous volumes of ash and pyroclasts. In volcanic systems, a wide range of processes are at work which may interact in different ways and in different combinations. Thus, many different lines of research are open. At present, however, two outstanding problems seem to have priority: the physics of the fragmentation process and the factors governing the evolution of conduit shape and size during eruption.

Acknowledgements. I thank all the members of the Geology Department, Bristol University, for their kindness and hospitality. George Hellfrich and John Brodholt introduced me to the computer system. Steve Sparks improved the manuscript with his comments and criticisms. Colin Wilson shared his ideas and gave the photograph appearing in Plate 1 (see Plates section).

TABLE 1

LIST OF SYMBOLS

SYMBOL	DEFINITION
a	Radius of volcanic conduit
α	Void fraction (volume fraction of gas bubbles)
α_c	Critical void fraction for viscosity equation (2.7)
β	Compressibility
γ	Coefficient for the variation of viscosity as a function of dissolved volatile content (equation 5.7)
C	Dimensionless number to characterize the importance of compressibility
e	Strain rate
G	Mass flux per unit area
h	Thickness of lava flow or dome
H	Height of volcanic conduit (also depth to the magma chamber)
η	Similarity variable
η_n	Value of dimensionless similarity variable at flow front
K	Dilatational viscosity
λ	Second coefficient of viscosity for lava
M	Molar mass of gas species
μ	First coefficient of viscosity for lava
μ_1	Newtonian viscosity of silicate melt
P	Pressure in vesicular magma rising towards the Earth's surface
P_a	Pressure at the conduit walls (function of height z)
P_f	Pressure at the eruptive vent
P_i	Pressure at the bottom of the volcanic conduit
P_L	Lithostatic pressure (function of height z)
ΔP	Overpressure in magma chamber
Q	Mass flux of volcanic eruption
R	Perfect gas constant
ρ	Lava density
ρ_a	Density of lava at pressure P_a
ρ_o	Density of compressible lava at atmospheric pressure
ρ_g	Gas phase density
ρ_l	Liquid density
ρ_r	Density of country rock
τ	Stress
v	Vertical velocity outside the vent
w	Vertical velocity in the conduit
x	Solubility of volatiles in silicate melt

REFERENCES

[1] L. WILSON, *Explosive eruption columns, III, Plinian eruption columns*, Geophys. J.R. Astr. Soc., 45 (1976), pp. 543–556.

[2] R.S.J. SPARKS and L. WILSON, *A model for the formation of ignimbrite by gravitational column collapse*, J. Geol. Soc. London, 132 (1976), pp. 441–451.

[3] L. WILSON, R.S.J. SPARKS and G.P.L. WALKER, *Explosive volcanic eruptions, IV, The control of magma properties and conduit geometry on eruption column behaviour*, Geophys. J.R. Astr. Soc., 63 (1980), pp. 117–148.

[4] S.W. KIEFFER, *Blast dynamics at Mount St Helens on 18 May 1980*, Nature, 291 (1981), pp. 568–570.

[5] K.A. WOHLETZ, T.R. MCGETCHIN, M.T. SANDFORD and M.E. JONES, *Hydrodynamics of caldera-forming eruptions: numerical models*, J. Geophys. Res., 89 (1984), pp. 8269–8285.

[6] A.W. WOODS, *The fluid dynamics and thermodynamics of eruption columns*, Bull. Volcanol., 50 (1988), pp. 169–193.

[7] S.N. CAREY, H. SIGURDSSON and R.S.J. SPARKS, *Experimental studies of particle-laden plumes*, J. Geophys. Res., 93 (1989), pp. 15314–15328.

[8] P.M. BRUCE and H.E. HUPPERT, *Thermal control of basaltic fissure eruptions*, Nature, 342 (1989), pp. 665–667.

[9] H.E. HUPPERT, J.B. SHEPHERD, H. SIGURDSSON and R.S.J. SPARKS, *On lava dome growth, with application to the 1979 lava extrusion of the Soufriere de Saint Vincent*, J. Volcanol. Geotherm. Res., 14 (1982), pp. 199–222.

[10] J.H. FINK and R.W. GRIFFITHS, *Radial spreading of viscous gravity currents with solidifying crust*, J. Fluid Mech. (1990) in press.

[11] C.R. ILLINGSWORTH, *Some solutions of the equations of flow of a viscous compressible fluid*, Proc. Camb. Phil. Soc., 46 (1950), pp. 469–478.

[12] F.M. WHITE, *Viscous Fluid Flow*, McGraw Hill, New York, 1972.

[13] G.I. TAYLOR, *The two coefficients of viscosity for an incompressible fluid containing air bubbles*, Proc. R. Soc. London Ser.A226 (1954), pp. 34–39.

[14] R.K. PRUD'HOMME and R.B. BIRD, *The dilatational properties of suspensions of gas bubbles in incompressible Newtonian and non-Newtonian fluids*, J. non-Newtonian Fluid Mech., 3 (1977/78), pp. 261–279.

[15] H.R. SHAW, T.L. WRIGHT, D.L. PECK and R. OKAMURA, *The viscosity of basaltic magma: an analysis of field measurements in Makaopuhi lava lake, Hawaii*, Am. J. Sci., 266 (1968), pp. 225–264.

[16] F.J. SPERA, A. BORGIA and J. STRIMPLE, *Rheology of melts and magmatic suspensions*, J. Geophys. Res., 93 (1988), pp. 10273–10294.

[17] M.P. RYAN and J.Y.K. BLEVINS, *The viscosity of synthetic and natural silicate melts and glasses at high temperature and 1 bar pressure and at higher pressures*, U.S, Geol. Surv. Bull., 1764 (1987).

[18] C.W. BURNHAM, *The importance of volatile constituents*, in: The Evolution of Igneous Rocks, ed. by H.S. YODER, Princeton Univ. Press, Princeton, New Jersey, 1979, pp. 439–482.

[19] E. STOLPER and J.R. HOLLOWAY, *Experimental determination of the solubility of carbon dioxide in molten basalt at low pressure*, Earth Planet. Sci. Lett., 87 (1988), pp. 397–408.

[20] R.S.J. SPARKS, *The dynamics of bubble formation and growth in magmas: a review and analysis*, J. Volcanol. Geotherm. Res., 3 (1978), pp. 1–37.

[21] B.F. HOUGHTON and C.J.N. WILSON, *A vesicularity index for pyroclastic deposits*, Bull. Volcanol., 51 (1989), pp. 451–462.

[22] H.M. PRINCEN, M.P. ARONSON and J.C. MOSER, *Highly concentrated emulsions, II, Real systems. The effect of film thickness and contact angle on the volume fraction of creamed emulsions*, J. Colloid Interface Sci., 75 (1980), pp. 246–270.

[23] A.M. KRAYNICK, *Foam flows*, Ann. Rev. Fluid Mech., 20 (1988), pp. 325–357.

[24] W.R. SCHOWALTER, *Mechanics of non-Newtonian fluids*, Pergamon, 1978, pp. 300.

[25] J.O. SIBREE, *The viscosity of froth*, Trans. Faraday Soc., 30 (1934), pp. 325–331.

[26] L.W. SCHWARTZ and H.M. PRINCEN, *A theory of extensional viscosity for flowing foams and concentrated emulsions*, J. Colloid Interface Sci., 75 (1987), pp. 246–270.

[27] H.M. PRINCEN, *Rheology of foams and high concentrated emulsions. II. Experimental study of the yield stress and wall effects for concentrated oil-in-water emulsions*, J. Colloid Interface Sci., 105 (1985), pp. 150–171.

[28] D.A. SWANSON and R.T. HOLCOMB, *Regularities in growth of the Mount St Helens dacite dome 1980–86*, Proc. Volcanol., 2 (1990), pp. 1–24.

[29] S.R. TAIT, C. JAUPART and S. VERGNIOLLE, *Pressure, gas content and eruption periodicity of a shallow crystallizing magma chamber*, Earth Planet. Sci. Lett., 92 (1989), pp. 107–123.

[30] D.A. SWANSON, *Pahoehoe flows from the 1969–1971 Mauna-Ulu eruption, Kilauea volcano, Hawaii*, Geol. Soc. Am. Bull., 84 (1973), pp. 615–626.

[31] D.A. SWANSON, *Magma supply rate at Kilauea volcano, Hawaii*, Science, 175 (1972), pp. 169–170.

[32] A. LACROIX, *La Montagne Pelee et ses eruptions*, Masson, Paris, 1904.

[33] H.E. HUPPERT, *The propagation of two-dimensional and axisymmetric viscous gravity currents over a rigid horizontal surface*, J. Fluid Mech., 121 (1982), pp. 43–58.

[34] J.C. EICHELBERGER, C.R. CARRIGAN, H.R. WESTRICH and R.H. PRICE, *Non-explosive silicic volcanism*, Nature, 323 (1986), pp. 598–602.

[35] T.M. GERLACH AND E. GRAEBER, *Volatile budget of Kilauea volcano*, Nature, 313 (1985), pp. 273–277.

[36] L.P. GREENLAND, W.I. ROSE and J.B. STOKES, *An estimate of gas emissions and magmatic gas content from Kilauea volcano*, Geochim. Cosmochim. Acta, 49 (1985), pp. 125–129.

[37] G. HEIKEN, K.A. WOHLETZ and J.C. EICHELBERGER, *Fracture fillings and intrusive pyroclasts, Inyo domes, California*, J. Geophys. Res. 93 (1988), pp. 4335–4350.

[38] O.L. ANDERSON and P.C. GREW, *Stress corrosion theory of crack propagation with applications to geophysics*, Rev. Geophys. Space Phys., 17 (1977), pp. 77–104.

[39] H.L. WEISSBERG, *Laminar flow in the entrance region of a porous pipe*, Phys. Fluids, 2 (1959), pp. 510–516.

MODELING TRANSPORT PROCESSES IN NONLINEAR SYSTEMS: THE EXAMPLE OF SOLIDIFICATION AND CONVECTION

CURTIS M. OLDENBURG†* AND FRANK J. SPERA†

Abstract. Although convection is commonly associated with phase change in the earth, numerical modeling of phase change and convection has not received as much attention as the modeling of either process alone. Recently formulated continuum models of phase change and convection are readily amenable to numerical solution. In the continuum models, one set of equations is used over the entire solution domain. Iteration within each time step accommodates the nonlinearity and strong coupling of the transport equations and the supplementary equations which relate enthalpy to temperature, fraction solid, and composition under the assumption of local thermodynamic equilibrium. We present a comparison of a set of laboratory experiments of the solidification and convection of aqueous ammonium chloride solution with two-dimensional numerical experiments using the continuum model. The numerical experiments reproduce the behavior observed in the laboratory experiments and show the profound effects of solidification on convective systems.

I. Introduction. Considering the importance of convective heat transfer throughout the solid earth as well as in its oceans and atmosphere, it is not surprising that the equations describing natural convection comprise one of the more well-studied sets of nonlinear partial differential equations in geophysics. What is surprising is that phase change coupled to convection has received so little attention. In the earth, convection commonly occurs together with phase change. For example in the inner and outer core, both phase change and convection of iron are important. In the convecting mantle, partial melting of silicates occurs giving rise to magma which erupts at the surface or accumulates and eventually solidifies in magma chambers. In crustal hydrothermal systems, evaporation and condensation of hydrothermal fluids are coupled to natural convection. In the oceans and atmosphere, many kinds of phase change processes occur coupled to natural convection.

Motivated by the diversity of igneous rocks observed at the earth's surface, our interest is primarily in understanding the process of solidification and convection in crustal magma bodies. Magmatic solidification is complicated by the fact that natural magmas are generally multicomponent silicate melts. As such, phase change occurs over a range of temperatures for any given melt with a solid-liquid mush region occurring between the liquidus and solidus temperatures (called the melting interval). In addition, the density of magma is a strong function of composition as well as temperature. Thus the thermal and compositional gradients set up by the solidification process cause convection, which in turn affects the thermal and compositional variations in the melt. In this way, the process of solidification is strongly coupled to natural convection in magma.

Our approach consists of two-dimensional numerical modeling of solidification and convection where we consider explicitly the complicated coupling between the processes of solidification and the nonlinear convection equations. The continuum

*current address: Earth Sciences Division, Lawrence Berkeley Laboratory, Berkeley, CA 94720
†Department of Geological Sciences and Institute for Crustal Studies, University of California, Santa Barbara, CA 93106

model for phase change and convection developed for problems in metallurgy (Bennon and Incropera, 1987a) is sufficiently general to provide a starting point for modeling phase change and convection in magmatic systems. In this report, we present a model for the solidification and convection of binary eutectic systems and directly compare model calculations to laboratory experiments in systems where the fundamental coupling between solidification and convection is similar to that in magma. The primary goal is to demonstrate that solidification and convection can be accurately modeled by numerical calculation. To this end, we compare the laboratory results of the solidification and convection of ammonium chloride solution (Cao and Poulikakos, 1990) with two-dimensional numerical calculations. Within the theme of nonlinear systems in geoscience, the secondary goal is to show that numerical modeling of complicated transport processes, such as phase change coupled to the nonlinear convection equations, is a viable technique. An underlying point of the report is that the behavior of convecting systems is profoundly affected by solidification.

We present first the model equations and a result for a single component pure metal (molten tin) which verifies the general continuum approach without the complications of compositional buoyancy and the melting interval. Then we summarize the experimental results of Cao and Poulikakos (1990) and present our results of numerical experiments of the same system where compositional buoyancy and the presence of a mush region are primary features. Finally, we discuss the interpretations of the numerical and laboratory results.

II. Continuum model

a. Differential Equations

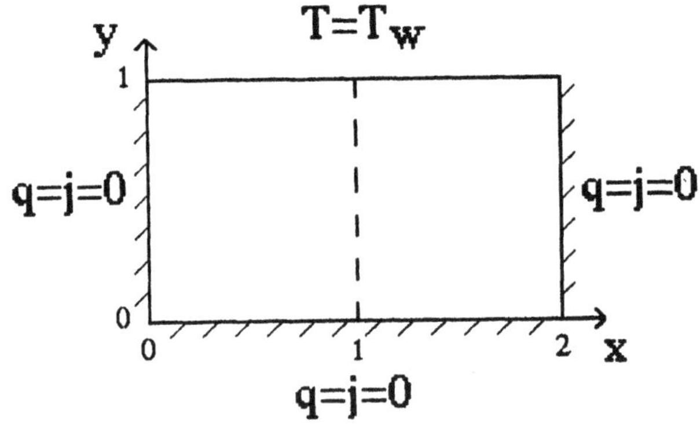

Fig. 1. Computational domain for solidification from above convection calculations. The sidewalls, bottom, and top are rigid ($u = v = 0$) and impermeable ($j = 0$) boundaries. The sidewalls and bottom are insulated ($q = 0$) while the top is held at sub-solidus temperature, T_W. Initially, the melt is motionless and held at $T_0 > T_{liq}$. Because of the symmetry of the problem, calculations are done in the left half of the domain with the line at $x = 1$ a mirror plane.

In continuum models of phase change (often called single-region models), one set of equations is valid over the entire solution domain (Fig. 1) regardless of what phases are present. The dimensionless, primitive variable, continuum conservation equations for solidification and convection in a two-dimensional domain such as shown in Fig. 1 are as follows:

(1) $$\nabla \cdot V = 0$$

(2) $$\frac{1}{Pr}\left(St\frac{\partial u}{\partial t} + (V \cdot \nabla u)\right) = -\frac{\partial p}{\partial x} + \nabla \cdot (\mu_r \nabla u) - \frac{\mu_r K_0}{Da\ K_x}(u - u_s)$$

(3) $$\frac{1}{Pr}\left(St\frac{\partial v}{\partial t} + (V \cdot \nabla v)\right) = -\frac{\partial p}{\partial y} + \nabla \cdot (\mu_r \nabla v) + Ra \cdot T + Rs \cdot C - \frac{\mu_r\ K_0}{Da\ K_y}(v - v_s)$$

(4) $$St\frac{\partial h}{\partial t} + (V \cdot \nabla h) = \nabla^2 h + \nabla^2 (h_s - h) - (V - V_s) \cdot \nabla(h_l - h)$$

(5) $$St\frac{\partial C}{\partial t} + (V \cdot \nabla C) = \nabla \cdot \frac{1}{Le}\frac{D}{D_l}\nabla C + \nabla \cdot \frac{1}{Le}\frac{D}{D_l}\nabla(C_l - C) - (V - V_s) \cdot \nabla(C_l - C)$$

Table 1. Scales and dimensionless groups appearing in Eqs. 1–5. Note that the hats (^) have been suppressed in (1)–(5).

coordinates	dependent variables	parameters
$\hat{x} = \frac{x}{L}$	$\hat{V} = \frac{VL}{\kappa}$	$Pr = \frac{\nu_0}{\kappa}$
$\hat{y} = \frac{y}{L}$	$\hat{p} = \frac{L^2}{\mu_0 \kappa} p$	$Ra = \frac{g\alpha(T_0 - T_{sol})L^3}{\kappa \nu_0}$
$\hat{t} = t\frac{\kappa}{L^2}\ St$	$\hat{h} = \frac{(h - h_{sol})}{(h^*_{liq} - h_{sol})}$	$Rs = \frac{g\beta(C_{s2} - C_{s1})L^3}{\kappa \nu_0}$
	$\hat{T} = \frac{(T - T_{sol})}{(T_0 - T_{sol})}$	$St = \frac{c_p(T_0 - T_{sol})}{(h^*_{liq} - h_{sol})}$
	$\hat{C} = \frac{(C - C_{s1})}{(C_{s2} - C_{s1})}$	$Le = \frac{\kappa}{D_l}$
		$Da = \frac{K_0}{L^2}$

(Bennon and Incropera, 1987a). The scales and dimensionless groups used to non-dimensionalize the continuum equations are given in Table 1.

In the above equations, density is assumed to obey the relation

$$\rho = \rho_0(1 - \alpha(T - T_{sol}) - \beta(C - C_{s1})) \tag{6}$$

where α and β are constants and T_{sol} is the solidus temperature and C_{s1} is a pure endmember. It is further assumed that density changes are important only in the body force term (the Boussinesq approximation). Also, the density of the solid is assumed equal to the density of the liquid if both phases are at the same temperature and composition. In the numerical experiments presented here, the solid phase velocity (V_s) is equal to zero which models the case where solid always adheres to other solid and to the rigid wall.

The last terms on the right-hand side of (2) and (3) are the Darcy flow terms; the last terms on the right-hand side of (4) and (5) are the interphase transport terms. The equations (1)–(5) are appropriate for systems in which the liquid phase can flow through the porous mush in the two-phase solid-liquid (mush) region and constitute the relative motion (RM) model. In mush regions, the permeability is isotropic and related to fs through the Blake–Kozeny–Carman relation:

$$K_x = K_y = K_0 \left(\frac{(1 - fs)^3}{fs^2} \right) . \tag{7}$$

b. Additional Equations

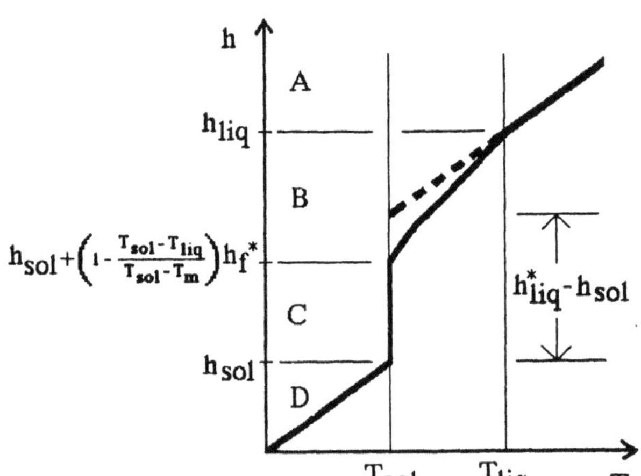

Fig. 2. Enthalpy-temperature diagram showing the 4 regions (A, B, C, D) of different $h-fs-T$ relations. h^*_{liq} is the projection of the superliquidus enthalpy down to T_{sol}. h_f^* is the mass-weighted heat of fusion of the mixed solid of eutectic composition. $h^*_{liq} - h_{sol}$ is the enthalpy scale in the Stefan number. Between T_{sol} and T_{liq}, one uses h_f (the heat of fusion of the first solid formed) for the source term which accounts for release of latent heat. At T_{sol}, the mass-weighted heat of fusion (h_f^*) is used to account for enthalpy produced during the formation of the first solid and the other end-member which solidify together in eutectic proportions.

The latent heat release of the second solid phase which may occur at T_{sol} during cooling in a pure binary eutectic system (Fig. 2) is treated by defining a mixed solid of eutectic composition with a corresponding mass weighted effective heat of fusion equal to

$$h_f^* = C_{eut}\, h_{fs2} + (1 - C_{eut}) h_{fs1} \; . \tag{8}$$

This approximation is valid for systems such as the silicates diopside and anorthite (Robie et al, 1979) where the differences in heat capacities between the two end-members can be neglected. We further assume that the heat capacity of the liquid is equal to the heat capacity of the solid. The different heats of fusion of the solid phases below T_{sol} are considered explicitly by mass averaging the heats of fusion of the two end members. Accounting for two solid phases is a departure from prior studies in metallurgical systems where solid solution occurs. In the metallic systems, there is a single solid phase at subsolidus temperatures, thus only one heat of fusion.

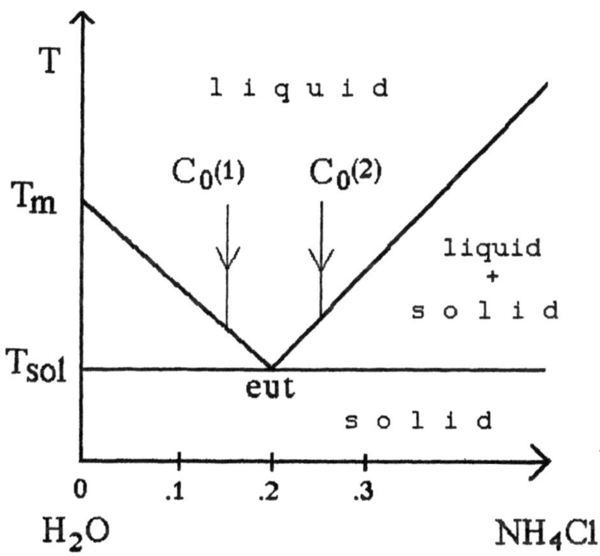

Fig. 3. Binary eutectic phase diagram for water-ammonium chloride (H_2O-NH_4Cl) system. The liquidi are linearized for simplifying the lever-rule relations. The two initial compositions correspond to experiments on the water-side and ammonium chloride-side of the eutectic, respectively. Some thermophysical properties of aqueous ammonium chloride are given in Table 1.

Supplementary equations relate the enthalpy to temperature, the temperature to fraction solid, and the phase compositions to fraction solid. With reference to the generalized enthalpy-temperature diagram (Fig. 3), and assuming local thermodynamic equilibrium, the dimensional equations in Table 2 relate the temperature (T) and fraction solid (fs) to enthalpy (h) depending on the enthalpy at any given location in the pure binary eutectic model.

Table 2. Relations between enthalpy (h), fraction solid (fs), and temperature (T) in the pure binary eutectic model in the four different regions of the enthalpy–temperature diagram.

Region	fs	T
A	0	$\dfrac{h-(h^*_{liq}-h_{sol})}{c_p}$
B	root of Eq. 13	$\dfrac{T_{liq}-T_m fs}{1-fs}$
C	$1-\dfrac{h-h_{sol}}{h^*_f}$	T_{sol}
D	1	$\dfrac{h}{c_p}$

In general, the enthalpy of the mixture (h) is given in terms of the enthalpy of the solid (h_s), the enthalpy of the liquid (h_l), and the fraction solid (fs) by

$$h = fs\ h_s + (1-fs)h_l \tag{9}$$

where

$$h_s = c_p T \tag{10}$$

and

$$h_l = c_p T + h^*_{liq} - h_{sol} \tag{11}$$

where it is assumed that there is only one heat capacity (c_p) in the system. The term h^*_{liq} is simply the projection of the superliquidus enthalpy curve through the two-phase region to T_{sol}.

The temperature in subsolidus and superliquidus regions (D and A, respectively) is linearly related to h through the heat capacity (c_p) (Fig. 3). In regions D and A, fs is unity and zero, respectively.

In supersolidus and subliquidus region B of Figure 3, fs is given by

$$fs = \frac{T-T_{liq}}{T-T_m}\ . \tag{12}$$

Substituting (10–12) into (9) and simplifying, we arrive at the equation giving the enthalpy of the mixture in terms of fs for region B:

$$h = \left(\frac{A}{1-fs}\right) fs^2 - \left(\frac{2\ A - c_p T_m}{1-fs}\right) fs + \left(\frac{c_p T_{liq} + A}{1-fs}\right) \tag{13}$$

where

$$A = h^*_{liq} - h_{sol}\ . \tag{14}$$

Equation (13) is valid as long as the local fraction solid (fs) does not exceed the value obtained when T just reaches T_{sol} (ie., $fs < (T_{liq} - T_{sol})/(T_m - T_{sol})$) during cooling in region B. Newton–Raphson iteration is used to solve (13) for fraction solid given the mixture enthalpy. The $T - fs$ relation in region B is given by (12) or, equivalently, as shown in Table 2.

In supersolidus region C where $1 > fs \geq (T_{liq} - T_{sol})/(T_m - T_{sol})$, a simple linear relation using the enthalpy of fusion of the eutectic mixed solid (h^*_f) is used to calculate fs.

Additional equations relate phase compositions and mixture properties to those of the particular phases in the pure binary eutectic model with an immobile solid phase:

$$\text{for} \quad fs < \frac{T_{liq} - T_{sol}}{T_m - T_{sol}}\,, \qquad C = (1 - fs)C_l \tag{15}$$

$$\text{for} \quad fs \geq \frac{T_{liq} - T_{sol}}{T_m - T_{sol}}\,, \qquad C = fs\left(fs - \frac{T_{liq} - T_{sol}}{T_m - T_{sol}}\right)C_{eut} + (1 - fs)C_{eut} \tag{16}$$

$$V = (1 - fs)V_l \tag{17}$$

$$D = (1 - fs)D_l \tag{18}$$

The derivation of (16) is presented in Oldenburg and Spera (1990a).

c. Significance of dimensionless groups

Although several classical dimensionless numbers appear in (1)–(5) above, their significance is somewhat different in phase-change systems from that in single-phase systems. For example, the thermal Rayleigh number (Ra) indicates the importance of thermal buoyancy effects relative to viscous effects in natural convection; the compositional Rayleigh number (Rs) is analogous for compositional buoyancy. Nevertheless, the convective heat transfer cannot be accurately related to Ra in a scaling relation (as done by Roberts (1977) for single phase convection) when Ra is based on the width of the domain because for much of the solidification evolution only a small part of the domain contains convecting liquid. Momentum transport in two-phase mush regions is unlike convection in single-phase liquid. The Lewis number (Le) is the ratio of thermal to chemical diffusivity and thus provides an indication of the ratio of thermal boundary layer thickness to chemical boundary layer thickness. However, since chemical diffusivity is a function of the fraction solid in this continuum model (ie., diffusion in the solid is negligible compared to diffusion in the melt), scaling relations for boundary layer thicknesses using Le can only be applied in pure single-phase liquid regions. The Darcy number (Da) is only relevant in regions where solid is present and plays no role in the single-phase liquid region. It should further be noted that the thermophysical quantities and the form of the phase diagram also enter into the specification of phase change problems.

The dimensional analysis shows that even in a simple binary system, the number of parameters (both dimensionless parameters and significant thermophysical parameters) which affect the solution of the phase change and convection equations is large. Consequently, simple scaling analyses are much less useful in solidification and convection calculations than in single-phase convection problems.

III. Solution methodology. Each of the differential equations (1)–(5) can be cast in the general form

$$(19) \qquad \frac{\partial(\rho\phi)}{\partial t} + u\frac{\partial(\rho\phi)}{\partial x} + v\frac{\partial(\rho\phi)}{\partial y} = \frac{\partial}{\partial x}\left(\Gamma_\phi \frac{\partial\phi}{\partial x}\right) + \frac{\partial}{\partial y}\left(\Gamma_\phi \frac{\partial\phi}{\partial y}\right) + S_\phi$$

where ϕ denotes any one of the dependent variables of the mixture and Γ is the general diffusion coefficient. The source term S accounts for the Darcy flow and body force terms in the momentum equations, the source of latent heat and relative phase motion in the energy equation, and the source of species and relative phase motion in the species equation. Once in this general form, the equations are amenable to solution by control-volume finite difference methods. We developed a two-dimensional code along the lines of the iterative, primitive variable, SIMPLER algorithm (Patankar, 1980; Fletcher, 1988) to solve the system of equations describing solidification and convection in a binary eutectic system. The use of primitive variables allows the direct extension to three-dimensions.

The continuum equations were solved on a uniform rectangular, staggered mesh. Strong bi-directional coupling between the energy and momentum equations at large Pr slows convergence of the iterative scheme (Galpin and Raithby, 1986). In order to speed convergence, we have used ramp under-relaxation in the momentum equations wherein the under-relaxation parameter is progressively decreased with iteration. The additional equations and Newton–Raphson iteration for finding the fraction solid from the enthalpy, where necessary, are performed at each iteration. More detail on our computational methods and example calculations can be found in Oldenburg and Spera (1990a,b).

The convergence of the momentum equations is governed by the satisfaction of the continuity equation (conservation of mass, (1)) to one part in 2×10^5 in each control volume. Other convergence criteria are that local fractional changes in enthalpy, composition, and fraction solid and the average heat extracted from the cold wall were all less than 1×10^{-5} ($10^{-3}\%$) from one iteration to the next. The number of iterations to convergence varies from about 30 to 90 for a time-step size of 5×10^{-4} and is limited by the momentum equations, requiring more iterations when velocities were high.

IV. Solidification and convection of a pure metal

a. Pure tin

A multi-region solution for a problem of the two-dimensional solidification and convection of pure tin has been presented by Ramachdran et al. (1982). The problem considers the freezing of pure molten tin along the vertical walls of a box and includes the effects of convection. The parameters and thermophysical properties used in our simulation are given in Table 3. Our numerical experiment was set up to be identical to that of Ramachdran et al. (1982). Because of differences in the definitions of the parameters between our model and theirs, some dimensionless parameters may have different numerical values. For example, their Ra is defined on the basis of the difference between the initial temperature and the cold wall temperature whereas our Ra is defined on the basis of the difference between the initial temperature and the solidus temperature.

Table 3. Thermophysical properties and dimensionless groups for the molten tin numerical experiment.

Property	units	value
T_{sol}	(K)	505*
T_0	(K)	745
T_w	(K)	265
ρ	(kg m^{-3})	6950*
μ	(kg m^{-1} s^{-1})	1.8×10^{-3}*
β	(K^{-1})	1.6×10^{-4}*
c_p	(J kg^{-1} K^{-1})	250*
h_{sol}	(J kg^{-1})	1.262×10^5
h^*_{liq}	(J kg^{-1})	1.862×10^5
h_{f^*}	(J kg^{-1})	6.0×10^4
h_f	(J kg^{-1})	6.0×10^4*
Dimensionless parameters		
Ra		2.5×10^3 [1]
Pr		0.98 [2]
Da		1×10^{-10}
St		1.

*data from Brent (1989).

[1] corresponds to the $Ra = 5000$ used by Ramachdran et al. (1982).

[2] Ramachdran et al. (1982).

Ramachdran et al. (1982) used a multi-region approach with separate equations for the melt and solid regions. The continuum model presented here was applied to the same problem they solved and the results compared. Although the problem is a single component problem and does not entail any species redistribution, it does involve important thermal energy transfer associated with convection and solidification. The continuum approach agreed very closely with the multi-region model. We refer the reader to their paper for details of their results and to the color plate (see Plates section) for a summary of our results. The color plate is a view of the stream function, temperature and fraction solid fields at $t = .02, .14$, and .27. The final time ($t = .27$), is the same time as was shown in Ramachdran et al. (1982). Our calculation used a 15×31 grid and time step size of 5×10^{-4} in a box of width 1 and height 2.2.

V. A laboratory experiment of solidification and convection

a. Experiment

The laboratory solidification and convection experiments of Cao and Poulikakos (1990) were done with the binary system ammonium chloride (NH_4Cl)–water (H_2O) for which the water-rich side of the phase diagram is shown in Fig. 3. Cao and Poulikakos (1990) conducted experiments of the solidification from above of aqueous solutions on both sides of the eutectic, which is at $C = .20$. Their experimental apparatus consisted of a tank of internal dimensions 48.3 cm long by 25.4 cm tall by 12.7 cm deep. The sides were insulated while the top and bottom were made of aluminum for specifying top cooling and bottom heating boundary conditions by means of embedded circulating constant temperature baths. Thermocouples were placed in the top and bottom and along two rods in the cavity to monitor internal temperature.

b. Results

The following summarizes the results of Cao and Poulikakos (1990) for their experiments in which the system was cooled from above with no bottom heating. The top boundary of the cavity of the system was held at a temperature of -25 C while the initial temperature of the solution was 25 C. Briefly, the behavior of the system is as follows. At early times, a mush forms along the top while bi-cellular convection occurs in the liquid. Some time after the mush region is established, they observe the formation of a pure solid region. As time goes on both the solid and mush regions grow, with the mush region about twice the thickness of the solid region.

Cao and Poulikakos (1990) observed that as the concentration is increased from pure water ($C = 0$) toward the eutectic ($C = .20$), the bottom temperature decreases noticeably with higher concentration. This indicates that the solidification near the top is occurring at lower and lower temperatures as the concentration increases, a result of the solutal freezing point depression effect. Specifically, the liquid which arrives at the bottom of the cavity by natural convection is expelled during the lower-temperature solidification occurring near the top for higher ammonium chloride concentrations. The interfaces between the three regions (solid, mush, and liquid) are reported to be relatively smooth indicating that convection in the mushy zone is not vigorous. As the concentration of ammonium chloride is increased, the thickness of the mush zone decreases. However, as the initial composition is increased to $C = .25$ which is on the ammonium chloride side of the eutectic, they were surprised to observe that no mushy zone exists at all after 15 hours. They suggest this occurs because convection in the liquid region inhibits the creation of a permeable mush. They observed fingers of solid-laden density currents of ammonium chloride precipitate descending from the solid-liquid interface which apparently enhance the convective velocities near the solidification front.

VI. Numerical experiments

a. Results

In a series of numerical experiments, we have modeled the solidification from above of aqueous ammonium chloride (NH_4Cl)solution. The conditions of the experiments are chosen to be the same as the conditions in the laboratory experiments of Cao and Poulikakos (1990). The parameters and thermophysical properties we used are given in Table 4. A 31×31 uniform grid and time step of 5×10^{-4} were used in the calculations. The tank used by Cao and Poulikakos was larger than the effective tank used in our numerical experiment by a factor of 100. Nevertheless, the calculations reproduce the significant features of the laboratory results.

Table 4. Thermophysical properties and dimensionless groups for the ammonium chloride numerical experiment.

Property	units	value
T_{sol}	(K)	258
T_0	(K)	298
T_w	(K)	248
ρ	($kg\ m^{-3}$)	1000
μ	($kg\ m^{-1}\ s^{-1}$)	1.3×10^{-3}
α	(K^{-1})	3.8×10^{-4}
β	(−)	0.257
c_p	($J\ kg^{-1}K^{-1}$)	2560
h_{sol}	($J\ kg^{-1}$)	6.605×10^5
h^*_{liq}	($J\ kg^{-1}$)	9.743×10^5
h_{f^*}	($J\ kg^{-1}$)	3.138×10^5
h_f	($J\ kg^{-1}$)	3.138×10^5
Dimensionless parameters		
Ra		1×10^4
Rs		5×10^4 *
Pr		9
Le		33.
Da		1×10^{-6}
St		.3263

* can be positive or negative depending on initial composition.

Two features distinguish the evolution of the system when the initial composition is on the water side and ammonium chloride side of the eutectic. Briefly, for initial solution compositions on the water side of the eutectic, the first solid formed is pure ice (H_2O) and the residual liquid is more enriched in the solute and therefore denser. Because the solidification occurs at the top, this denser liquid produced at the solidification front readily flows downward and tends to enhance convective velocities. For initial compositions on the ammonium chloride side of the eutectic, the first solid to form is (NH_4Cl)and the rejected liquid is less enriched in solute and less dense. This liquid tends to rise and inhibits convection since cooling is from above. The other feature which distinguishes the water and ammonium chloride sides of the eutectic is the evolution of the fs field. The two initial conditions shown in this report are $C_0 = .15$ and $C_0 = .25$. Both of these initial conditions are equidistant from the eutectic at $C = .20$ (Fig. 2). Assuming local thermodynamic equilibrium, simple lever rule relations dictate that as the temperature decreases through T_{sol} for $C_0 = .15$, the fraction solid will jump from .25(= .05/.20) to 1. This means that a significant mush region with fs between 0.0 and .25 exists in regions where the temperature decreases from its high values to T_{sol}. For $C_0 = .25$, the lever rule implies that the fraction solid jumps from .0625 (=.05/.8) to 1 at T_{sol}. This means that the two-phase solid-liquid mush region which exists at temperatures above T_{sol} is restricted to values between 0.0 and .0625. This mush region will be very small for typical temperature fields. These results will be shown in detail below.

b. Water side

Shown in Figs. 4 and 5 are the stream function (ψ), temperature (T), fraction solid (fs), and normalized composition of the mixture (C^*) fields at two times in the evolution for an initial composition on the water side of the eutectic. With such an initial composition, the first solid formed is pure ice (H_2O) and the remaining liquid is enriched in the solute, ammonium chloride (NH_4Cl), which increases the density of the residual liquid. This enhances convective velocities in the cooling from above geometry. In Fig. 4a, the stream function field shows a clockwise circulation with a strong downwelling along the right-hand side at early time ($t = .10$). Fig. 4b shows the corresponding temperature field. Note the conductive temperature gradient across the upper solid region and the convective isotherms in the lower part of the domain. Fig. 4c shows the fraction solid field with the top-most region of the tank frozen and the rest liquid. In Fig. 4d, the C^* field shows the strong downwelling flow and the tongue of solute-enriched liquid which is enhancing downward flow on the right-hand side of the domain. The solid-mush interface is characterized by a depression in C^* which corresponds to more water-rich mixture compositions in the neighborhood of the solidification front.

Later in the evolution, one sees the solidification has taken up almost one-half of the domain. Fig. 5a shows the stream function field and the very weak convection present at $t = .50$. The temperature field (Fig. 5b) shows largely conductive profiles again indicating the sluggish convection present. Average dimensionless temperature at this time is -.015. Fig. 5c shows that mush takes up the entire domain and accounts for the sluggist convection. This thick two-phase solid-liquid mush region is just what was predicted from the lever rule relations and it acts to strongly inhibit convection. The C^* field still shows the water-rich region (low point in C^*) at the solidification front and a largely diffusive profile elsewhere.

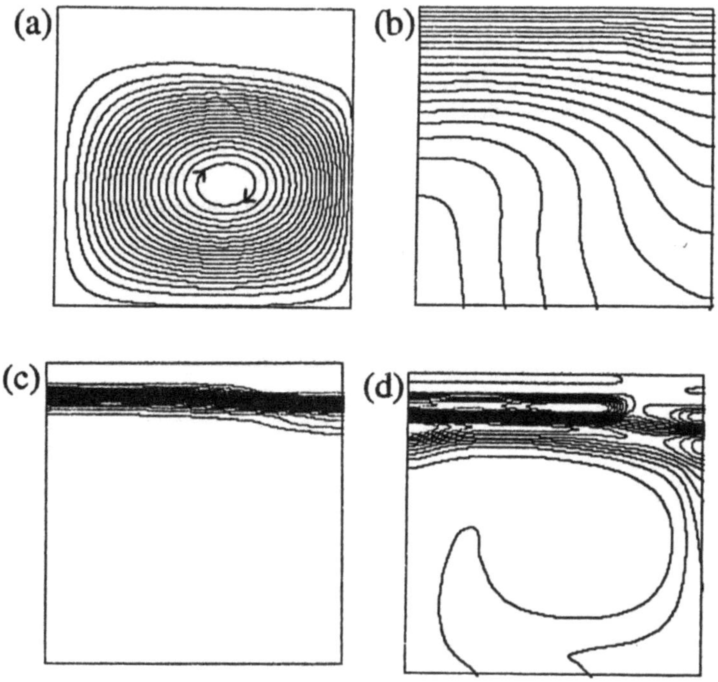

Fig. 4. Results for water-side of the eutectic initial composition numerical experiment at $t = .10$. a) stream function of the mixture ($\psi_{min} = -2.3$, $\psi_{max} = .02$) shows a clockwise circulation with stagnant top region. b) temperature ($T_{max} = .44$, $T_{min} = -.25$) shows convective flow in the lower part of the domain. c) fraction solid ($fs_{max} = 1, fs_{min} = 0$) shows solid region along the top, the mush region and the liquid region in most of the domain. d) normalized composition of the mixture ($C^* = (C - C_{min})/(C_{max} - C_{min})$). The minimum value is in the low region which extends three-quarters of the way from left to right near the top (${C^*}_{min} = .136$). The maximum value is in the high region just below the minimum (${C^*}_{max} = .170$). In the low region the mixture is H_2O-rich. The high region is enriched in solute and is at the solidification front. The enriched mixture composition drives convection downward in the domain and this is seen in the tongue of solute-enriched fluid extending around in the lower half of the domain.

c. *Ammonium chloride side*

The previous evolution contrasts with that for solutions on the ammonium chloride side of the eutectic. In Figs. 6 and 7 are shown the numerical results for solidification and convection of ammonium chloride solution with initial composition on the ammonium chloride side of the eutectic ($C_0 = .25$). Fig. 6a is at early time ($t = .10$) and shows the stream function field with a weak clockwise circulation. The convective velocities are slower here than in the previous experiment because the residual liquid at the solidification front is water-rich and therefore less dense. This liquid resists moving downward and thereby inhibits convection. Associated with this is some weak, transient layered convection which can be seen in Fig. 6a.

The temperature field (Fig. 6b) shows roughly parallel isotherms indicative of the weaker convection (cf. Fig. 4b). The fraction solid field (Fig. 6c) looks roughly the same at early time as in Fig. 4c. The C^* field (Fig. 4d) shows the same main depleted region at the solidification front, but in this case the low points indicate NH_4Cl-rich mixture compositions. On the right-hand side is a region of high C^* which corresponds to the small recirculation seen in the stream function plot and which represents H_2O-rich liquid which is sufficiently buoyant to remain in the upper region of the container.

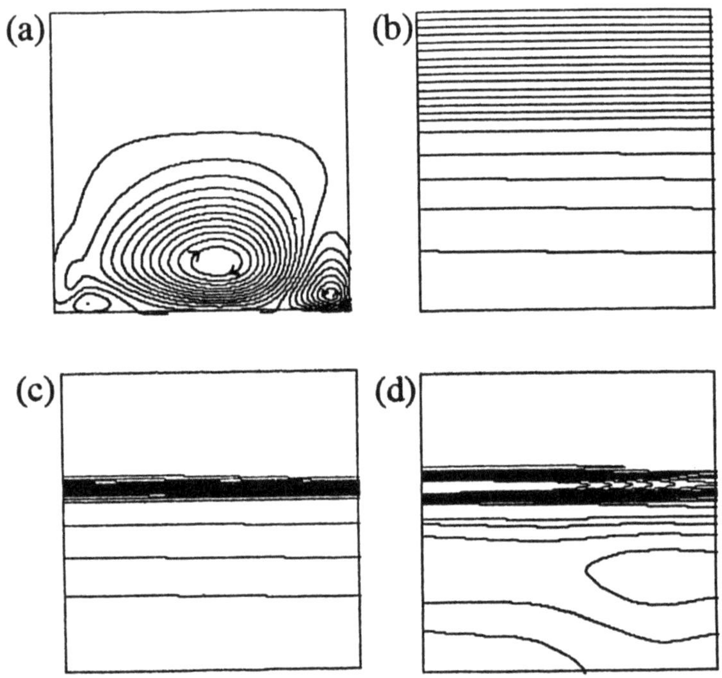

Fig. 5. Results for water-side experiment at $t = .50$. a) stream function of the mixture ($\psi_{min} = -.0006$, $\psi_{max} = .0004$) shows a very weak clockwise circulation with completely stagnant top region and two weak recirculations. b) temperature ($T_{max} = .08$, $T_{min} = -.25$) shows vigor of convective flow throughout the domain is very small and heat transport is mostly by conduction. c) fraction solid ($fs_{max} = 1, fs_{min} = .04$) shows solid region in the upper third of the domain, and the broad mush region in the rest of the domain. The thick mush region extending across the lower part of the domain leads to very weak convection. d) normalized composition of the mixture $C^* = (C - C_{min})/(C_{max} - C_{min})$. The minimum value is in the low region which extends all of the way from left to right two-thirds of the way up from the bottom (${C^*}_{min} = .137$). The maximum value is on the broad rise on the right-hand side about one-third of the way up from the bottom (${C^*}_{max} = .160$). The low and high concentrations are dependent upon diffusion for their mixing at this point when the convection has almost stopped.

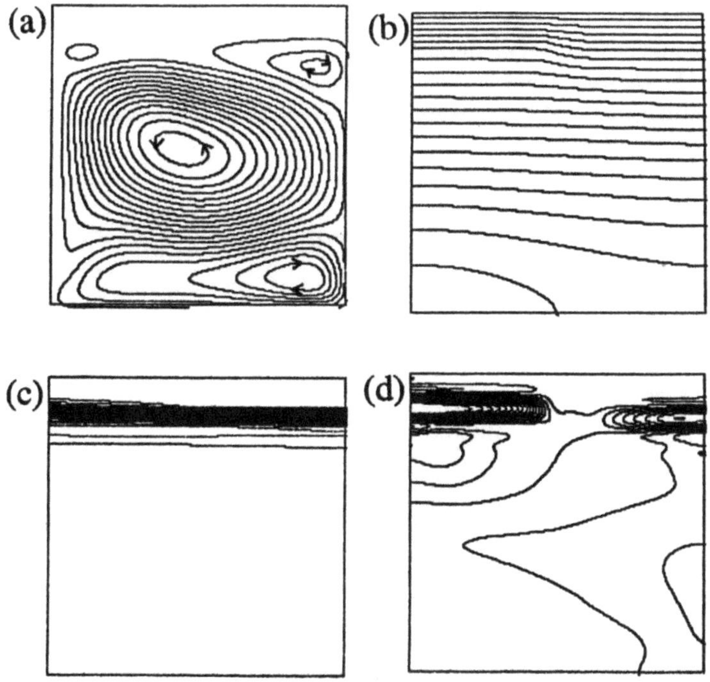

Fig. 6. Results for ammonium chloride-side of the eutectic initial composition numerical experiment at $t = .10$. a) stream function of the mixture ($\psi_{\text{min}} = -.08$, $\psi_{\text{max}} = .17$) shows a large counter-clockwise circulation with subordinate recirculations. Convection here is less vigorous than in the water-side experiment at the same time. b) temperature ($T_{\text{max}} = .56$, $T_{\text{min}} = -.25$) shows the weak convective flow in the lower part of the domain. c) fraction solid ($fs_{\text{max}} = 1, fs_{\text{min}} = 0$) shows solid region along the top, the mush region and the liquid region in most of the domain. d) normalized composition of the mixture ($C^* = (C - C_{\text{min}})/(C_{\text{max}} - C_{\text{min}})$) (in this case, $C_{eut} = .80$, and $C_o = .75$). The minimum value is in the region which extends almost one-half of the way from left to right near the top (${C^*}_{\text{min}} = .666$). The maximum value is from the low region on the right-hand side (${C^*}_{\text{max}} = .762$). The higher values of C^* represent a region where the mixture is H_2O-rich. This buoyant liquid is pooled near the top of the domain and the overall effect is to decrease convective vigor.

Later in the evolution ($t = .50$), the stream function (Fig. 7a) reveals a complicated flow which is stronger than in the previous experiment due to differences in the fs field. The temperature field (Fig. 7b) shows the conductive profile through the solid region at the top and the well-mixed and isothermal lower liquid-dominated region. The average dimensionless temperature is -.049, colder than in the water side experiment. In Fig. 7c one sees the significant result that fs drops from 1 to some small value over a very narrow interface. The thickness of the mush region which inhibits convection is much narrower at the same time in the evolution in the ammonium chloride-side initial composition run than in the water-side initial composition run. The model has reproduced the lever-rule prediction that the mush region should be narrower on the ammonium chloride-side of the eutectic ($C_0 = .25$). Finally, the C^* field shows the region of ammonium chloride-rich mixture near the solidification front and the relatively isochemical lower region where convection is mixing the solution.

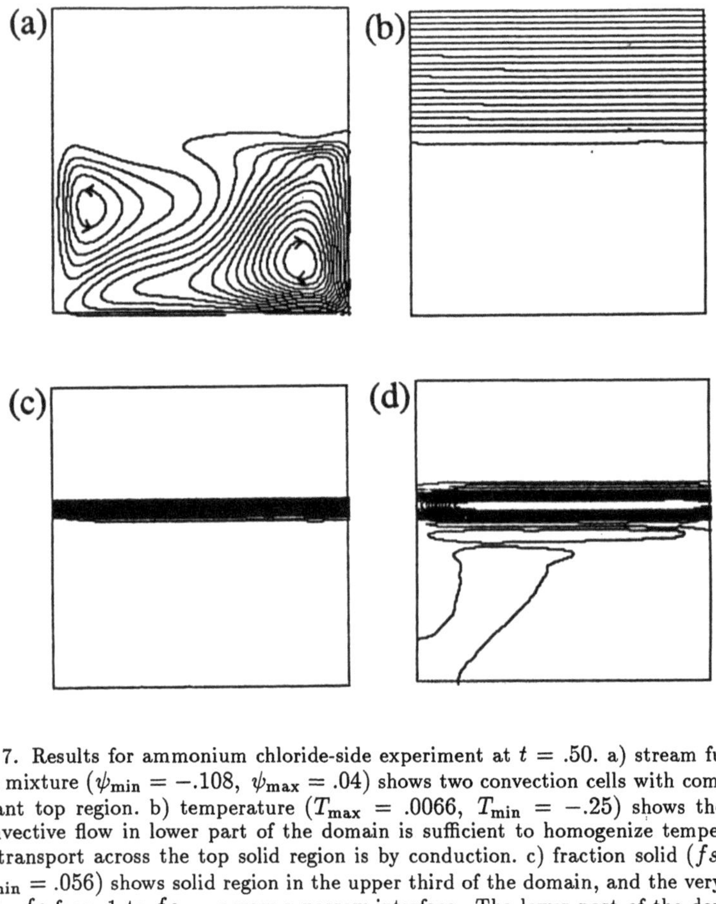

Fig. 7. Results for ammonium chloride-side experiment at $t = .50$. a) stream function of the mixture ($\psi_{\min} = -.108$, $\psi_{\max} = .04$) shows two convection cells with completely stagnant top region. b) temperature ($T_{\max} = .0066$, $T_{\min} = -.25$) shows the vigor of convective flow in lower part of the domain is sufficient to homogenize temperature. Heat transport across the top solid region is by conduction. c) fraction solid ($fs_{\max} = 1, fs_{\min} = .056$) shows solid region in the upper third of the domain, and the very sharp drop in fs from 1 to $fs_{\min}$ across a narrow interface. The lower part of the domain is relatively homogeneous with fs sufficiently small that convection still occurs in the lower part. d) normalized composition of the mixture ($C^* = (C - C_{\min})/(C_{\max} - C_{\min})$). The minimum value is in the low region which extends all of the way from left to right across the domain ($C^* = .670$). The maximum value is in the high region just below the low region (${C^*}_{\max} = .763$).

The globally-averaged kinetic energy elucidates the differences in vigor of convection between the two numerical experiments. Shown in Fig. 8 is the kinetic energy for both water-side and ammonium chloride-side experiments. Both initial compositions produce maxima between $t = 0$ and $t = .1$, after which the evolutions diverge. Whereas the water-side system becomes filled with mush and convection becomes very sluggish, more of the domain in the ammonium chloride-side experiment contains mush of very small fs which is still capable of convection. In fact, the kinetic energy is seen to be slightly increasing out to a time of $t = .5$ for the ammonium chloride-side experiment. At later times, the convection in both systems completely ceases.

VII. Discussion. The main observations of the study of Cao and Poulikakos (1990) are that (1) the mush region is often significantly larger than the solid region, (2) buoyancy driven flow is weak for the cooling from above geometry, (3) convection in the mush is very weak, (4) the average temperature of the cavity decreases as the initial composition becomes more ammonium chloride-rich, and (5) there is no mush region for $C_0 = .25$. As seen in Figures 4–7 and discussed above, the numerical experiments capture these features extremely well. We concentrate on observation (5) below and provide a simple alternative explanation for it.

Whereas Cao and Poulikakos (1990) call on convection in the lower region to prevent the formation of a mush region, the lever rule and the assumption of local thermodynamic equilibrium provide another explanation for the lack of mush in the ammonium chloride-side experiment. Simply put, the jump in fs which occurs across T_{sol} is smaller for the $C_0 = .15$ (water-side) experiment than for $C_0 = .25$ (ammonium chloride-side) experiment. This means that over the temperature range from the hotter regions to T_{sol}, there will be more mush for the $C_0 = .15$ experiment than for $C_0 = .25$ experiment. In other words, the mush zone is thicker in the former case than in the latter. The mush region may be negligibly thin in the ammonium chloride side laboratory experiment. Now it is also true that convection is more vigorous in the latter case because the mush region is small. This was observed in our numerical experiments (Fig. 8, and cf. Fig 5a and Fig. 7a) and may enhance the effect seen in the laboratory experiments. This is a good example of a new nonlinearity entering the system when solidification and convection are coupled.

We emphasize the huge effect on the system of including phase change in convective systems. For example, during solidification, much of the domain becomes stationary solid and no longer participates in convective heat transfer. Nevertheless, the solid region buffers convection in the remaining liquid region by controlling the conductive heat transfer, an effect which has already been noted previously (Carrigan, 1987; Marsh, 1990). In the convection equations alone, the advection of energy and species coupled with the body force terms is the fundamental nonlinearity. When solidification occurs, other nonlinearities enter the problem. At the same time though, the relative importance of the various nonlinear effects changes. For example, at the cold top boundary in an isoviscous single-phase system, negatively buoyant plumes would form and break off and make the top be a region of complicated flow. In the solidification system with the same geometry, however, solid forms at the top and these negatively buoyant plumes form only ahead of the solidification front and rely on conduction across the solid region for generation of negative buoyancy. This tends to decrease the importance of this effect and simplify the convective flow.

Despite the difference in the effective physical domain size used and the use of some specific approximations, the numerical results agree very well with the laboratory experiments. Good agreement between laboratory and numerical experiments with the continuum formulation has been noted by other researchers as well (Christenson et al., 1989). The parameters chosen for our numerical experiments are limited by the numerical method. For example, the Ra in the laboratory experiments was about 10^{10} whereas the Ra in the numerical experiments was, for practical reasons, equal to 10^4. Furthermore, several approximations have gone into the numerical model, including the assumption of local thermodynamic equilibrium, the restriction that the velocity of solid is zero, the relation between permeability and fraction solid, and the approximation that the heat capacity difference between the solid endmembers could be neglected. Nevertheless, the numerical experiments were able to capture the important transport phenomena.

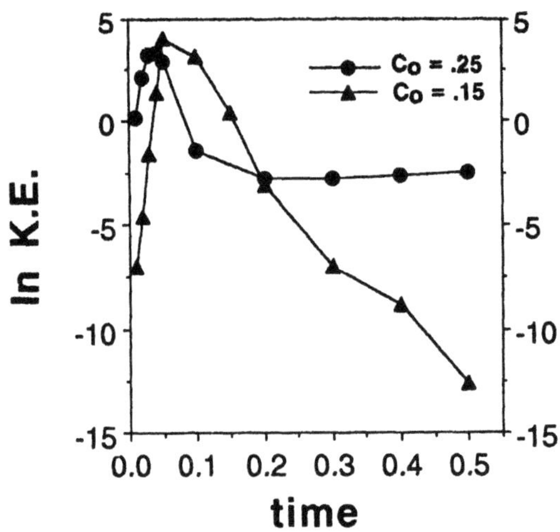

Fig. 8. Log of the globally-averaged kinetic energy $K.E. = \frac{1}{2}\int_0^1\int_0^1 (u^2+v^2)dx\,dy$ versus dimensionless time (t). Figure shows the decreasing vigor for the water-side ($C_0 = .15$) experiment and the more steady vigor of the ammonium chloride-side ($C_o = .25$) experiment. Mush inhibits convection in the water-side experiment.

VIII. Conclusions. Numerical experiments using a continuum model for solidification and convection have replicated the results of a multi-region model and a laboratory experiment. The successful modeling of solidification and convection shows that even quite complicated processes in nonlinear systems can be modeled numerically. Finally, we note that phase-change has a profound effect on convective systems in general. Future work will focus on the magmatic environment in particular.

Acknowledgements

This work was supported by NSF (OCE89–11396) and NSF (EAR90–18128).

Nomenclature

A	difference $(h^*_{liq} - h_{sol})$.
C	mixture composition.
c_p	heat capacity.
d	depth of domain.
D	chemical diffusivity.
Da	Darcy number.
eut	eutectic.
g	acceleration of gravity.
h	enthalpy.
h^*_{liq}	projection of h_{liq} to T_{sol}.
j	species flux.
J	Joules.
k	thermal conductivity.
K	permeability.
L	width of domain.
Le	Lewis number.
p	pressure.
q	heat flux.
Ra	Thermal Rayleigh number.
Rs	Compositional Rayleigh number.
S	source term.
s_1	solid 1.
s_2	solid 2.
St	Stefan number.
t	time.
T	temperature.
u	horizontal velocity.
v	vertical velocity.
V	velocity vector.
x	horizontal coordinate.
y	vertical coordinate.

Greek symbols

α	coefficient of thermal expansion.
β	coefficient of compositional expansion.
Γ	general diffusion coefficient.
ϕ	general field variable.
κ	thermal diffusivity.
μ	dynamic viscosity.
ρ	density.
ν	kinematic viscosity.
ψ	stream function.

Subscripts

eut	eutectic.
f	fusion.
l	liquid.
liq	liquidus.
m	pure end-member.
r	ratio.
s	solid.
sol	solidus.
w	wall.
o	initial or reference value.

Accented Characters

Circumflex ($\widehat{x}$) is used to indicate dimensionless variables.

REFERENCES

W.D. Bennon and F.P. Incropera, *A continuum model for momentum, heat and species transport in binary solid-liquid phase change systems–I. Model formulation*, Int. J. Heat Mass Trans. 30 (1987), 2161–2170.

W.D. Bennon and F.P. Incropera, *A continuum model for momentum, heat and species transport in binary solid-liquid phase change systems–II. Application to solidification in a rectangular cavity*, Int. J. Heat Mass Trans. 30 (1987), 2171–2187.

A.D. Brent, *Numerical simulation of tin solidification under the influence of natural convection in a thick walled annular crucible*, Int. Commun. Heat Mass Transfer 16 (1989), 693–701.

W.-Z. Cao and D. Poulikakos, *Solidification of an alloy in a cavity cooled through its top surface*, Int. J. Heat Mass Trans. 33 (1987), 427–434.

C.R. Carrigan, *The magmatic Rayleigh number and time-dependent convection in cooling lava lakes*, Geophys. Res. Lett. 14 (1987), 915–918.

M.S. Christenson, W.D. Bennon and F.P. Incropera, *Solidification of an aqueous ammonium chloride solution in a rectangular cavity–II. Comparison of predicted and measured results*, Int. J. Heat Mass Trans. 32 (1989), 69–79.

C.A.J. Fletcher, *Computational Techniques for Fluid Dynamics*, vol. II, Springer (1988), 486 pp.

P.F. Galpin and G.D. Raithby, *Numerical solution of problems in incompressible fluid flow: Treatment of the temperature-velocity coupling*, Num. Heat Trans. 10 (1986), 105–129.

B.D. Marsh, *On convective style and vigor in sheet-like magma chambers*, J. Petrology 30 (1989), 479–530.

C.M. Oldenburg and F.J. Spera, *Numerical modeling of solidification and convection in a viscous pure binary eutectic system*, Int. J. Heat Mass Trans., in press.

C.M. Oldenburg and F.J. Spera, *Simulation of phase change and convection in magma bodies*, C. Carrigan and T.Y. Chu (editors), Heat Transfer in Earth Science Studies ASME-HTD 149 (1990), 35–42.

S.V. Patankar, *Numerical Heat Transfer and Fluid Flow*, Hemisphere (1980), 197 pp.

N. Ramachdran, J.P. Gupta and Y. Jaluria, *Thermal and fluid flow effects during solidification in a rectangular enclosure*, Int. J. Heat Mass Trans. 25 (1982), 187–194.

G.O. Roberts, *Fast viscous Benard convection*, Geophys. Fluid Dynam. 12 (1979), 235–272.

R.A. Robie, B.S. Hemingway and J.R. Fisher, *Thermodynamic properties of minerals and related substances at 298.15 K and 1 bar (10^5 Pascals) pressure and at higher temperatures*, U.S. Geol. Surv. Bull. 1452, (1979), 456 pp.

CONVECTION IN POROUS MEDIA WITH THERMAL AND CHEMICAL BUOYANCY: A COMPARISON OF TWO MODELS FOR SOLUTE DISPERSION

N.D. ROSENBERG* AND F.J. SPERA*

Abstract. Results of a study of the role of solute dispersion model on convection in porous media heated and salted from below are given. In the first model, use is made of an effective dispersion coefficient that is taken as a constant scalar. In the second, a velocity-dependent dispersivity term that explicitly considers both molecular and mechanical dispersion is used. The results show that for the particular cases studied ($Ra = 600, \phi/\sigma = 1$, aspect ratio $= 1$, $R\rho = 0$ and $R\rho = 0.5$), although the choice of dispersion model does not affect the basic character of the fluid dynamics of the system, the details of the solute fields are somewhat different. In the case of the passive tracer ($R\rho = 0$), the solute fields for both models evolve to a steady-state which is very similar. The main difference between the evolution of the solute fields for $R\rho = 0$ is that in the case of the simple model, the solute field mixes more quickly. For $R\rho = 0.5$, chaotic flow characterizes convection for both dispersion models and regions of sharp solute gradients are highly mobile in time and space. The main difference between the two solute dispersion models for $R\rho = 0.5$ is that the interior of the solute field is less homogeneous and more saline in the case of the simple dispersion model.

Introduction. A number of geophysical, geochemical and engineering problems require an understanding of convection phenomena in fractured or granular porous media in which two sources of buoyancy (e.g., salt and heat) are present. Traditionally, two regimes of double-diffusive convection are distinguished. When the faster diffusing component (heat) is destablizing, the system is in the diffusive regime. When the slower diffusing component (salt) is destabilizing, the system is in the finger regime. In this study, attention is focused on double-diffusive convection in the diffusive regime, specifically the case of porous media heated and salted from below. Such would be the case, for example, in a hydrothermal system in which a hot, saline brine was overlain by relatively cold and fresh fluids. Convection can occur in the diffusive regime even when the overall density gradient is stable and linear stability analysis predicts that infinitesimal disturbances will decay. This subcritical, finite-amplitude convection occurs because the diffusivity of the destabilizing component, in this case heat, is greater than the diffusivity of the stabilizing component, salt.

Double-diffusive convection in viscous fluid systems has received a great deal of attention (Turner, 1985; Huppert and Moore, 1976). Oscillatory and chaotic flows are characteristic of many of these systems. Much less work has been done, however, on double-diffusive convection in porous media. Most of the published work on this subject concerns linear stability analyses and steady-state flows. Trevisan and Bejan (1990) recently reviewed the literature on double-diffusion convection in porous media. The time-dependent dynamics of convection in porous media heated and salted from below as a function of chemical to thermal buoyancy ratio in the system ($R\rho$) was recently studied by Rosenberg and Spera (1990a,b; in press). They

*Department of Geological Sciences, University of California, Santa Barbara, CA 93106

reported that as $R\rho$ increases, the dynamics change from a system which evolves to a convective steady-state, to one in which flow is chaotic, and finally to one which evolves to a conductive steady-state. A simple Fickian solute dispersion model in which a constant scalar represents an effective dispersion coefficient was used in this study. The most appropriate way to model solute dispersion in porous media is, however, the subject of much controversy (Dagan, 1990), Tompson and Gray, 1989, Gist et al., 1990) and is critically important in many practical and environmental problems.

In the present contribution, time-dependent simulations of convection in porous media heated and salted from below are discussed for two different Fickian solute dispersion models. In the first model, use is made of an effective dispersion coefficient that is taken as a constant scalar, as in the earlier study by Rosenberg and Spera (1990a,b; in press). In the second, a velocity-dependent dispersivity coefficient that explicitly considers both molecular and mechanical dispersion is opted. In the first part of this paper, the equations governing the transport of heat, momentum and solute are given for the two-dimensional case investigated here. A brief description of the numerical methods used is then presented. Results are reported and discussed for both solute dispersion models for the passive tracer case ($R\rho = 0$) and for the case where the ratio of chemical to thermal buoyancy is one-half ($R\rho = 0.5$).

Mathematical Formulation. The equations governing double-diffusive convection in porous media include equations for the conservation of mass, momentum, energy, and species and an equation of state for the fluid (Bear, 1979). Conservation of fluid mass, assuming an incompressible fluid and no sources or sinks, can be expressed as:

(1) $$\nabla \cdot \mathbf{q} = 0 \; .$$

Conservation of fluid momentum is expressed by Darcy's law, providing the fluid moves slowly so that inertial effects are negligible ($Re < 1$), as:

(2) $$\mathbf{q} = -\frac{k}{\mu}(\nabla p - \rho \mathbf{g})$$

where k, the permeability, is isotropic and spatially invariant, and μ, the fluid viscosity, is taken as a constant.

Conservation of energy can be expressed as:

(3) $$\sigma \frac{\partial T}{\partial t} + \mathbf{q} \cdot \nabla T = \kappa \nabla^2 T$$

where it is assumed that the fluid and matrix are in thermal equilibrium and κ, the effective thermal diffusivity of the saturated medium, is constant.

Conservation of species can be expressed as:

(4) $$\phi \frac{\partial C}{\partial t} + \mathbf{q} \cdot \nabla C = \phi \mathbf{D_h} \nabla^2 C$$

where $\mathbf{D_h}$ is the coefficient of hydrodynamic dispersion. (All symbols are defined in a separate notation section.)

In writing equation (4), it is assumed that the dispersive flux of solute can be expressed in Fickian form, in accordance with classical models of hydrodynamic dispersion. Field data, however, show that dispersion depends on the size of the initial anomaly and on travel-time, a clear violation of Fickian behavior (Gelhar, 1986; Dagan, 1990). Interestingly, in heterogeneous (e.g., layered) media, Fickian behavior is found asymptotically in the limit of large field-scale distances or for long times (Matheron and Marsily, 1980). Controversy surrounds the issue of how best to model solute dispersion in porous and further work must be accomplished before the issue is considered resolved.

In this contribution, two simple Fickian models of dispersion have been considered. The simplest model assumes that the dispersion term can be represented as a constant scalar, D_{eff}. In this case, the conservation of species is expressed as:

$$\phi\frac{\partial C}{\partial t} + \mathbf{q}\cdot\nabla C = D_{\text{eff}}\nabla^2 C \ . \tag{5}$$

A second and somewhat more refined treatment of dispersion recognizes that the coefficient of hydrodynamic dispersion, $\mathbf{D_h}$, is the sum of the coefficient of molecular diffusion in a porous medium, $\mathbf{D^*_{mol}}$, and the coefficient of mechanical dispersion, $\mathbf{D_{mech}}$. $\mathbf{D^*_{mol}}$ is equal to the molecular diffusivity of solute in the fluid multiplied by the tortuosity of the medium. $\mathbf{D_{mech}}$ is a function of fluid velocity and medium characteristics and is usually much greater than $\mathbf{D^*_{mol}}$ except for very slow flows. For this second case, the conservation of species can be expressed as:

$$\phi\frac{\partial C}{\partial t} + \mathbf{q}\cdot\nabla C = [\phi\mathbf{D^*_{mol}} + \mathbf{D_{mech}}]\nabla^2 C \tag{6}$$

where

$$\mathbf{D_{mech}} = (a_L - a_T)\frac{(q_x q_z)}{|\mathbf{q}|} + a_T|\mathbf{q}|\delta_{ij} \tag{7}$$

Here, a_L and a_T represent the longitudinal and transversal dispersivities and δ_{ij} is the Kronecker delta.

It is possible to express the conservation of energy equation (4) in a form analogous to equation (6) with a coefficient of thermal dispersion similar to $\mathbf{D_h}$. However, since the effective thermal diffusivity is about 100 times greater than the molecular diffusivity of a solute, the thermal dispersion term can be neglected over a much wider range of velocities. For the velocity range for which the momentum equation (3) is valid, the thermal dispersion term can be neglected (Bear, 1979).

For the equation of state of the fluid, it is assumed that fluid density is a function of temperature and concentration only:

$$\rho = \rho_0[1 - \alpha(T - T_0) + \beta(C - C_0)] \ . \tag{8}$$

All fluid and transport properties are assumed to be constant except for density in the body force term of the momentum equation (the Boussinesq approximation).

The governing equations may be written in the following dimensionless form:

$$\nabla^2 \widehat{\psi} = -Ra \left(\frac{\partial \widehat{T}}{\partial \widehat{x}} - R\rho \frac{\partial \widehat{C}}{\partial \widehat{x}} \right) \tag{9}$$

$$\frac{\partial \widehat{T}}{\partial \widehat{t}} + \left(\widehat{q}_x \frac{\partial \widehat{T}}{\partial \widehat{x}} + \widehat{q}_z \frac{\partial \widehat{T}}{\partial \widehat{z}} \right) = \nabla^2 \widehat{T} \tag{10}$$

and in the case of the simplest dispersion model

$$\phi^* \frac{\partial \widehat{C}}{\partial \widehat{t}} + \left(\widehat{q}_x \frac{\partial \widehat{C}}{\partial \widehat{x}} + \widehat{q}_z \frac{\partial \widehat{C}}{\partial \widehat{z}} \right) = \frac{1}{Le_{\text{eff}}} \nabla^2 \widehat{C} \,. \tag{11}$$

In the above equations, ψ is the streamfunction ($q_x = \partial\psi/\partial z$, $q_z = -\partial\psi/\partial x$),

$$\begin{array}{llll} \widehat{x} = x/H & \widehat{z} = z/H & \widehat{t} = t\kappa/H^2\sigma & \widehat{q} = qH/\kappa \\ \widehat{\psi} = \psi/\kappa & \widehat{T} = (T - T_0)/\Delta T & \widehat{C} = (C - C_0)/\Delta C & \end{array} \tag{12}$$

and the four dimensionless parameters governing the convective dynamics are the Rayleigh number (Ra), an effective Lewis number (Le_{eff}), ϕ^*, and the buoyancy ratio ($R\rho$) where:

$$Ra = \frac{k\rho_0 g\alpha\Delta T H}{\mu\kappa} \qquad Le_{\text{eff}} = \frac{\kappa}{D_{\text{eff}}} \qquad \phi^* = \phi/\sigma \qquad R\rho = \frac{\beta\Delta C}{\alpha\Delta T} \,. \tag{13}$$

In the case of the more complete dispersion model, equation (11) is replaced by equation (14):

$$\phi^* \frac{\partial \widehat{C}}{\partial \widehat{t}} + \left(\widehat{q}_x \frac{\partial \widehat{C}}{\partial \widehat{x}} + \widehat{q}_z \frac{\partial \widehat{C}}{\partial \widehat{z}} \right) = \frac{\partial}{\partial \widehat{x}} \left(\left(f + \frac{1}{Le_{\text{mol}}} \right) \frac{\partial \widehat{C}}{\partial \widehat{x}} + h \frac{\partial \widehat{C}}{\partial \widehat{z}} \right) + \frac{\partial}{\partial \widehat{z}} \left(h \frac{\partial \widehat{C}}{\partial \widehat{x}} + \left(g + \frac{1}{Le_{\text{mol}}} \right) \frac{\partial \widehat{C}}{\partial \widehat{z}} \right) \tag{14}$$

where

$$\begin{array}{lll} f = \dfrac{\widehat{a}_T (a_r \widehat{q}_z^2 + \widehat{q}_x^2)}{|\mathbf{q}|} & g = \dfrac{\widehat{a}_T (a_r \widehat{q}_x^2 + \widehat{q}_z^2)}{|\mathbf{q}|} & h = \dfrac{\widehat{a}_T ((a_r - 1)\widehat{q}_x \widehat{q}_z)}{|\mathbf{q}|} \\ \widehat{a}_T = a_T/H \quad a_r = a_L/a_T & Le_{\text{mol}} = \dfrac{\kappa}{D_{\text{mol}} {}^* \phi} & \end{array} \tag{15}$$

In this case, there are six dimensionless parameters governing the convective dynamics: the Rayleigh number (Ra), a molecular Lewis number, ϕ^*, the buoyancy ratio ($R\rho$), the dimensionless transversal dispersivity ($\widehat{a}_T$) and the ratio of the longitudinal to transversal dispersivities (a_r).

The nondimensional computational domain consists of a square box (Figure 1). The box has a hot bottom wall with a cold top wall and adiabatic sides. All walls are impermeable to flow and the fluid is initially cold and motionless. The system is salted from below (i.e., $\widehat{C} = 1$ along the bottom) and the fluid is initially at the same concentration as the top wall ($\widehat{C} = 0$).

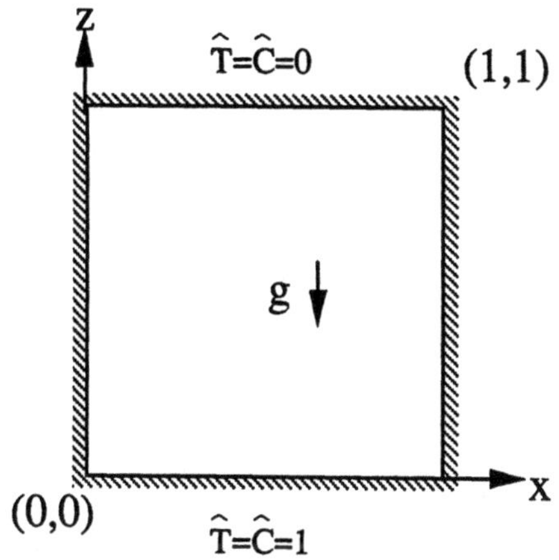

Figure 1. Domain and boundary conditions. All walls are impermeable to flow and no heat or solute leaves the domain through the side walls. The system is heated and salted from below. Initially, the fluid is motionless and $\widehat{T} = \widehat{C} = 0$ everywhere.

The flow, temperature and solute fields are determined at each time step. In addition, the spatially-averaged kinetic energy per unit mass of fluid (KE) is also computed. KE is a convenient measure of the vigor of the flow and is defined for a 1×1 box as:

$$KE = \frac{1}{2} \int_0^1 \int_0^1 [\widehat{q}_x^2 + \widehat{q}_z^2] \, d\widehat{x} d\widehat{z} \,. \tag{16}$$

Nu, the ratio of the total heat transferred to the heat transferred by conduction alone for the imposed ΔT, is defined as:

$$Nu = - \int_0^1 \left(\frac{\partial \widehat{T}}{\partial \widehat{z}} \right)_{\widehat{z}=1} d\widehat{x} \,. \tag{17}$$

Finally, Sh, the ratio of the total solute transferred to the solute transferred by diffusion alone for the imposed ΔC, is defined as:

$$Sh = - \int_0^1 \left(\frac{\partial \widehat{C}}{\partial \widehat{z}} \right)_{\widehat{z}=1} d\widehat{x} \,. \tag{18}$$

The governing equations are solved using a Galerkin finite element code used previously for other convection calculations (Oldenburg, et al, 1989; Rosenberg and Spera, in press); details of the method are given in Sewell (1985). Briefly, triangular finite elements with quadratic basis functions are used to discretize the conservation equations and the resulting set of algebraic equations is solved by Newton–Raphson iteration. A fully implicit scheme is used to advance the energy and species equations forward in time. The results were obtained using a graded grid of 1120 elements and a time step of $\sim 10^{-4}$. Additional calculations with finer spatial and temporal resolution were performed to verify the accuracy of the solutions. The resolution question is more difficult for chaotic flow solutions because, by definition, chaotic flows are highly sensitive to small differences in initial conditions and therefore to small differences in resolution. Simulations of flows performed at a variety of spatial and temporal resolutions suggest that the chaotic flows reported here are indeed chaotic (i.e., running average KE and the broadband character of the corresponding power spectra are preserved).

Results. The time-dependent solutions were obtained for $R\rho = 0$ and 0.5. A Rayleigh number of 600, a reasonable value for hydrothermal systems, is used. ϕ^* is set equal to 1. In the simple solute dispersion case, Le is taken as 20. For the second solute dispersion case, $\widehat{a}_T = 10^{-4}$, $a_r = 10$, and $Le_{\text{mol}} = 10^4$.

For $R\rho = 0$, corresponding to the problem of pure thermal convection with a passive tracer, the flow and temperature fields rapidly reach a convective steady-state consisting of two side-by-side convection cells elongated in the vertical direction ($KE = 1690$, $Nu = 6.6$). The flow consists of rising hot fluid along the adiabatic and impermeable side walls and colder fluid sinking in the center of the domain. Thin temperature boundary layers are found at the top and bottom of the domain and the interior is well-mixed (Figure 2). For a passive tracer, the evolution of the flow and temperature fields is of course the same for both dispersion models.

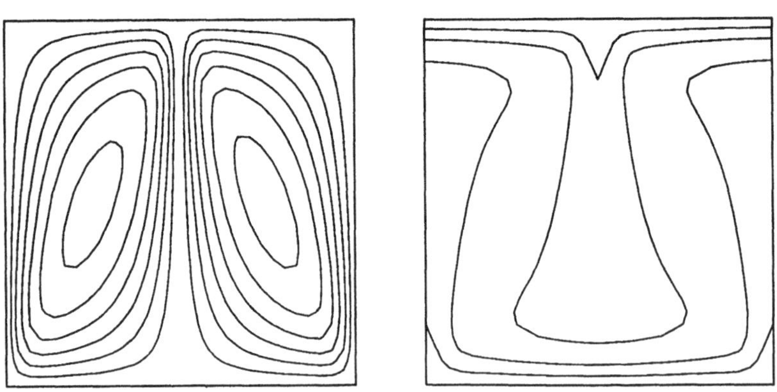

(a) (b)

Figure 2. (a) Flow and (b) temperature fields for passive tracer case. $\psi_{\max} = 12$, $\Delta\psi = 2$; $T_{\max} = 1, \Delta T = 0.2$.

The nature of the solute fields for these two models is, however, different. Figure 3 (see Plates section) shows the solute field for both dispersion models at two early times and at steady-state. The steady-state solute fields for both dispersion models are similar. In both cases, fluid with a high solute concentration rises along the side walls and fluid with a low solute concentration sinks in the center of the domain. The solute fields are similar to the temperature field, only with much thinner boundary layers. For the second dispersion model, the interior is almost completely homogeneous and the chemical boundary layers are even thinner than in the first dispersion model. This difference is reflected in the Sherwood numbers. For the simple dispersion model, $Sh \approx 30$. Rates of solute transport are larger by about a factor of two in the velocity-dependent dispersion case.

The evolution of the solute fields in the two $R\rho = 0$ cases is also different. The solute field for the case of the simple dispersion model mixes and evolves to a steady-state more quickly. This is illustrated in the solute fields shown in Figure 3 and in the plot of spatially-averaged solute concentrations for both cases (Figure 4) which show that solute concentration is always greater at any given time for the simple dispersion model before steady-state is reached. In the second dispersion model, advective transport of solute results in the isopleths being more concentrated along lines of constant streamfunction. In the first dispersion model, the evolution of the solute field is more similar to the evolution of the temperature field.

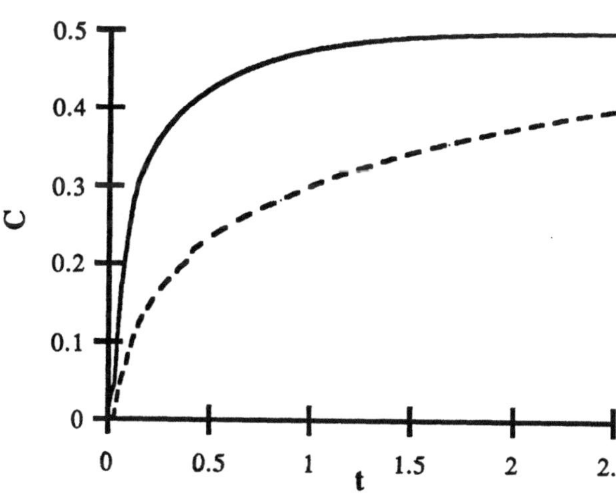

Figure 4. Spatially-averaged solute concentration versus time for passive tracer case. Solid line is for simple dispersion model and dashed line is for complex dispersion model.

For $R\rho = 0.5$, corresponding to a system in which chemical buoyancy is one-half the thermal buoyancy, convection is chaotic. This is true for both dispersion models. Plots of the solute field for both dispersion models are shown in Figure 5 (see Plates section). In both cases, thin solute boundary layers are found on the top and bottom walls of the system and plumes of solute arise from two small areas on the bottom, carrying solute upwards. These source areas move aperiodically to the left and to the right with time, resulting in regions of sharp solute gradients that are highly mobile in time and space. The general nature of chaotic flow is the same for both dispersion models at $R\rho = 0.5$. The main difference between the two cases is that the interior of the system is more homogeneous and less salty for the second dispersion model.

Plots for both dispersion models for $R\rho = 0.5$ of KE time series (Figure 6), phase diagrams of KE versus $\dot{KE}$ (Figure 7) and the power spectrum of KE (Figure 8) clearly exhibit the features of chaotic flow. The $KE - \dot{KE}$ trajectories follow irregular orbits in a loosely defined region of phase space, never reaching a fixed point and the KE power spectra exhibit broadband structure.

Conclusions. Results of a study of the role of solute dispersion model on double-diffusive convection in porous media are presented. In the first model, use is made of an effective dispersion coefficient that is taken as a constant scalar. In the second, a velocity-dependent dispersivity term that explicitly considers both molecular and mechanical dispersion is used. The results show that for the particular cases studied ($Ra = 600$, $\phi^* = 1$, aspect ratio $= 1$, $R\rho = 0$ and $R\rho = 0.5$), the choice of dispersion model does not affect the first order convective dynamics of the system; the details of the solute fields are however different. In the case of the passive tracer ($R\rho = 0$), the solute fields for both models evolve to a steady-state which is very similar. The chemical boundary layers in the case of the more complex dispersion model are however a bit thinner and the interior is slightly more homogeneous. The main difference between the evolution of the solute fields is that in the case of the simple model, the solute field mixes more quickly than in the case of the more complex dispersion coefficient. For the simple dispersion model, the evolution of the solute field is very similar in character to the evolution of the temperature field, whereas in the case of the more complex dispersion coefficient, the isopleths more closely follow lines of constant streamfunction. For $R\rho = 0.5$, chaotic flow characterizes convection for both dispersion models and regions of sharp solute gradients are highly mobile in time and space. The main difference between the two solute dispersion models for $R\rho = 0.5$ is that the interior of the solute field is more homogeneous and less salty in the case of the complex dispersion model.

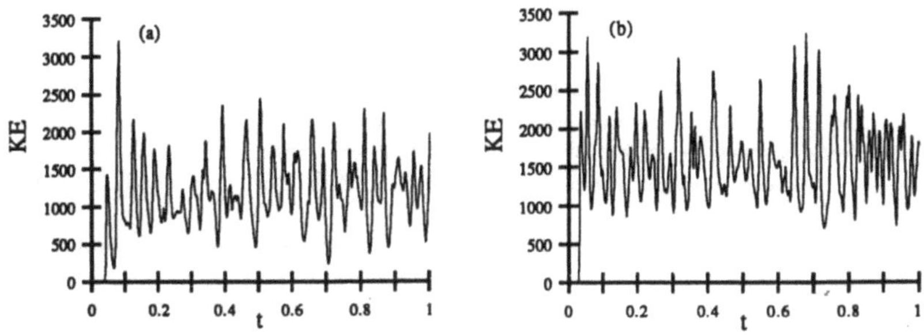

Figure 6. KE time series plots for (a) simple dispersion model and (b) complex dispersion model for $R\rho = 0.5$. These time series show that the flows are unsteady and aperiodic.

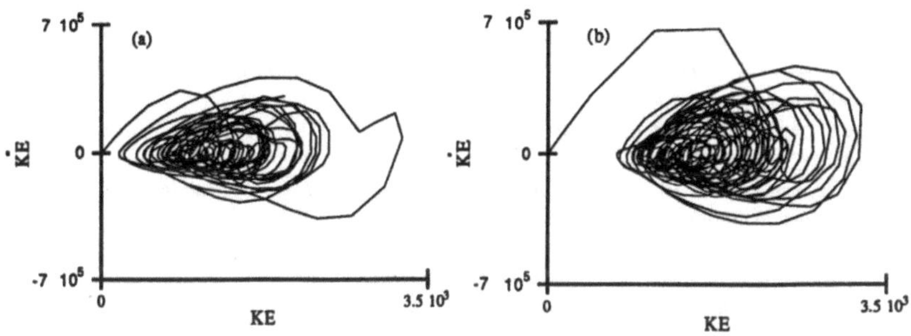

Figure 7. $KE - \dot{KE}$ plots for (a) simple dispersion model and (b) complex dispersion model for $R\rho = 0.5$. The $KE - \dot{KE}$ trajectory follows an irregular orbit in a loosely defined region on phase space and never reaches a fixed point. This is characteristic of chaotic flow.

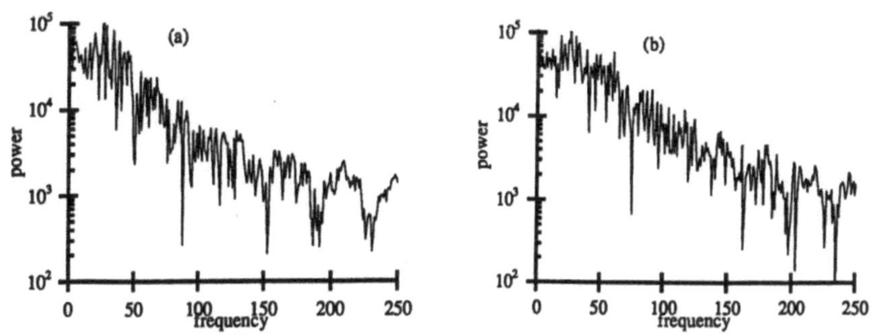

Figure 8. KE power spectra plots for (a) simple dispersion model and (b) complex dispersion model for $R\rho = 0.5$. Note the broadband nature of the power distribution, a characteristic of chaotic flows.

REFERENCES

J. Bear, *The Dynamics of fluids in porous media*, Dover Publications, New York, 1972.

G. Dagan, *Transport in heterogeneous porous formations: spatial moments, ergodicity, and effective dispersion*, Water Res. Research **26** (1990), 1281–1290.

————, *Theory of solute transport by groundwater*, Ann. Rev. Fluid Mech. **19** (1987), 183–215.

L.W. Gelhar, *Subsurface hydrology from theory to applications*, Water Res. Research **22** (1986), 135S–145S.

G.A. Gist, A.H. Thompson, A.J. Katz and R.L. Higgins, *Hydrodynamic dispersion and pore geometry in consolidated rock*, Phys. Fluids A **2** (1990), 1533–1544.

H.E. Huppert and D.R. Moore, *Nonlinear double-diffusive convection*, J. Fluid Mech. **78** (1976), 821–854.

G. Matheron and G. de Marsily, *Is transport in porous media always diffusive? a counter example*, Water Res. Research **16** (1980), 901–917.

C.M. Oldenburg, F.J. Spera, D.A. Yuen and G. Sewell, *Dynamic mixing in magma bodies: theory simulations, and implications*, J. Geophys. Res. **94** (1989), 9215–9236.

N.D. Rosenberg and F.J. Spera, *Role of anisotropic and/or layered permeability in hydrothermal systems*, Geophys. Res. Let. **17** (1990a), 235–238.

————, *Simulations of fluid flow in fractured porous media with thermal and chemical buoyancy*, EOS Transactions, American Geophysical Union **71** (1990b), 1687.

————, *Thermohaline convection in a porous medium heated from below*, Int. J. Heat Mass Trans., in Press.

G. Sewell, *Analysis of a finite element method: PDE/PROTRAN*, Springer–Verlag, New York (1985).

A.F.B. Tompson and W.G. Gray, *A second-order approach for the modeling of dispersive transport in porous media: 1. Theoretical development*, Water Res. Research **22** (1986), 591–600.

O.V. Trevisan and A. Bejan, *Combined heat and mass transfer by natural convection in a porous medium*, Adv. in Heat Trans. **20** (1990), 314–352.

J.S. Turner, *Multicomponent convection*, Ann. Rev. Fluid Mech. **17** (1985), 11–44.

Notation

a_L	longitudinal dispersivity
a_T	transversal dispersivity
a_r	ratio a_L/a_T
C	concentration (mass fraction solute in fluid)
c_p	heat capacity
$\mathbf{D_h}$	hydrodynamic dispersion coefficient
$\mathbf{D^*_{mol}}$	molecular dispersion coefficient
$\mathbf{D_{eff}}$	effective dispersion coefficient
$\mathbf{D_{mech}}$	mechanical dispersion coefficient
g	gravitational constant
H	height of porous medium
k	permeability
KE	kinetic energy per unit mass fluid
$\dot{KE}$	time rate-of-change of KE
Le_{eff}	effective Lewis number (κ/D_{eff})
Le_{mol}	molecular Lewis number $(\kappa/\phi D^*_{\text{mol}})$
Nu	Nusselt number
p	pressure
$\mathbf{q}$	Darcy velocity
q_x	Darcy velocity in x-direction
q_z	Darcy velocity in z-direction
Ra	Rayleigh number
Re	Reynolds number
$R\rho$	chemical/thermal buoyancy ratio
Sh	Sherwood number
T	temperature
t	time
$\mathbf{v}$	fluid velocity $(\mathbf{q}/\phi)$
x, z	Cartesian coordinate

Greek:

α	coefficient of thermal expansivity
β	coefficient of chemical expansivity
δ_{ij}	Kronecker delta
$\phi^* = \phi/\sigma$	
ϕ	porosity
κ	thermal diffusivity of saturated porous medium
μ	viscosity
ρ	fluid density
σ	heat capacity ratio, $\phi + (1-\phi)(\rho c_p)_{\text{solid}}/(\rho c_p)_{\text{fluid}}$
ψ	streamfunction $(\partial\psi/\partial z = q_x,\ -\partial\psi/\partial x = q_z)$
ΔC	concentration difference in vertical direction
ΔT	temperature difference in vertical direction

Subscripts and superscripts

$\hat{}$	dimensionless variables
0	reference value

DYNAMO THEORY

PAUL H. ROBERTS*

Abstract. Many astrophysical bodies possess magnetic fields. The Earth has one, the geomagnetic field, and it is apparently as old as the Earth. Since the temperature throughout most of the Earth is above the Curie points of all known materials, the Earth is not a permanent magnet. The geomagnetic field must be produced by electric currents flowing mainly within its core, which is a good electrical conductor. Lacking sources, such currents would die out in about 10,000 years. It is now believed that the currents are maintained by electromagnetic induction, as in a commercial dynamo, the moving parts of the geodynamo being motions in the outer core, which is fluid.

The first question in dynamo theory is whether self-excitation is possible in a body having such a low degree of symmetry as the Earth's core. This well-studied "kinematic dynamo problem" defines a nontrivial non-self adjoint eigenvalue problem, which will be discussed after the basic electrodynamic theory has been summarized. The second question is whether motions of sufficient vigor, and which have the "helicity" required for field generation, would arise naturally in the Earth's outer core. This nonlinear "magnetohydrodynamic dynamo problem" raises some interesting questions in rotating magnetoconvection. The third question is whether the required energy source is available to power the dynamo. For bodies such as the solar convection zone, energy is available in abundance, but the geodynamo's needs are less easily met.

Other topics will be touched on, including fast dynamos, auto-reversing dynamos, and chaotic dynamos.

1. Introduction. Where and when the discovery was made that the Earth is magnetic is lost in the mists of time. Historical research points to China as the place and perhaps more than 4000 years ago as the time. As exploration of the Earth's surface blossomed, so did knowledge of the direction that a freely suspended magnetic needle would point at different geographic locations. It became known that the source of the Earth's magnetism lay within it. Figure 1 shows Gilbert's famous "terrella" or little Earth. The direction taken up by spikes of iron placed on this spherical loadstone resembled the "inclination", I, to the horizontal of a freely suspended magnetic needle on the Earth's surface.

The Earth was, however, not as axisymmetric as Figure 1 might suggest. There were no magnetic meridian planes in which the magnetic needle would lie; there was an angle of "declination", D, between true North and magnetic North, a fact that became well known to makers of portable sundials (Figure 2) long before Gilbert. Columbus noted in the log of his famous voyage of discovery that D varies over the Earth's surface, and Halley (of comet fame) realized that maps giving D as a function of latitude and longitude (ϕ) would be useful to mariners (Figure 3).

Figure 3 gives some idea of how "the magnetic elements" like D and I vary over continental scales. Figure 4, a more recent (1983) map of declination, shows that D varies in time, t, as well as space. In this slow "secular variation" the isolines of D – compare particularly the $D = 0$ isolines in Figures 3 and 4 – have moved from the mid-Atlantic to the continental United States, a phenomenon known as "westward drift", a motion opposite to that of the world's weather patterns.

*Department of Mathematics, University of California, Los Angeles, California 90024

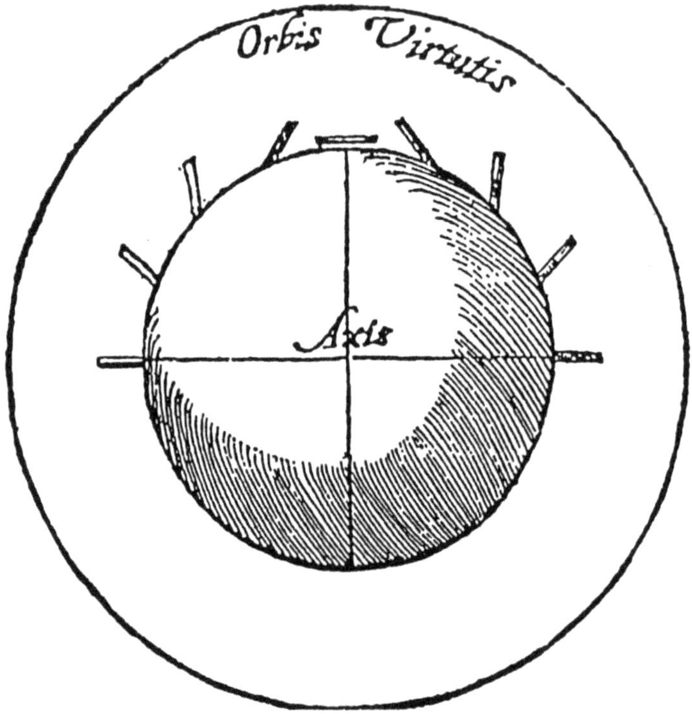

Figure 1. Gilbert's terrella, showing the inclinations taken up by iron spikes placed on its surface at different latitudes.

Figure 2. A portable sundial in the author's possession.

When examined in more detail, the Earth's field exhibits a very wide spectrum

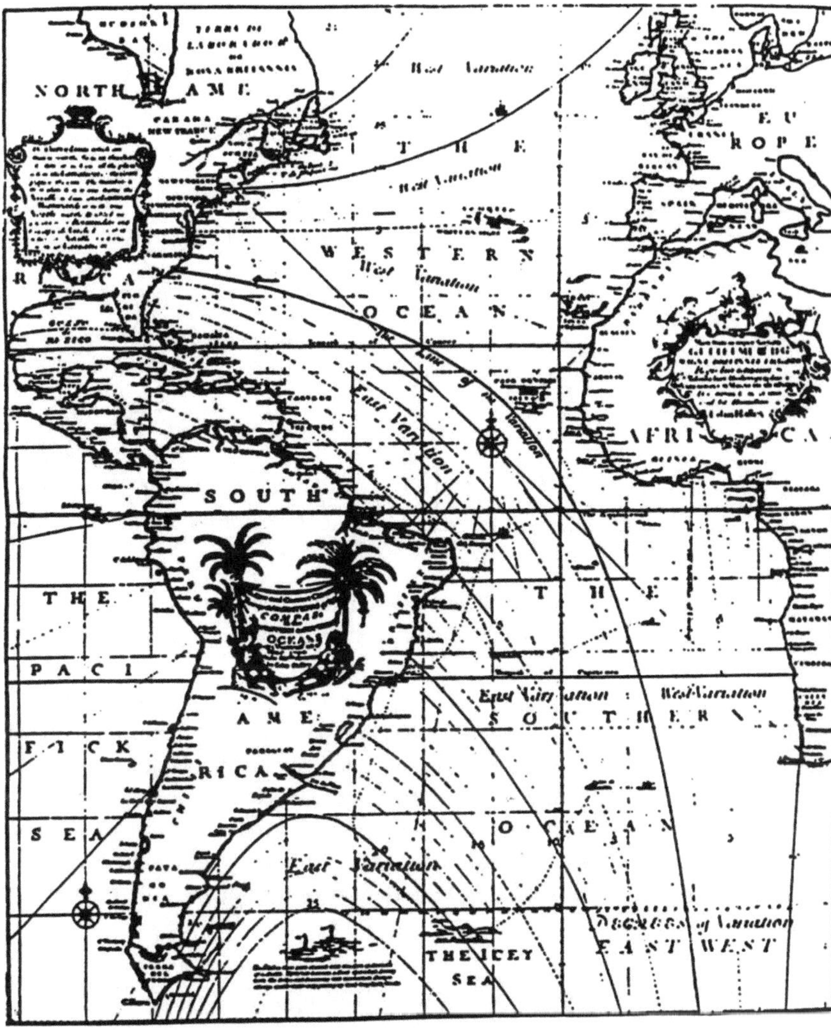

Figure 3. Reproduction of Halley's 1701 map of D. Note the position of the "Line of no variation", $D = 0$. (Figure courtesy of R.A. Langel.)

of time scales, even after the changes created from outside the Earth (e.g. in the ionosphere) are removed. Figure 5 shows a phenomenon that has attracted much attention during the past decade: secular change "impulses", or "jerks" as they are sometimes called. Abrupt changes in $\dot{X}$, $\dot{Y}$, $\dot{Z}$ (the rates of change of the Eastward, Southward and downward components of the magnetic field, $\mathbf{B}$) apparently occured all over the Earth's surface in 1969, and were completed in a very short time, perhaps less than a year.

Before describing the lower frequencies, we should note that the geomagnetic

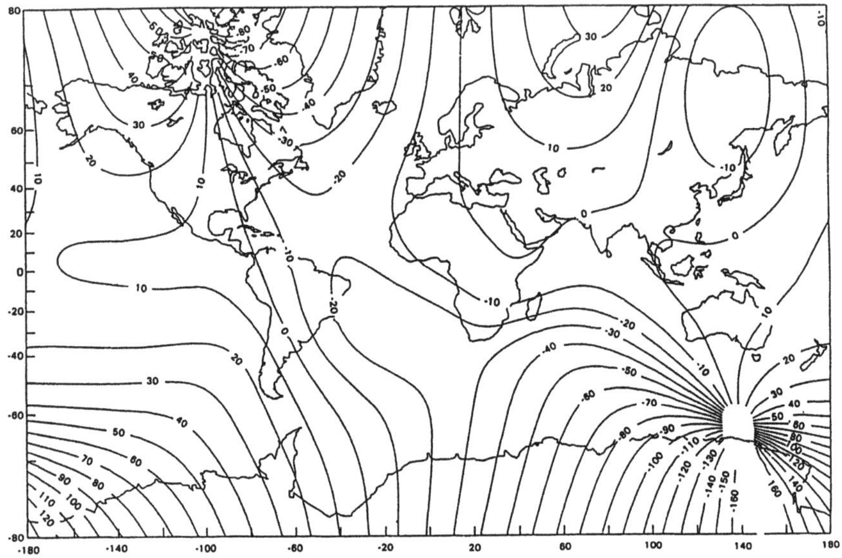

Figure 4. Isolines of D for epoch 1983. Note how far the contour $D = 0$ has move West relative to Figure 3. (Figure courtesy of R.A. Langel.)

field can be divided into dipole, quadrupole, octupole... parts, of which the dipole dominates at the Earth's surface. The direction of this dipole, $\mathbf{m}$, defines the "geomagnetic axis" which currently deviates from the geographic axis, Oz, by about $10°$. Thus, the "axial dipole", m_z, i.e. the component of $\mathbf{m}$ in the z-direction greatly exceeds the "equatorial dipole", $\mathbf{m}_E = \mathbf{m} - m_z\mathbf{1}_z$ (where $\mathbf{1}_q$ is the unit vector, the direction of increasing q). With $x = 0$ the Greenwich meridian $\phi = 0$, it is usual to write

$$\mathbf{m} = \frac{4\pi b^3}{\mu_0}\left(g_1^1\mathbf{1}_x + h_1^1\mathbf{1}_y + g_1^0\mathbf{1}_z\right), \tag{1.1}$$

where b is the radius of the Earth ($= 20{,}000\text{km}/\pi$), μ_0 is the permeability of free space, and the g and h constants are coefficients in the expansion,

$$V = b\sum_{n=1}^{\infty}\sum_{m=0}^{n}\left[g_n^m \cos m\phi + h_n^m \sin m\phi\right]\left(\frac{b}{r}\right)^{n+1} P_n^m(\theta), \tag{1.2}$$

of the geopotential; here θ is colatitude and P_n^m are Legendre functions. We use the fact here that the sources of $\mathbf{B}$ lie inside the Earth [hence the absence of terms proportional to $(r/b)^n$] and that no currents flow at the Earth's surface, so that

$$\mathbf{B} = -\nabla V\ , \tag{1.3}$$

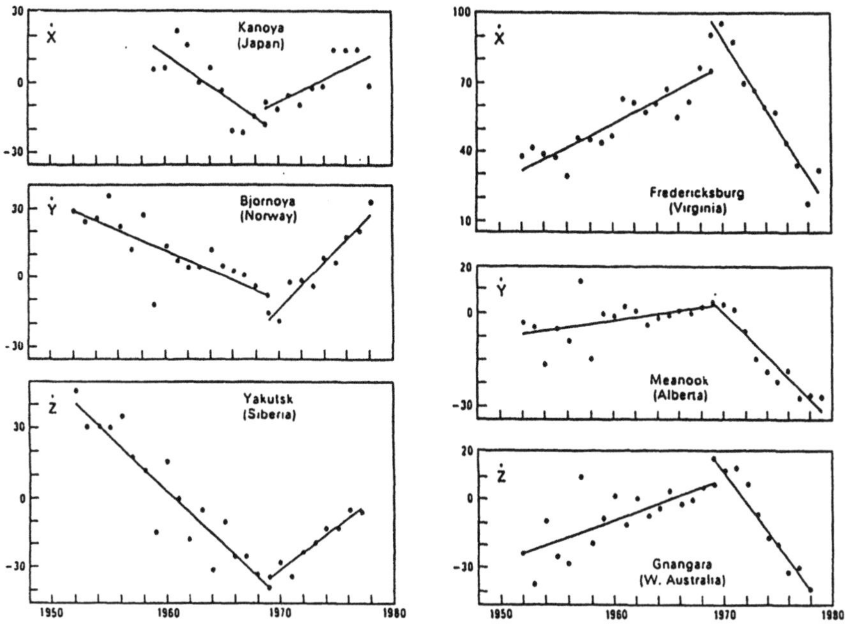

Figure 5. Secular change impulses, indicating their worldwide character. (After Madden and Le Mouël, 1982.)

with

$$\nabla^2 V = 0. \tag{1.4}$$

[There is no 4π in (1.2), so the units are unrationalized, g_n^m and h_n^m being in Tesla (T); the $(4\pi/\mu_0)$ factor in (1.1) makes $\mathbf{m}$ rationalized and a source of $\mathbf{H}(=\mathbf{B}/\mu)$, whose units are Am^2.]

The dipole is constantly changing in magnitude and direction. The rate of change of the axial dipole moment during this century is given in Figure 6; the recent history of $\dot{D}$ and of the length of day (l.o.d.) is displayed in Figure 7. To discuss the latter, which is apparently related to coupling between core and mantle, would take us too far from our theme. It is however clear from Figures 6 and 7 that changes in $\mathbf{B}$ occur on the decade timescale.

Moving to still lower frequencies, consider Figures 8, which show archeomagnetic data drawn from many sites over the globe; $F = |\mathbf{B}|$, I and D are displayed, and clearly exhibit variations on time scales of 100 yrs and longer. Figure 9 summarizes archeomagnetic data from the last 12,000 years. There is some suggestion of an oscillation period of about 8000 years, sometimes called "the fundamental period" of the geomagnetic field.

Figure 10 brings us to one of the most exciting properties of the geomagnetic field: polarity reversals. Here I, obtained from the magnetism of segments of a deep-sea core that acquired their magnetism when they were created at different

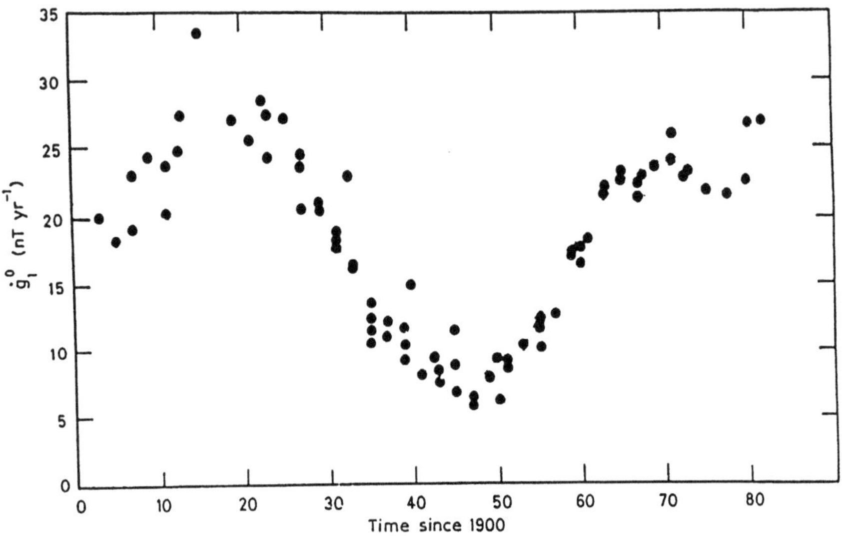

Figure 6. Rate of change, $\dot{m}$, of axial dipole moment since 1900, indicating a periodicity near 60 yr. (After Langel, 1987.)

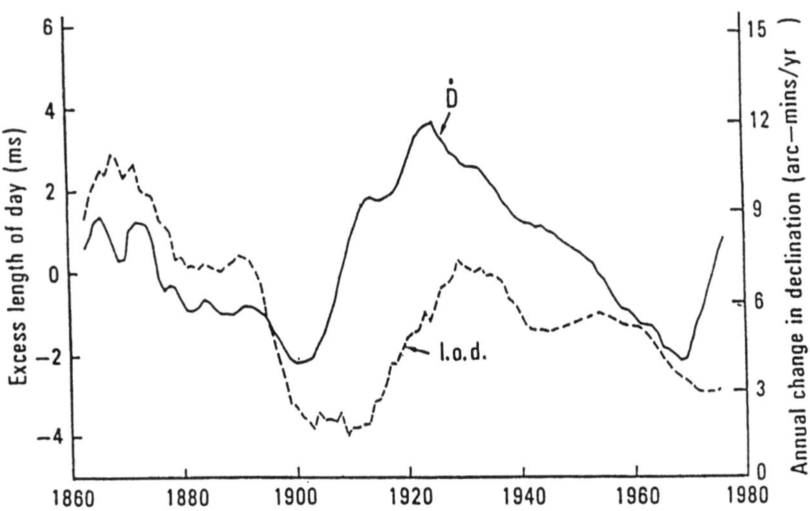

Figure 7. Comparison of 5 yr means of $\dot{D}$ at Paris with the variation in the length of the day ("l.o.d.") (After Le Mouël and Courtillot, 1981.)

times in the past, is shown. The column on the right indicates when the field had today's polarity (with geomagnetic and geographic axes almost antiparallel). These periods are indicated in black and the reversed polarity (in which $\mathbf{m}$ and $\mathbf{\Omega}$, the angular velocity of the Earth, are almost parallel) are displayed in white. The same

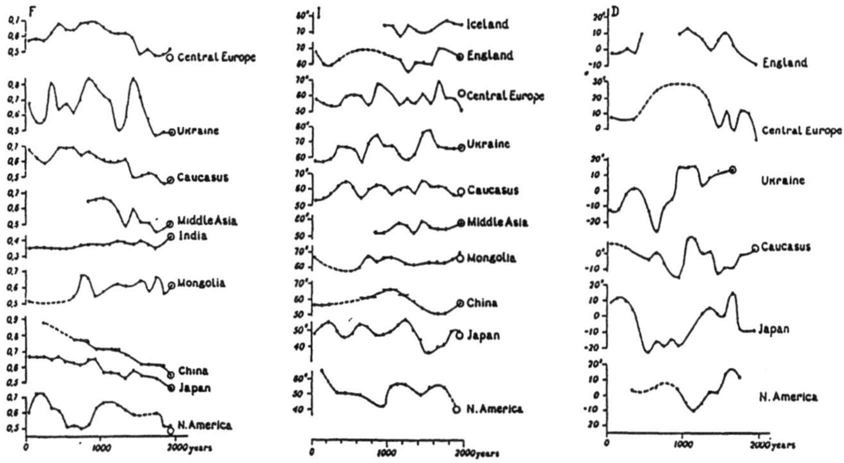

Figure 8. Variation on F, I and D over the past 2000 years at a number of sites. (After Bucha, 1983.)

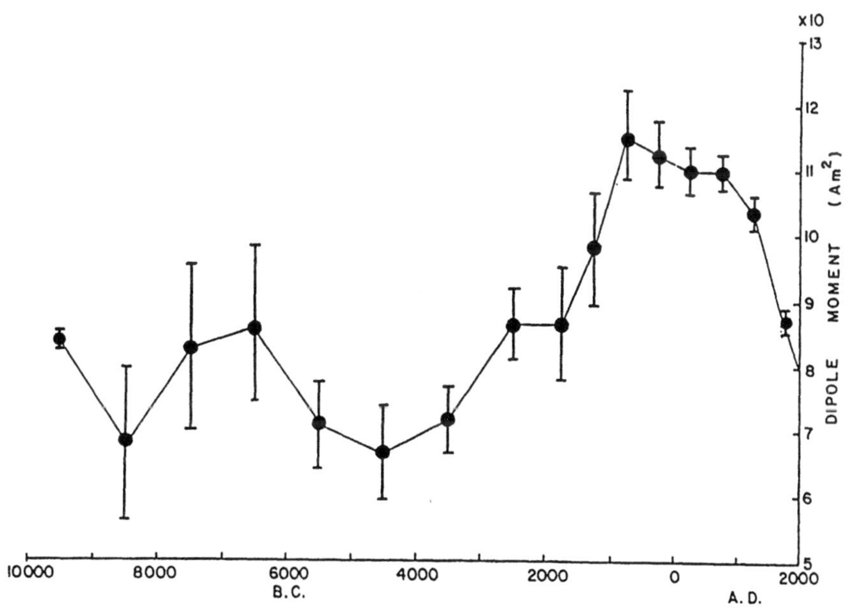

Figure 9. Variation of the dipole moment, m, over the last 12,000 years obtained from archeomagnetic data; error bars at 95%. (After McElhinny and Senanayake, 1982.)

code is used in Figure 11, the result of fitting a timescale to the variations of B during the part 5 My, using radioactive dating techniques. It seems from Figures 10 and 11 that there are "polarity epochs," during which the sign of m_z is stable, but also "polarity excursions" in which the polarity changes over a geologically short time and then returns to its stable polarity.

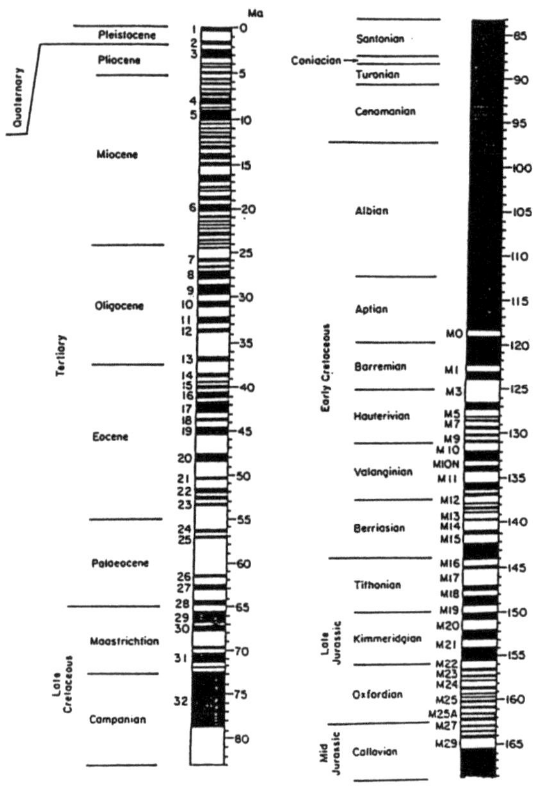

Figure 10. Summary of recent polarities of the geomagnetic field: black, normal polarity; white, reversed polarity. (After Merrill and McElhinny, 1983.)

The frequency of reversals has changed markedly over geological time, as is indicated in Figure 12. Currently we live during a period of frequent polarity change, but during the late Cretaceous and Permian the reversal frequency fell close to zero.

No very definite pattern has yet emerged of the way that $\mathbf{B}$ varies during a reversal. It is thought that the decrease in m_z as it passes through zero is not compensated by an increase in $|\mathbf{m}_E|$, i.e. the dipole $\mathbf{m}$ does not simply rotate about its axis, while preserving its strength. The reversal process appears to take a time of the order of the fundamental period (8,000 yrs) to complete, but there are several reports that the transition takes place more quickly, and in a series of jerks rather than through a smooth transition; see Figure 13.

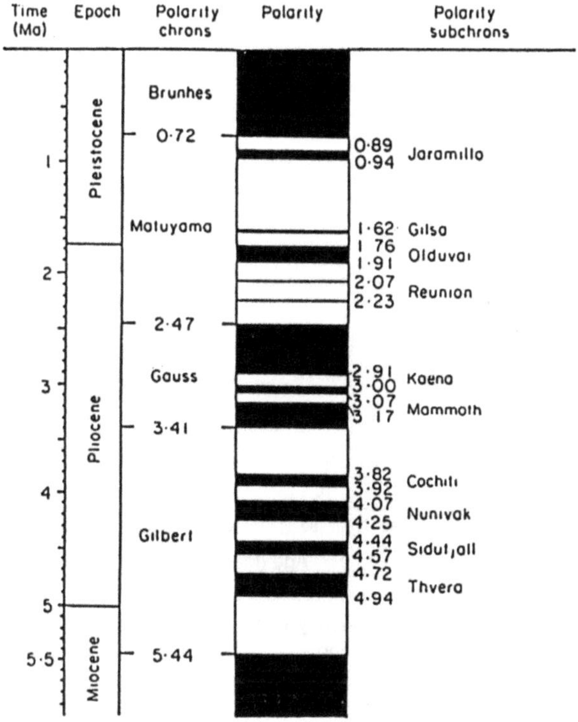

Figure 11. Reversals over the past 5 Myr. (After McElhinny, 1973.)

Finally there is paleomagnetic evidence that the field has existed throughout geological time, or at least until well into the pre-Cambrian, and its strength (although varying) appears to have been for much of that time within a factor of 2 of its strength today. Figure 14 gives an idea how little, over the past 350 Myr, the scatter of field directions has changed.

2. Some questions. Early "theories" were preposterous, e.g. that the directional property of the compass needle proved its affinity to the pole star. Even Gilbert's idea, "The Earth is a great magnet", began to seem very unlikely by 1635 when Gellibrand announced the existence of a secular variation; how could a permanent magnet change in such a way? By the turn of the eighteenth century, the lack of an explanation, and the lack of interest in finding an explanation, were beginning to be frustrating, as this near poetic declamation of Christopher Hansteen illustrates:

"The mathematicians of Europe since the times of Kepler and Newton have all turned their eyes to the heavens, to follow the planets in their finest motions and mutual perturbations: it is now to be wished that for a time they would turn their

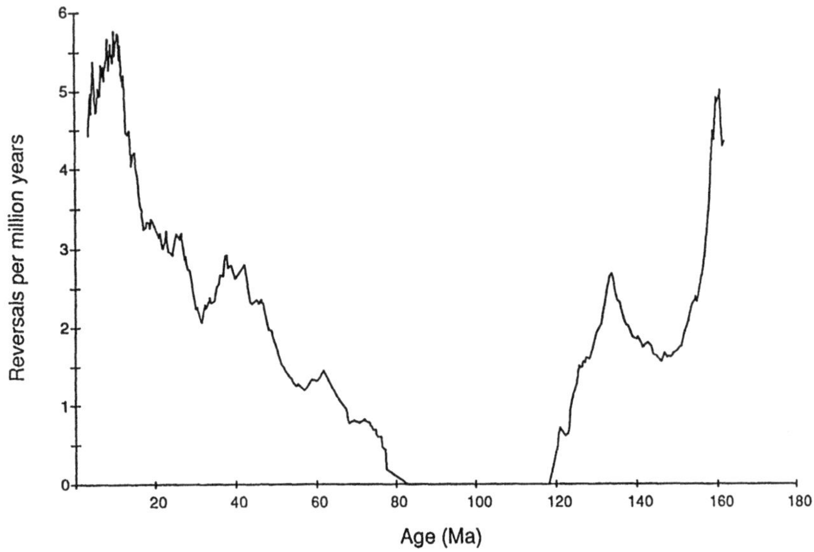

Figure 12. Reversal frequency over the past 165 Myr. (After Merrill and McFadden, 1990.)

gaze downwards towards the Earth's center where also there are marvels to be seen; the Earth speaks of its internal movements through the silent voice of the magnetic needle."

The final sentence appears today as a prophetic allusion to the fluid core of the Earth, whose existence was established seismologically only in 1906. Indeed today we know a great deal about the internal structure of the Earth. Moreover we have a far better understanding of fluid mechanics, electrodynamics and their union magnetohydrodynamics ("MHD"). We have, then, a far better chance than those Hansteen castigated of understanding the geomagnetic field. Also there is much more to explain than in Hansteen's day. We have a far more detailed picture of the Earth during recent, and particularly during post-satellite, times. Archeomagnetism has brought us data during historic and prehistoric times, and paleomagnetism has built up an impressive record of geomagnetism over geological time. What are the questions we should attempt to answer? Here are a few:

Q1. Why is the Earth magnetic? and

Q2. Why has its magnetic field persisted throughout geological time?

Q3. Why is it predominantly dipolar?

Q4. Why has its strength apparently varied so little, only by an order of magnitude over geological time? and

Q5. What determines that strength?

Q6. Why does the compass needle point North? (i.e. Why do geomagnetic and geographic axes so nearly coincide?) and

Q7. Why when averaged over geological time are they apparently coincident?

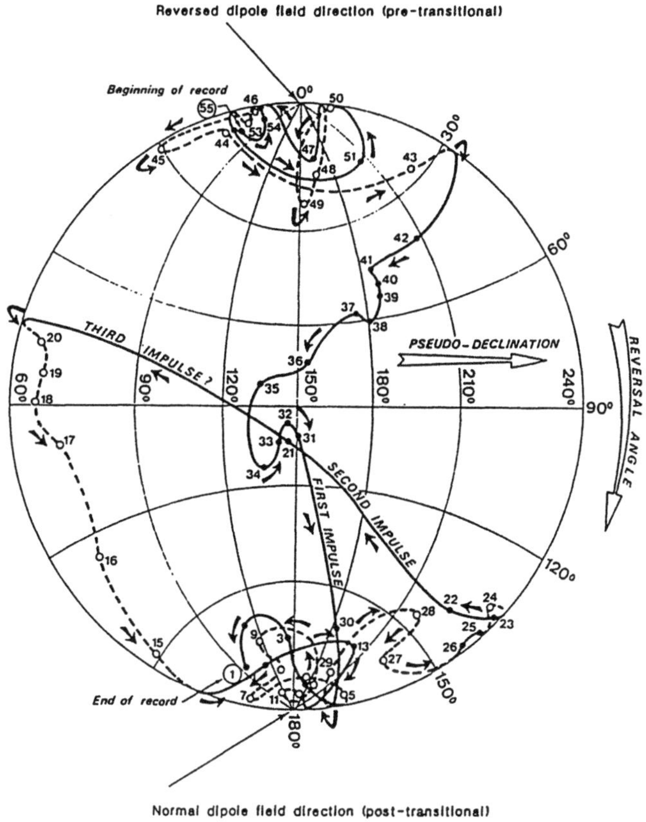

Figure 13. One report of the path of the pole during a reversal. (After Prévot *et al*, 1985.)

Q8. What causes the irregular polarity reversals of the field?

Q9. What happens to the geomagnetic field during a reversal? And

Q10. Why does the frequency of reversals vary so greatly over geological time?

Q11. What causes the slow "secular" changes in the field (with timescales of – say – 200 to 2000 yrs)?

Q12. What significance does the slow westward drift of geomagnetic field patterns have?

Q13. What causes the short period variations in field (with timescales of – say – 10 to 200 yrs)?

Q14. What causes the secular variation "impulses"?

Q15. Can the Earth be fitted into a general scheme into which the magnetism or other planets also fits?

Many theories of geomagnetism have been advanced but all except one have foundered because of their utter incapability of answering one or more of these questions. The idea of Blackett, that all bodies acquire magnetism when they rotate has, like the idea of a permanently magnetized Earth, had to be abandoned.

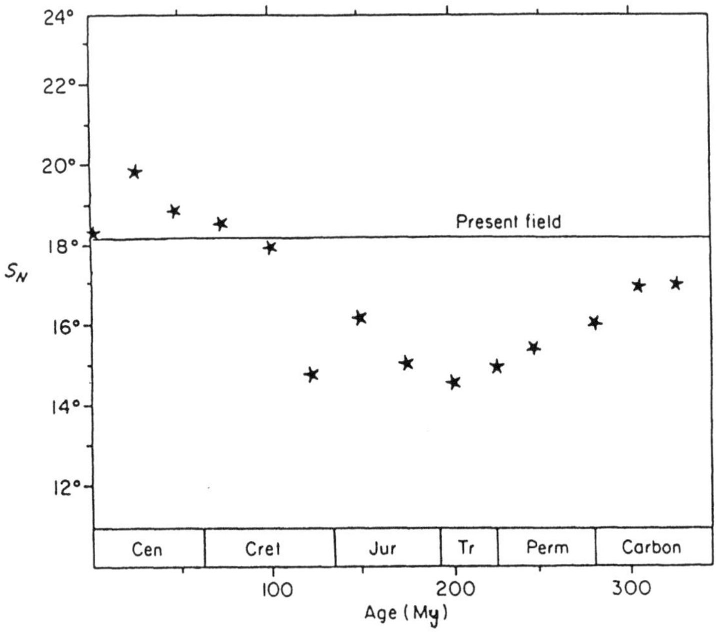

Figure 14. Angular variance of field directions over geological time. (After Irving and Pullaiah, 1976.) Geological epochs are indicated by the abbreviations at the bottom.

In the absence of intrinsic magnetization one must suppose that the geomagnetic field is produced by currents flowing within the Earth. But once one has currents, one also has ohmic dissipation. We are led therefore to ask further questions:

Q16. What kind of "battery" maintains the electric currents? and

Q17. Can that battery supply energy at the necessary rate over geological time?

The one theory that has survived is the dynamo theory of geomagnetism, which attributes the Earth's field to the currents induced when a fluid conductor (the Earth's core) moves in the presence of a magnetic field, namely the very same magnetic field those that the currents produce. In other words, the dynamo works on the same principle as, but in a more complicated way than, the simple homopolar dynamo illustrated in Figure 15. It is worth stressing that the process of self-excitation of fields is in no sense " hoisting oneself up by his own bootstraps". The currents, **J**, flowing between the axle and rim of the rotating disk have to cross the magnetic field that they produce. The resulting "Lorentz force", $\mathbf{J} \times \mathbf{B}$, is in the opposite direction to the velocity, **V**, of the disk. Thus work must be done to maintain the rotation of the disk against the Lorentz force. The device transforms kinetic energy into magnetic energy, and the electrical resistance of the circuitry converts the magnetic energy to heat.

In view of what follows, it may also be worth noting that the homopolar dynamo is asymmetric, it would not work if the sliding brush, S_1, were spread with axially symmetry around the rim of the disk, and were joined symmetrically to a complete encircling wire loop. If that loop encircled the rim in the opposite direction, or if the disk turned in the opposite sense, the device would destroy rather than generate field! Man creates this asymmetry deliberately, but God created an Earth that is predominantly axisymmetric. It is not at all obvious how the fluid core can compensate for the lack of structural asymmetry. For example, if one plunged the device shown in Figure 15 into an (axisymmetric!) bucket of fluid of the same conductivity, the electric currents would be short-circuited and the device would stop producing field, even though the disk were kept spinning! It is unclear how a "homogeneous dynamo" can work.

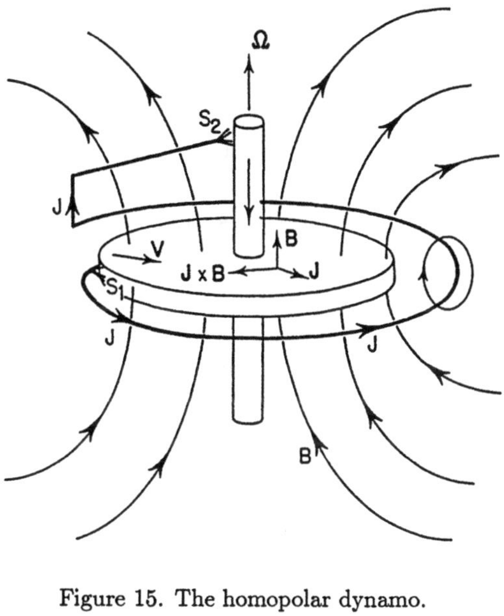

Figure 15. The homopolar dynamo.

It is the object of this paper to review the status of geomagnetic dynamo theory today. We will return in §6 to the seventeen questions above,[1] and we will find that dynamo theory has so far been unable to provide satisfactory answers to several of these. But this seems to be due not to its intrinsic incapability of providing those answers but to an insufficient commitment by the scientific community in finding those answers. The main geomagnetic field has no impact on the environment and

[1]Concerning Q3, we may note that, as r increases, the series (1.2) is increasingly dominated by the first dipolar terms $n = 1$. Even if all coefficients g and h in (1.2) contributed equally at the CMB, they would not do so at the Earth's surface. The fact that the centered dipole already dominates (1.2) at $r = b$ suggests that the sources of **B** are deep within the Earth, within the core in fact. There is something for dynamo theory to answer, however: "Should the asymmetric ($m \neq 0$) and the symmetric ($m = 0$) parts of **B** be of the same order of magnitude at the CMB?"; see §3. Observationally the former is perhaps of order 5 times smaller than the latter. Braginsky (1964a) used this as an argument in favor of the strong field dynamo discussed below. We have not had space in this review to describe his fascinating theory.

has no bearing on global change. To be sure, by learning how the geomagnetic field behaved in the remote past, geoscientists have transformed it into a valuable tool, which has enabled them to determine the history of the sea floor, and discover how the continents have drifted. But one does not need a *theory* of geomagnetism to do that! Dynamo theory rests on well-understood nineteenth century physics, and is therefore boring to today's physicists. It is perhaps not surprising that advancing geodynamo theory is the esoteric occupation of a few applied mathematicians and (increasingly) computational fluid dynamicists. The band, though dedicated, is small. Progress is slow. Much remains to be done.

3. Kinematic theory. It is helpful to think of the Earth in the zero approximation as a well-mixed body in hydrostatic equilibrium (Figure 16): in other words

$$0 = -\nabla p_0 + \rho_0 \mathbf{g}_0, \tag{3.1}$$

$$\mathbf{g}_0 = \nabla \left[\Phi_0 + \frac{1}{2}(\mathbf{\Omega} \times \mathbf{r})^2 \right], \tag{3.2}$$

$$\nabla^2 \Phi_0 = -4\pi G \rho_0, \tag{3.3}$$

$$S_0 = \text{constant}. \tag{3.4}$$

Here ρ is density, p is pressure, S is specific entropy, Φ is gravitational potential, G is the universal constant of gravitation, $\mathbf{g}$ is effective gravity; the suffix 0 distinguishes values in this hydrostatic state. The constants in (3.4) are not meant to be identical in the inner core, outer core and mantle; indeed, the constancy of S over the solid inner core and over parts of the mantle is far from certain. We shall however focus on the fluid outer core, which is well mixed by convection currents (except possibly in boundary layers).

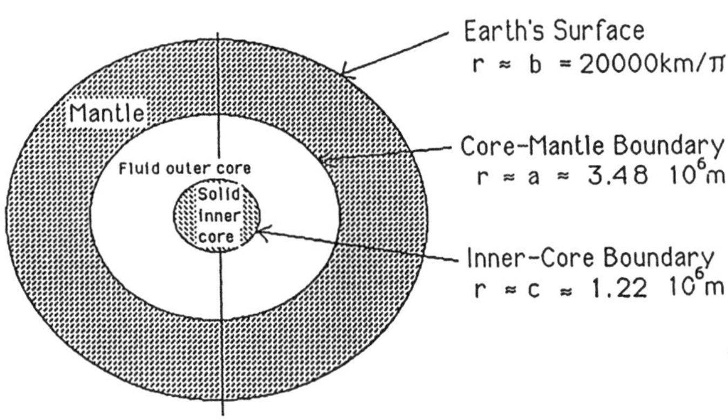

Figure 16. General structure of the Earth (ellipticity greatly exaggerated!)

All hydrostatic and thermodynamic departures of the Earth from the axisymmetry defined by (3.1) – (3.4) are now considered to be small perturbations. These include the departures $\mathbf{g}', \rho', p', S'$ and T' in $\mathbf{g}, \rho, p, S$ and the temperature T from $\mathbf{g}_0, \rho_0, p_0, S_0$ and T_0. We shall set $\mathbf{g}' = 0$, preferring to exclude the inessential complications of a small effect and to regard $\mathbf{g}_0$ as given. The linearization of the thermodynamics about the reference state (3.1) – (3.4) is usually termed "the anelastic approximation". We shall however assume more: we shall adopt a "Boussinesq" model. This seems reasonable for the Earth's core; ρ_0 increases by only about 10% from the CMB (core-mantle boundary) to the ICB (inner core boundary). The urge to include it in simulations of the geodynamics may be irresistible to some, but by involving it in geomagnetic theory today adds complications without enlightenment, darkness without light.

As a further simplification, we shall assume that the CMB is the sphere $r = a$. This seems reasonable from the smallness of $\Omega^2 a/g$; the oblateness ($\sim 1/340$) of the CMB shown in Figure 16, is greatly exaggerated! After omitting the final term in (3.2), we obtain a radial $\mathbf{g}$ in other words, $\mathbf{g} = -g\mathbf{1}_r$. The inner core does not occupy much of the volume of the whole core and, again for simplicity, we shall risk the wrath of geophysists by supposing fluid fills the entire core. In short, our starting point will be the simplest possible; see Figure 17.

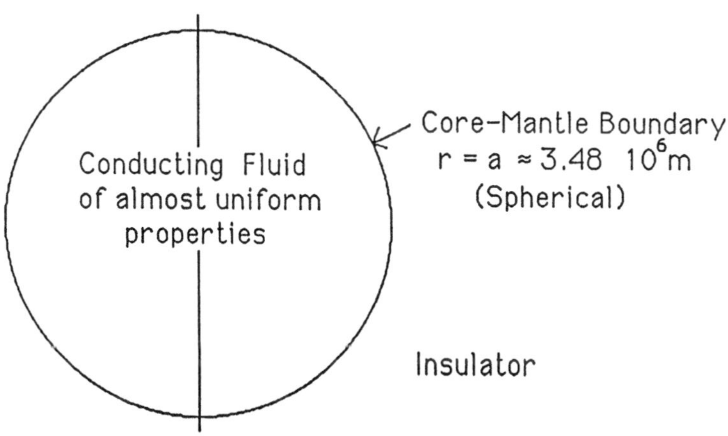

Figure 17. Theoretician's idealization of the core (uniform density, spherical CMB, electrically insulating mantle, no inner core).

Dynamo theory is called "kinematic" when it focusses entirely on the electrodynamics, and supposes that the motions are given. It is called "dynamic", or "magnetohydrodynamic", or "fully self-consistent" when the fluid dynamics is considered along with the electrodynamics. In this section we study only kinematics.

This rests on the pre-Maxwell equations:

$$\nabla \cdot \mathbf{B} = 0, \tag{3.5}$$

$$\nabla \times \mathbf{B} = \mu_0 \boldsymbol{J}, \tag{3.6}$$

$$\nabla \times \mathbf{E} = -\frac{\partial \mathbf{B}}{\partial t}, \tag{3.7}$$

and Ohm's law for a moving conductor:

$$\mathbf{J} = \sigma(\mathbf{E} + \mathbf{V} \times \mathbf{B}). \tag{3.8}$$

Here **J** is electric current density, **E** is electric field, μ_0 is the permeability, and σ is the electrical conductivity, which we usually replace by the magnetic diffusivity $\eta = 1/\mu\sigma$. From (3.5) – (3.8) follows the equation of electromagnetic induction

$$\frac{\partial \mathbf{B}}{\partial t} = \nabla \times (\mathbf{V} \times \mathbf{B}) + \eta \nabla^2 \mathbf{B}. \tag{3.9}$$

Labeling the terms in (3.9) in order as (a), (b) and (c), we may quickly compare their relative sizes:

Case 1: (a) versus (c), as when $\mathbf{V} = 0$.

We have

$$\frac{(a)}{(c)} \sim \frac{B/T}{\eta B/L^2} \sim \frac{L^2/\eta}{T}.$$

We here locate the characteristic "electromagnetic diffusion time" of a field of scale L:

$$\tau_\eta = L^2/\eta\,. \tag{3.10}$$

Since (3.9) is a diffusion equation when $\mathbf{V} = 0$ we may correctly surmise that, when there are no external sources of **B**, τ_η gives the lifetime of a field in a stationary conductor. More precisely the e-folding time of the current system in a sphere of radius L is τ_η/π^2. In the case of the Earth's core, $\eta = 3\text{m}^2/\text{s}$ gives $\tau_\eta/\pi^2 \sim 14{,}000$yr. Since the Earth's field has existed for more than 10^8 yr, the need to answer Q16 and Q17 is urgent.

Case 2: (a) versus (b), as when $\eta = 0$.

We have

$$\frac{(a)}{(b)} \sim \frac{B/T}{UB/L} \sim \frac{L/U}{T}.$$

We see here a "kinematic timescale",

$$\tau_K = L/U, \tag{3.11}$$

where U is a characteristic flow speed. Mental picture building is greatly assisted by Alfvén's "frozen flux theorem," which states that magnetic flux tubes are material

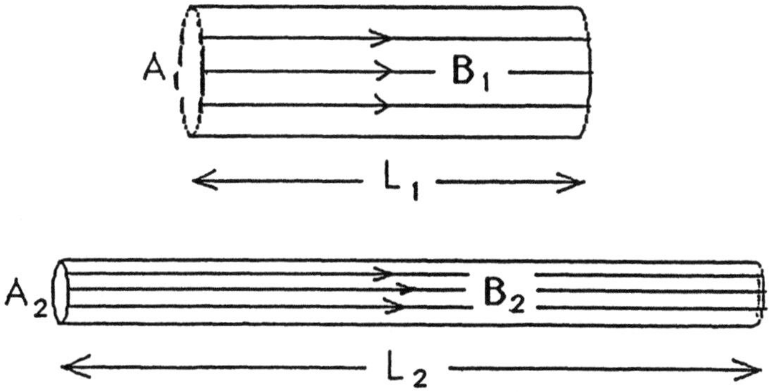

Figure 18. The stretching of a magnetic flux tube.

volumes, i.e. the fluid within them carries the magnetic lines of force with it in its motion, just as though those lines were frozen to the medium.

This theorem is no surprise to the experts on vorticity,

$$\boldsymbol{\omega} = \nabla \times \mathbf{V}, \tag{3.12}$$

present at this workshop. In an inviscid incompressible (or barotropic) fluid $\boldsymbol{\omega}$ obeys

$$\frac{\partial \omega}{\partial t} = \nabla \times (\mathbf{V} \times \boldsymbol{\omega}), \tag{3.13}$$

which is the form taken by (3.9) when $\eta = 0$

$$\frac{\partial \mathbf{B}}{\partial t} = \nabla \times (\mathbf{V} \times \mathbf{B}). \tag{3.14}$$

By analogy with Kelvin's theorem, **B** is frozen to the fluid. Actually (3.14) is simpler conceptually than (3.13). In (3.14), **B** is a "passive" vector but, in (3.13), $\boldsymbol{\omega}$ has to maintain consistency with (3.12), i.e. (3.13) is nonlinear.

Although energetics are more appropriately left until §4, we may note here how motion is able to create magnetic energy. Consider a segment of length L of a flux tube of cross-sectional area A, in a field of strength B_1; see Figure 18. Fluid motions can stretch that out into a segment of length L_2 of a flux tube of cross-sectional area A_2 in a field of strength B_2. Since by (3.14) the total flux across each cross-section is conserved,

$$A_1 B_1 = A_2 B_2, \tag{3.15}$$

so that the magnetic energy density increases from $B_1^2/2\mu_0$ to

$$\frac{B_2^2}{2\mu_0} = \left(\frac{A_1}{A_2}\right)^2 \frac{B_1^2}{2\mu_0} > \frac{B_1^2}{2\mu_0}. \tag{3.16}$$

If the fluid is incompressible, this new energy density fills the same volume of space, i.e. the net magnetic energy of the tube has increased through the stretching. This energy has been extracted from the kinetic energy of motion.

Case 3: (b) versus (c), as in a real fluid.

We have

$$\frac{(b)}{(c)} \sim \frac{UB/L}{\eta B/L^2} = R \equiv \frac{UL}{\eta} . \tag{3.17}$$

The ratio of induction to diffusion is measured by the "magnetic Reynolds number", R. Obviously there is no dynamo problem if $R = \infty$; Case 2. By Alfvén's theorem the fluid retains the magnetic field for ever, although how the field entered the conductor in the first place is open to question! When, as in the case of Earth, the field is old compared with τ_η, Case 2 is irrelevant. We may not assume $\eta = 0$; we may not consider sheet currents. At the other extreme, if $\mathbf{V} = 0$, i.e., if $R = 0$, Case 1 applies and the field disappears in on time of order τ_η. By continuity, it will also disappear when R is "small". Hopefully, when R is sufficiently large, the creation of magnetic energy by (b) can more than offset its destruction by (c), and the dynamo will be "self-excited". In other words: *a necessary condition for geodynamo action* is

$$R = \mathrm{O}(1). \tag{3.18}$$

If $\eta = 3 \ \mathrm{m}^2/\mathrm{s}$, $b = 3.48 \ 10^6 \mathrm{m}$ and U is a typical speed of westward drift (as extrapolated back to the core surface), i.e. $U \sim 10^{-4}$ m/s, then $R = UL/\eta \approx 100$. Apparently (3.18) is well satisfied in the core.

In dimensionless variables, (3.5) and (3.9) may be written

$$\frac{\partial \mathbf{B}}{\partial t} = R\nabla \times (\mathbf{V} \times \mathbf{B}) + \nabla^2 \mathbf{B}, \qquad \nabla \cdot \mathbf{B} = 0. \tag{3.19, 3.20}$$

These hold in $r < 1$. In the surrounding insulator, $r > 1$, we have $\mathbf{J} = 0$, so that (3.6) implies (1.3), and (3.5) then gives (1.4):

$$\mathbf{B} = -\nabla V, \qquad \nabla^2 V = 0. \tag{3.21,3.22}$$

The dynamo must be self-excited; there must be no external sources of $\mathbf{B}$ even at $r = \infty$. Thus the expansion (1.2) of V contains no terms that increase with r, and by (1.2) we see that

$$V = \mathrm{O}(r^{-2}), \qquad r \to \infty. \tag{3.23}$$

Finally, since there can be no surface currents when $R \neq \infty$ ($\eta \neq 0$), all components of $\mathbf{B}$ must be continuous as the CMB:

$$[\mathbf{B}] = 0 \quad \text{on} \quad r = 1. \tag{3.24}$$

Equations (3.19) – (3.24) summarize the kinematic dynamo problem which is a *linear* system for the unknown $\mathbf{B}$, with the coefficient $R\mathbf{V}$ assumed given. If $\mathbf{V}$ is independent of t, this is a non-self-adjoint eigenvalue problem for the growthrate, s, of the field, where

$$\mathbf{B} \propto e^{st}. \tag{3.25}$$

We may regard $\mathbf{V}$ as fixed, R as variable, and $s = s(R)$; in general, $s(-R) \neq s(R)$ as the homopolar dynamo shows. Without less of generally, we may suppose that $R > 0$. (If not, reverse $\mathbf{V}$!).

The spectrum of s is discrete with limit point at $s = -\infty$. As for any linear stability problem, interest centers on when and how the dominant eigenvalue, s_1, the one (or ones) having largest real parts, acquires a positive real part as R is increased from zero. In some cases $\Re(s_1) < 0$ for all R, but in others a marginal R_c can be found such that $\Re(s_1)$ passes from negative to positive values as R increases through R_c. Of course if $R > R_c$, $\mathbf{B}$ will increase exponentially without bound. In reality the kinematic dynamo problem concerns infinitesimal B and, if ever B becomes large enough to modify $\mathbf{V}$ through the Lorentz force, we should consider the dynamic problem (§4).

It can be proved from (3.19)–(3.24) that $\Re(s_1)$ is bounded above by an $O(R)$ supremum. In dimensional terms, the growth rate cannot exceed $O(\tau_K^{-1})$. Recently, much theoretical interest has been focussed on "the fast dynamo", i.e. the search for flows in which $\Re(s_1) = O(\tau_k^{-1})$, independent of η! One approach to the fast dynamo problem has been described by Bayly during this workshop.

Much of the difficulty in both the electrodynamic and hydrodynamic problems centers on the axisymmetric parts, $\overline{\mathbf{V}}$ and $\overline{\mathbf{B}}$, of $\mathbf{V}$ and $\mathbf{B}$. These are sketched in Figure 19, where they are divided into the sum of zonal and meridional parts $\overline{\mathbf{B}}_T, \overline{\mathbf{B}}_P, \overline{\mathbf{V}}_T, \overline{\mathbf{V}}_P$. The suffices T and P here stand for 'toroidal' and 'poloidal'; e.g.

$$\mathbf{B} = \nabla\times(T\mathbf{r}) + \nabla\times\nabla\times(P\mathbf{r}). \tag{3.26}$$

All axisymmetric toroidal fields are zonal; all axisymmetric poloidal fields are meridional.

With the help of Alfvén's theorem, it is easy to understand how $\overline{\mathbf{V}}_T$ can create $\overline{\mathbf{B}}_T$ from $\overline{\mathbf{B}}_P$ by a process sometimes called "the omega effect" for no better reason than somebody's accidental choice of ω for the variable $\overline{V}_\phi/s$, where s is distance from the symmetry axis. The process depends on the shear in V_ϕ, i.e. on $\nabla\boldsymbol{\omega}$ being nonzero. [Constant ω means solid body rotation, i.e. to a state of rest in the co-rotating frame. Since the electrodynamic problem is unaffected by this rotation, obviously all $\mathbf{B}$ must decay; see Case (a) above.] Figure 20 (a) illustrates a zonal $\overline{\mathbf{V}}$ and meridional $\mathbf{B}$; in Figure 20 (b) the former has stretched the field lines of the latter along lines of latitude, and out of the meridian planes; when the original $\mathbf{B}_P$ is subtracted from this stretched field, a zonal $\overline{\mathbf{B}}_T$ remains, as shown in Figure 20 (c).

If we modify (3.17) slightly by writing

$$R_\omega = \tilde{\omega}L^2/\eta, \tag{3.27}$$

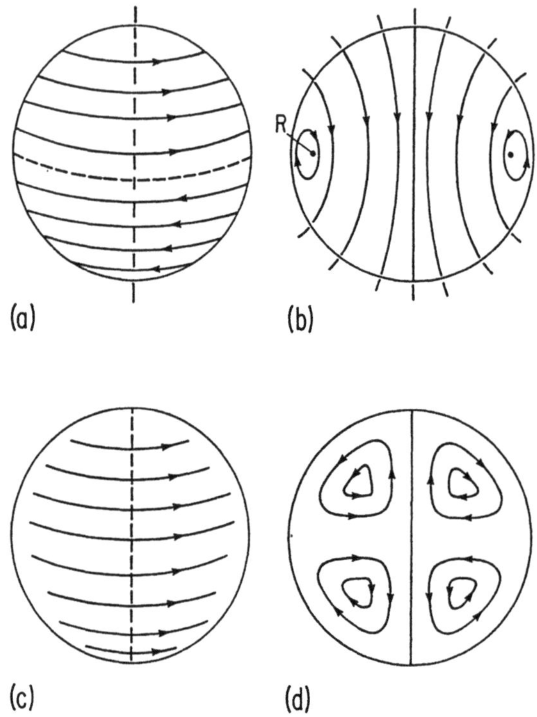

Figure 19. The axisymmetric parts of the field and the motion: (a) $\overline{\mathbf{B}}_T$, (b) $\overline{\mathbf{B}}_P$, (c) $\overline{\mathbf{V}}_T$, (d) $\overline{\mathbf{V}}_P$.

where $\tilde{\omega}$ is a typical value of ω, one can show that

(3.28) $$\overline{B}_T = O(R_\omega)\overline{B}_P.$$

If $R_\omega \approx 100$ (see above) the resulting toroidal field is about 100 times the poloidal field that gave it birth. But unless $\overline{\mathbf{B}}_P$ is itself replenished, it will die away, after which $\overline{\mathbf{B}}_T$ will follow suit. Before we consider how $\overline{\mathbf{B}}_P$ is maintained we should note that very probably

(3.29) $$\overline{V}_T >> \overline{V}_P, \quad \overline{B}_T >> \overline{B}_P,$$

in the core, the former because of the ease with which zonal motion, and especially geostrophic flow, are created in a rotating system (see §4). In a first approximation we have

(3.30) $$\mathbf{V} \approx V_\phi \mathbf{1}_\phi, \qquad \mathbf{B} = B_\phi \mathbf{1}_\phi;$$

see Figure 21.

Although $\overline{\mathbf{V}}_P$ induces a zonal emf from $\mathbf{B}_P$, this emf vanishes on the ring of O-type neutral points indicated by the Rs in Figure 19 (b). Thus $\overline{\mathbf{B}}_P$ cannot be

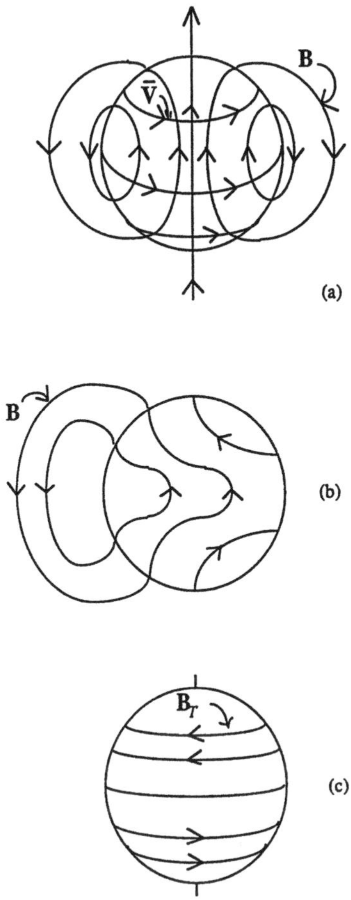

Figure 20. The ω-effect: (a) initial $\overline{\mathbf{V}}_T$ and $\overline{\mathbf{B}}_P$, (b) creation of $\overline{\mathbf{B}}_T$ from $\overline{\mathbf{B}}_P$ by $\overline{\mathbf{V}}_T$, (c) the difference between (b) and (a) — a zonal field has been created.

maintained in the neighborhood of that ring, and $\overline{\mathbf{B}}_P$ itself must therefore disappear in a time of order τ_η, since every meridional magnetic field created by currents flowing in $r \leq 1$ must possess at least one such ring within the fluid or on its surface. This is the basic idea behind Cowling's theorem, which states that axisymmetric $\mathbf{B}$ cannot be maintained by dynamo action.

To maintain $\overline{\mathbf{B}}_P$ the symmetry must be broken, i.e. the process must depend on the asymmetric parts, $\mathbf{V}'$ and $\mathbf{B}'$, of $\mathbf{V}$ and $\mathbf{B}$; i.e. by

$$\mathbf{V}' = \mathbf{V} - \overline{\mathbf{V}}, \qquad \mathbf{B}' = \mathbf{B} - \overline{\mathbf{B}}, \tag{3.31}$$

through the axisymmetric emf they create, namely

$$\mathcal{E} = \overline{\mathbf{V}' \times \mathbf{B}'}. \tag{3.32}$$

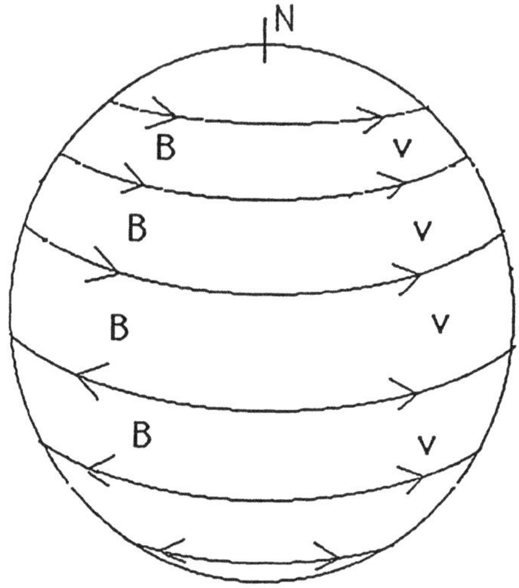

Figure 21. The "basic" MHD state of the core.

In the case of the Earth, it is thought that*

$$\boldsymbol{\mathcal{E}} \approx \alpha \overline{\mathbf{B}}_{\phi}, \tag{3.33}$$

and the process is called "the alpha effect" for no better reason than the accidental choice of α for the constant of proportionality in (3.33).

The existence of α is generally due to the helicity,

$$H = \mathbf{u} \cdot \boldsymbol{\omega}, \tag{3.34}$$

of the flow, and is most easily visualized by an example provided by Steenbeck, Krause and Rädler.

They imagined an eddy that lasted only a short time but which possesses helicity. Such a flow can be pictured as a wrench: a tug and a twist; see top of Figure 22. Figure 22 (a) shows a section of a $\overline{\mathbf{B}}_T$ flux tube. In Figure 22 (b) it has been tugged upwards into an Ω-shape by the eddy. The twist now turns that Ω out of the plane of the paper, as illustrated in Figure 22 (c). [Of course, in reality the effects of tug and twist happen simultaneously.] Diffusion reconnects the field lines in the region of high field gradient near the feet of the Ω, and a ring of flux separates as indicated in Figure 22 (d). The coalescence of many such rings, again through field line reconnections, creates a meridional field. Alternatively we may note that a ring is associated by (3.6) with a current antiparallel to $\overline{\mathbf{B}}_T$, and the current from all such rings has a nonzero mean component, $\overline{J}_{\phi}$, that by (3.6) generated $\overline{\mathbf{B}}_P$.

Steenbeck, Krause and Rädler concocted their example in the context of the solar dynamo, which operates in the highly turbulent solar convection zone. Although core flow is unlikely to be laminar (see §5), there is no evidence that it is

*This a slight oversimplification; see Braginsky (1964a) and his "effective" fields.

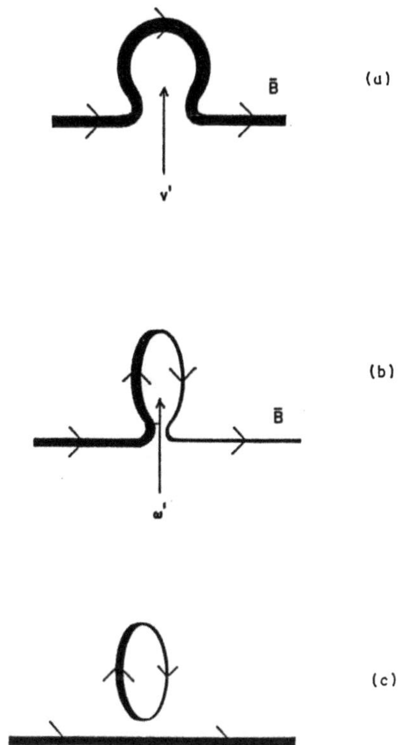

Figure 22. The α-effect in the Steenbeck, Krause, Rädler visualization. The effect of an eddy with positive helicity $(\overline{\mathbf{v}' \cdot \boldsymbol{\omega}'} > 0)$ on a field line of $\overline{\mathbf{B}}$ is thought of in three stages: (a) the $\mathbf{v}'$ creates an Ω-shaped dent in the field line, (b) the $\boldsymbol{\omega}$ twists that Ω out of its plane, (c) the ω detaches from the original line of force.

in a state of violent turbulence. It is improbable that the example of Figure 22 is relevant to the Earth, and currently favored is the idea that H and α are intrinsic to large-scale waves that 'ride' on the basic state (3.30), traveling along lines of latitude. But clearly a new set of questions arise:

Q18. What physical processes create asymmetry in the MHD state of the core?

Q19. What are the characteristic length and time scales of these asymmetries?

Q20. Can these be tied into the observed asymmetries of the geomagnetic field?

Q21. Do the asymmetries possess the required H and α?

The effectiveness of α is measured by an ω-effect Reynolds number

(3.35) $$R_\alpha = \tilde{\alpha} L / \eta,$$

where $\tilde{\alpha}$ is a typical value of α. One can show that

(3.36) $$\overline{B}_P = \mathrm{O}(R_\alpha) \overline{B}_T$$

Equations (3.28) and (3.36) close the dynamo's regeneration loop (see Figure 23); the axisymmetric field will be maintained if

(3.37) $$D = O(1),$$

where

(3.38) $$D = R_\alpha R_\omega$$

is the "dynamo number". Since (see above) $R_\omega \gg 1$, (3.37) and (3.38) show that $R_\alpha \ll 1$. Clearly (and symmetrically)

(3.39) $$\overline{B}_T = O(R_\omega/R_\alpha)^{1/2}\overline{B}_P \gg \overline{B}_P.$$

Because of (3.39), this "$\alpha\omega$-dynamo" is also often called the "strong field dynamo." Many such dynamos have been constructed numerically. [Indeed a few fully three-dimensional kinematic dynamos have been integrated successfully, but our object here has been to focus on the point of difficulty; the maintenance of the axisymmetric part of **B**.]

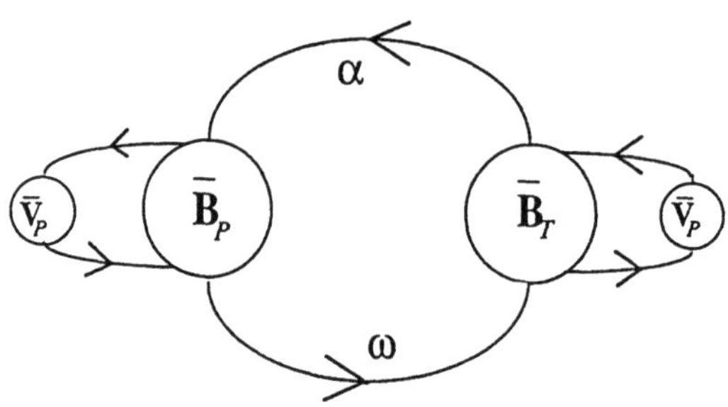

Figure 23. The interactions taking place in an $\alpha\omega$-dynamo.

The generic $\alpha\omega$-dynamo is oscillatory, i.e.

$$\Im(s_1) \neq 0 \quad \text{at} \quad R = R_c; \quad \text{in fact} \quad s_1 = O(\tau_\eta^{-1})$$

in dimensional units. Magnetic activity progresses from the equator to the poles or in the reverse directions, and the latter behavior so closely resembles that of the Sun that solar dynamo theory usually centers on $\alpha\omega$-models of the solar convection zone. The Sun has no free surface, but the existence of a CMB in the Earth to which the core is coupled gives rise to meridional circulations. It is interesting and possibly significant to observe that, when a suitable meridional flow is present, represented by the two ears of Figure 23, the oscillatory character of the $\alpha\omega$-dynamo is generally lost, i.e. $\Im(s_1) = 0$ at $R = R_c$. By "suitable" we mean "sufficiently large" [i.e. $O(\eta/b)$] and "in the right sense" of circulation. By such a choice, the critical Reynolds number, R_c, can also be substantially reduced [e.g. Roberts,

1972]. Meridional circulation can help the $\alpha\omega$-dynamo to function by transporting flux generated by α action to regions where the ω-process is especially effective, and *vice versa.*

We should remark that when $R_\omega \ll R_\alpha$, "weak field" dynamos exist in which (3.28) is replaced by

$$\overline{B}_T = O(\overline{B}_P). \tag{3.40}$$

The α-process depicted in Figure 22 then acts with equal efficacy on $\overline{\mathbf{B}}_P$, so that (3.33) is replaced by

$$\mathcal{E} = \alpha\overline{\mathbf{B}}, \tag{3.41}$$

and the α-effect, rather than the ω-effect, creates $\overline{\mathbf{B}}_T$ from $\overline{\mathbf{B}}_P$. The regeneration loop is closed as in Figure 23, but with ω replaced by α. The dual function of α has led to the alternative name "α^2-dynamos" for this type of mechanism.

4. MHD theory. The geodynamo dissipates electromagnetic energy at the rate J^2/σ per unit volume of core. In the strong field dynamo in which $\overline{\mathbf{B}}_T$ is the dominant part of $\mathbf{B}$ [see (3.30)], $\mathbf{J} \approx \overline{\mathbf{J}}_P = O(\overline{B}_T/\mu_0 a)$. The net ohmic dissipation is then

$$Q_\eta \equiv \int_{\text{core}} \frac{J^2}{\sigma} d^3x \approx \frac{4\pi a^3}{3} \cdot \frac{1}{\sigma}\left(\frac{\overline{B}_T}{\mu_0 a}\right)^2 = \frac{4\pi a}{3\mu_0}\eta \overline{B}_T^2 . \tag{4.1}$$

By extrapolating $\mathbf{B}_P$ from the Earth's surface to the CMB, we surmise that $\overline{B}_P \approx 5\ 10^{-4}$T in the core so that, by (3.28), $\overline{B}_T \approx 5\ 10^{-2}$T. Then (4.1) gives $Q_\eta \approx 10^{11}$W. If we use (4.1) for the weak field dynamo, using (3.40) for $\overline{B}_T \approx \overline{B}_P \approx 5\ 10^{-4}$T, we obtain $Q_\eta \approx 10^7$W. The question will arise below, "How seriously is 10^7W an underestimate for Q_η in a weak field model?" Meanwhile, we simply observe that $10^7 - 10^{11}$W of energy must be replaced every second by some driving force.

The favored power supply is buoyancy, but compositional rather than thermal. The Earth is becoming more centrally condensed as it cools. The outer core may be pictured as an alloy of iron and some lighter material (sulfur?), whereas the inner core is more nearly pure iron. The fluid core is cooled at the top but freezes from the bottom. (This is because of the relative slopes of the adiabat and the melting curve.) Phase equilibrium obtains at the ICB. An iron-rich fraction of the fluid alloy is deposited on the ICB as the Earth cools, and the light sulfur-rich residue rises to stir the core into motion. Assuming that the inner core is the result of this process operating since the birth of the Earth, and comparing the gravitational energy of a then homogeneous fluid core with the liquid-solid structure existing today, we can show that the energetic needs of the geodynamo can be met now and throughout geological time.

Although the theory of compositional convection is not dissimilar from that of thermal convection, it is less familiar. In what follows we shall suppose that the convection is thermally driven, and Boussinesq [see §3]. The rate at which buoyancy forces do work is then $-\gamma T\mathbf{g}\cdot\mathbf{V}$ per unit volume, where γ is the coefficient of volume expansion; T is most conveniently measured from the reference temperature of the Boussinesq state, which is then $T=0$. The integral

$$Q_g = \int_{\text{core}} g\gamma T V_r d^3x, \tag{4.2}$$

gives the input of buoyant power. It is hard to estimate Q_g but it is clear that a necessary condition for geodynamo action is

$$Q_g > Q_\eta. \tag{4.3}$$

It is easy to write down the equations governing the MHD dynamo in the Boussinesq approximation:

$$D_t\mathbf{B} = \mathbf{B}\cdot\nabla\mathbf{V} + \eta\nabla^2\mathbf{B}, \qquad \nabla\cdot\mathbf{B} = 0, \tag{4.4, 4.5}$$

$$D_t\mathbf{V} + 2\mathbf{\Omega}\times\mathbf{V} = -\nabla\Pi + \frac{1}{\rho_0}\mathbf{J}\times\mathbf{B} - \mathbf{g}\gamma T + \nu\nabla^2\mathbf{V}, \tag{4.6}$$

$$\nabla\cdot\mathbf{V} = 0, \qquad D_tT = \kappa\nabla^2T + S, \tag{4.7, 4.8}$$

together with (3.6), i.e. $\mathbf{J} = \nabla\times\mathbf{B}/\mu_0$, and appropriate boundary conditions. The kinematic dynamo problem is embedded in the MHD theory through (4.4) and (4.5); (4.6) is the Boussinesq form of the equation of motion with ν the kinematic viscosity; $\Pi = p/p_0$ where p contains the centrifugal force; (4.8) is the heat conduction equation with a source S, and a thermal diffusivity of κ; $D_t = \partial_t + \mathbf{V}\cdot\nabla$ is the motional derivative.

The system (4.4) – (4.8) poses a fascinating nonlinear stability problem. From the spherical symmetry assumed in §3, there exists a conduction solution in which $\mathbf{B}\equiv 0$ and $\mathbf{V}\equiv 0$ but $T\neq 0$. A *nonmagnetic state* in fact exists under *all* conditions, although $\mathbf{V}$ will be nonzero only if the Rayleigh number,

$$\widetilde{Ra} = \frac{g\beta\gamma L^4}{\nu\kappa}, \tag{4.9}$$

(where β is a characteristic temperature gradient and L is a typical length scale of the system, e.g. a) exceeds a critical value $\widetilde{Ra}_c$. If $\widetilde{Ra}$ is increased, the vigor of the convective motions in this non-magnetic state will increase, and a stage may be reached ($\widetilde{Ra} = \widetilde{Ra}_m$) at which the magnetic Reynolds number, R, attains the critical value, R_c, required for dynamo action. The non-magnetic solution is then linearly unstable; dynamo action is initiated.

If $\widetilde{Ra}$ is increased beyond $\widetilde{Ra}_m$, than R will exceed R_c and $\mathbf{B}$ will grow exponentially without bound according to kinematic theory. One might intuitively expect that instead $\mathbf{B}$ will grow until $\mathbf{J}\times\mathbf{B}$ in (4.5) reduces $\mathbf{V}$ by the amount necessary to make R for that $\mathbf{V}$ again marginal. This is tantamount to saying that the bifurcation at $\widetilde{Ra} = \widetilde{Ra}_m$ is supercritical, as in Figure 24(a).

This expectation is not necessarily justified: perhaps as $\mathbf{B}$ grows, $\mathbf{J}\times\mathbf{B}$ in (4.6) will *increase* $\mathbf{V}$! Perhaps the bifurcation at $\widetilde{Ra} = \widetilde{Ra}_m$ is subcritical, as in Figure

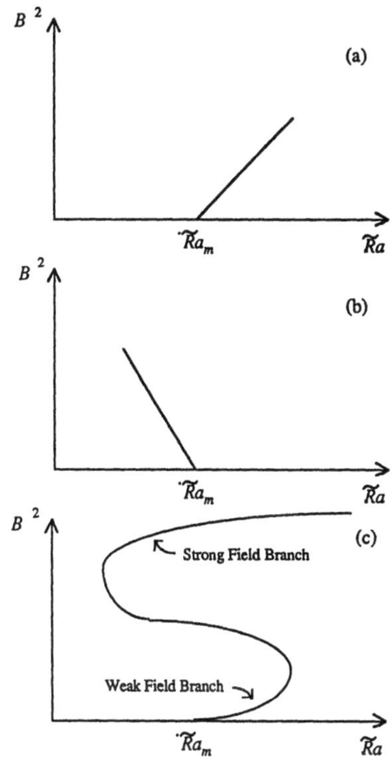

Figure 24. Types of bifurcation: (a) supercritical bifurcation, (b) subcritical bifurcation, (c) a more complicated bifurcation.

24(b). Why should this be so? We shall give reasons below. Meanwhile it is clear, particularly if $\widetilde{Ra}_m$ is a subcritical bifurcation, that there must be another branch (a strong field branch) "above" these weak field states to which the system will evolve, as in the notional bifurcation diagrams shown in Figure 24(c).

Perhaps the Earth can provide a clue about the existence and nature of the strong field branch? Let us compare the magnitude of the terms in the equation of motion (4.6). As in traditional geophysical fluid dynamics ("GFD"), we may introduce the Taylor number, Ekman number, and Rossby numbers by

$$Ta = \left(\frac{2\Omega L^2}{\nu}\right)^2, \quad E = \frac{\nu}{2\Omega L^2}, \quad Ro = \frac{U}{2\Omega L}, \tag{4.10}$$

the first two of which compare the viscous and Coriolis forces ($Ta = E^{-2}$) and the last of which compares the inertial force, $\rho_0 D\mathbf{V}/Dt$, with the Coriolis force, $2\rho_0 \mathbf{\Omega} \times \mathbf{V}$. In the core of the Earth,

$$Ta \approx 10^{30}, \quad E \approx 10^{15}, \quad Ro \approx 10^{-7}, \tag{4.11}$$

where we have taken $\nu \approx 10^{-6}\mathrm{m}^2/\mathrm{s}$. Viscosity seems extremely insignificant, and the inertial forces rather unimportant. It appears legitimate to replace (4.6) by the magnetostrophic equation.

$$(4.12) \qquad 2\boldsymbol{\Omega}\times\mathbf{V} = -\nabla\Pi + \frac{1}{\rho_0}\mathbf{J}\times\mathbf{B} - \mathbf{g}\gamma T,$$

although by omitting $D\mathbf{V}/Dt$ we eliminate torsional waves (see below). In this way, we transform (4.6) from a predictive into a diagnostic equation.

The decision to omit viscosity is reinforced when we compute the Chandrasekhar and Hartmann numbers,

$$(4.13) \qquad Q \approx \frac{B^2L^2}{\mu\rho\nu\eta}, \qquad M = \left(\frac{\sigma}{\rho\nu}\right)^{1/2} BL.$$

which compare the viscous and Lorentz forces ($Q = M^2$). Taking $B = 0.02$T we find

$$(4.14) \qquad Q \approx 4.10^{16}, \qquad M \approx 2.10^8\,.$$

Of course, viscous (Ekman) layers are needed at the CMB if no slip conditions are to hold there, but the layers are very thin, apparently of order $(\nu/2\Omega)^{1/2} \approx 0.1$m thick. Nevertheless when we wish to discuss core-mantle coupling below, viscous coupling provides a very simple theoretical model, even if it is geophysically unrealistic.

The ratio of magnetic and rotational terms in (4.6) and (4.12) is by (3.6)

$$(4.15) \qquad \frac{\text{Lorentz Force}}{\text{Coriolis Force}} = \frac{B^2/\mu_0\rho_0 L}{2\Omega U} = \frac{\Lambda}{R} \equiv C,$$

where C is the "magnetic coupling parameter" and

$$(4.16) \qquad \Lambda = \frac{\sigma B^2}{2\Omega\rho_0}$$

is the "Elsasser number". Taking $B = \overline{B}_T \approx 0.01$T, $\rho_0 = 10^4$kg/m^3 we obtain

$$(4.17) \qquad C \approx 8000, \qquad \Lambda \approx 80.$$

With $B = \overline{B}_P \approx \overline{B}_T/R$, these are reduced to

$$(4.18) \qquad C \approx 0.8. \qquad \Lambda = 8.10^{-3}.$$

The values (4.17) and (4.18) are far less extreme than (4.11) and (4.14). They suggest that the Coriolis force sets the "characteristic" scale, B_{ch}, of the magnetic field:

$$(4.19) \qquad B_{\text{ch}} = (2\Omega\rho_0/\sigma)^{1/2}.$$

For this B, we have $\Lambda = O(1)$. Numerically $B_{ch} \approx 2.10^{-3}$T, and by (3.28)

$$\overline{B}_T \sim B_{ch} R^{1/2}, \qquad \overline{B}_P \sim B_{ch} R^{-1/2}, \tag{4.20}$$

i.e. $\overline{B}_T \approx 0.02$T, $\overline{B}_P \approx 2.10^{-4}$T.

It is obviously theoretically important to understand why the core has apparently struck this dynamical balance rather than (say) made the kinematic and magnetic energy densities equal. In the inertial frame the kinetic energy density is enormously the larger; in the co-rotating frame the opposite is true:

$$\frac{\text{Kinetic Energy Density}}{\text{Magnetic Energy Density}} \approx \frac{\rho_0 U^2/2}{B^2/2\mu_0} = A^2, \tag{4.21}$$

where

$$A = \frac{U}{V_A}, \qquad V_A = \frac{B}{\sqrt{(\mu_0 \rho_0)}}, \tag{4.22}$$

are the "Alfvén number" (or "magnetic Mach number") and the Alfv́en velocity. Taking $B \approx \overline{B}_T$, we have $V_A \approx 0.2$ m/s, $A \approx 5.10^{-4}$.

The system (4.4) – (4.8) can be attacked by generalizing standard GFD techniques to include **B**. And this has been done by Gilman and Glatzmaier (though in the context of the solar dynamo), and more recently by Zhang and Busse. Since (4.11) and (4.14) are such extreme values, one must wonder if this is the right approach. Perhaps one should solve instead the unfamiliar system obtained by replacing (4.6) by (4.12)? This has been done for a few "intermediate models" (see below), but involves some unusual difficulties which are again associated with axisymmetric components, this time of (4.12).

Since **B** and **V** are zonal and axymmetric in the first approximation [see (3.30)], we may write (4.12) as

$$2\Omega V_* \mathbf{1}_s - \gamma T \mathbf{g} = \nabla\left(\Pi + \frac{\overline{B}_\phi^2}{2\mu_0\rho_0}\right), \tag{4.23}$$

where

$$V_* = V_\phi - \frac{\overline{B}_\phi^2}{2\Omega\mu_0\rho_0 s}, \tag{4.24}$$

and (s, ϕ, z) are cylindrical coordinates. Since $\nabla \times \mathbf{g} = 0$ and $\mathbf{g} = -g\mathbf{1}_r$, we see from $\nabla \times$ (4.23) that

$$2\Omega \frac{\partial V_*}{\partial z} = -\frac{g\gamma}{r}\frac{\partial T}{\partial \theta}. \tag{4.25}$$

This is a generalization of the usual "thermal wind equation". When integrated, (4.25) gives

$$\overline{V}_\phi = V_C + V_M + V_G(s,t). \tag{4.26}$$

Here V_C is the wind set up by the density difference, i.e. by the right-hand side of (4.25);

(4.27) $$V_M = \frac{B_\phi^2}{2\Omega\mu_0\rho_0 s}$$

is the "magnetic wind", and V_G is the "geostrophic flow" which is independent of z and is so far arbitrary.

It may be shown that, if $V_C = 0$, the $\alpha\omega$-dynamo described in §3 fails. Thus, our earlier estimate $U \approx 10^{-4}\text{ms}^{-1}$ must refer to V_C rather than more generally to $\overline{V}_\phi$.

Since V_C is so necessary, a mechanism is needed to maintain the pole-equator temperature differences appearing on the right-hand side of (4.25). One way is through meridional flows such as the $\mathbf{V}_P$ sketched in Figure 19(d). That such flows exist that are magnetically driven is obvious through the ϕ-component of (4.12). Since neither $\nabla\overline{\Pi}$ nor $-\gamma T\mathbf{g}$ have ϕ-components, this gives

(4.28) $$\overline{V}_s = \frac{1}{2\Omega\rho_0}\overline{(\mathbf{J}\times\mathbf{B})}_\phi.$$

Both the axisymmetric and the asymmetric fields contribute to (4.28) through $\overline{\mathbf{J}_P \times \mathbf{B}_P}$ and $\overline{(\mathbf{J}'\times\mathbf{B}')}_\phi$ respectively. Once $\overline{V}_s$ is obtained from (4.28), $\overline{V}_z$ can be derived by integrating (4.7):

(4.29) $$\overline{V}_z = -\int_0^z \frac{1}{s}\frac{\partial}{\partial s}(s\overline{V}_s)dz.$$

The only trouble is that the resulting $\overline{\mathbf{V}}_P$ is not confined to the core, i.e. the essential condition,

(4.30) $$V_r = 0, \quad \text{on} \quad r = a,$$

is not obeyed. Axisymmetry strikes again!

Two ways out of this difficulty have been proposed: the Taylor way and the Braginsky way. Taylor (1963) showed that (4.30) is satisfied if

(4.31) $$\langle(\mathbf{J}\times\mathbf{B})_\phi\rangle = 0,$$

where $\langle Q\rangle$ is the "cylinder average" of a field Q:

(4.32) $$\langle Q\rangle(s,t) = \frac{1}{4\pi s z_1}\int_0^{2\pi} d\phi \int_{-z_1}^{z_1} dz\; Q(s,\phi,z,t),$$

and $z_1(s) = \sqrt{(a^2 - s^2)}$. In other words, it is the average of Q over the cylinder shown in Figure 25. Taylor went further. He showed how $V_G(s,t)$ could evolve in

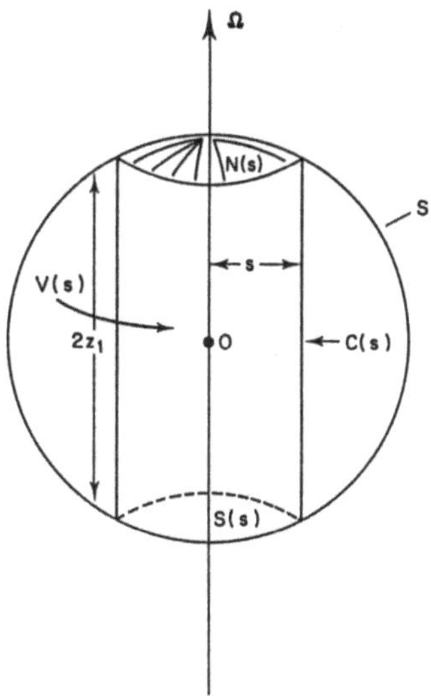

Figure 25. The Taylor cylinder $C(s)$, of radius s, capped on the CMB by $N(s)$ and $S(s)$.

such a way that (4.31), and therefore (4.30), are perpetually satisfied, if they were satisfied initially.

The physical idea behind Taylor's idea may be clarified by reference to torsional waves, which can bring about geostrophic adjustment. In a geostrophic motion, $V_G(s,t)$, the fluid on a cylinder such as that shown in Figure 25 turns about Oz in a rigid-body motion, but the set of all such cylinders move differentially. Because they are threaded together by the component B_s of $\mathbf{B}$, with which magnetic tension is associated, the cylinders are coupled together by the "elasticity" of $\mathbf{B}_P$, and can transmit waves from axis to core equator and back. These are the torsional waves. The geostrophic flow and torsional waves are unaffected by Coriolis forces which, for them, can be absorbed into the pressure gradient. Thus weaker forces, and particularly the inertial force, become dynamically significant; $D_t\mathbf{V}$ must be restored to (4.12). Since inertial forces replace Coriolis forces, the torsional wave is essentially an Alfvén wave, and it crosses the core with the Alfvén velocity, V_A. Its period is therefore of order $\tau_{\text{tor}} = a/V_A$. If we take 2.10^{-4}T as representative of B_s, we obtain $V_A \approx 2.10^{-3}$m/s and $\tau_{\text{tor}} \approx 50$yr. It is interesting that recently LeMouël *et al.* have been able to extract from geophysical data evidence of a torsional wave in the core. It is not clear how such waves are excited, but one must expect them to be damped not only by viscosity and resistivity but also by electrical conduction in the mantle.

If we start at $t = 0$ with an arbitrary $\mathbf{B}$, this will generally not obey (4.31), and the unbalanced torque will set the Taylor cylinders into motion. The resulting torsional wave will cross and recross the core, and be continually damped. Taylor visualized that, when it has disappeared, the core will have been brought into the state (4.31), and that subsequently the torsional waves will be able, because of their short timescale, to keep pace with other processes changing $\mathbf{B}$, in such a way that (4.31) is perpectually maintained.

This seemed so plausible in 1963 after publication of Taylor's insights that many geotheoreticians thought it was only a matter of time and faster computers before a working MHD model of the geodynamo would be constructed. Before this happened, however, Braginsky threw a monkey-wrench into the works. Independently of Roberts and Stewartson (1975), he gave reasons in 1975 why Taylor's concept was not necessarily correct for the Earth. He proposed "model-Z".

Braginsky's idea depends in an essential way on core-mantle coupling. To make the discussion as simple as possible we shall suppose that the coupling is viscous, even though in geophysical reality it is probably dominated by magnetic or topographic coupling.

Suppose that (4.30) is not obeyed. We recognize that $\overline{\mathbf{V}}_P$ was obtained by setting $\nu = 0$ [see (4.12)] and that an Ekman layer will be present at $r = a$ in which viscosity brings the core to rest relative to the mantle. Such a layer creates "Ekman pumping", a radial flow that must match to the mainstream flow, i.e. the flow obeying (4.12). The idea now is that the $\overline{V}_r$ that fails to satisfy (4.30) is precisely this Ekman flux.

This implies that V_G must be very large in the core, since a V_G pumps with radial velocity, $\overline{V}_r$, that is $O(\nu/\Omega a^2)^{1/2} V_G$, and we have already seen that $\overline{V}_r \equiv O(\eta/a)$. Thus $V_G = O(\eta(2\Omega/\nu)^{1/2}) \cong 60\text{m/s}$. [This preposterous value is a consequence of assuming viscous coupling, but it is also true that, when magnetic coupling acts instead, V_G remains embarrassingly large. We are, however, interested here only in the basic idea.] More precisely in the steady state.

$$V_G = \frac{1}{\rho_0}\left(\frac{z_1}{\nu\Omega a}\right)^{1/2} \langle(\mathbf{J}\times\mathbf{B})_\phi\rangle. \tag{4.33}$$

We see an essential difference from Taylor's concept. Here $V_G = O(\nu^{-1/2}) \to \infty$ as $\nu \to 0$; in Taylor's proposal $V_G = O(1)$ as $\nu \to 0$ (in fact for $\nu = 0$). This also means that the viscous energy expended by the geostrophic shear at the CMB is not negligible if Braginsky's proposal is correct.

As we have seen, a meridional flow is useful in creating pole-equator temperature differences, and may also (§3) improve the efficiency of the $\alpha\omega$-dynamo mechanism. For all its larger viscous dissipation, it is far from obvious that the meridional flow of model-Z is less effective than that in Taylor's state in promoting the efficiency of the dynamo mechanism and in generating favorable pole-equator temperature differences. Because V_G dominates $\overline{V}_\phi$ in this solution, the ω-effect of §3 tends to shear $\overline{\mathbf{B}}_P$ onto the Taylor cylinders, so that $\overline{B}_s$ becomes small compared with $\overline{B}_z$. It was because of this alignment of $\overline{\mathbf{B}}_P$ with the z-axis, that Braginsky christened his brainchild 'model-Z'.

The question of whether the geodynamo is of Taylor or Braginsky type is as yet unknown. To attempt to settle the question, a number of "intermediate" dynamo models have been integrated numerically. "Intermediate" signifies that the model has gone beyond kinematics, but has still not come to grips with the full three-dimensional MHD problem. Its aim is to elucidate the axisymmetric *dynamics*; the kinematics is regarded as essentially solved, and without compunction α and V_C are chosen arbitrarily. The aim is to understand the resulting axisymmetric state, and in particular whether the final state is of Taylor or Braginsky type.

The integrations of intermediate models performed by Braginsky (1978), Braginsky and Roberts (1988) and Roberts (1989), invariably equilibrated to model-Z form (Figure 26). Roberts argued that the magnetically-driven $\overline{\mathbf{V}}_p$ produced by these models-Z actually assists dynamo action, so that a subcritical bifurcation occurs; see Figure 27.

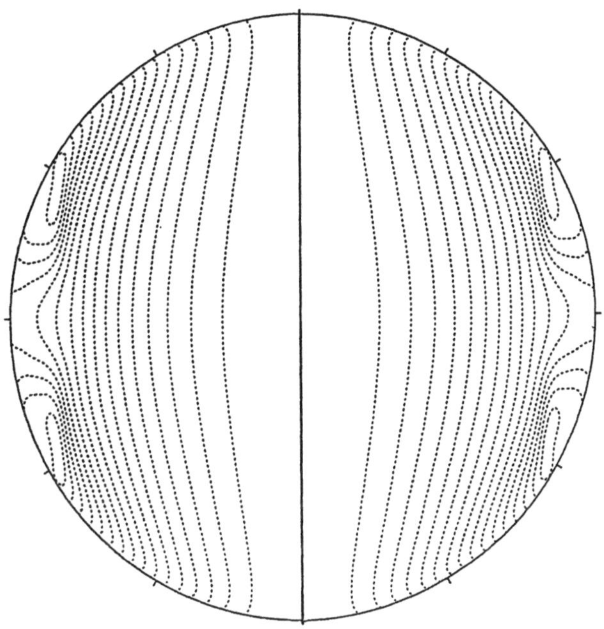

Figure 26. Lines of force of $\overline{\mathbf{B}}_P$ in a model-Z integrated by Braginsky and Roberts (1987).

It will be interesting to see if fully three-dimensional MHD dynamos follow this model-Z pattern, or something closer to Taylor's concept.

5. Waves, convection and turbulence. The system (4.4), (4.5), (4.7), (4.8) and (4.12) not only determine the basic axisymmetric state of the core (denoted now by $\mathbf{B}_0$, $\mathbf{V}_0$, etc); it also governs departures from that basic state (though not short period disturbances, such as the torsional waves, which are excluded by the magnetostrophic approximation). The most important deviations from the basic state are those that are asymmetric, for it is they that may be able to defeat Cowling's theorem (§3), provided that their amplitudes are sufficiently great. Although therefore interest centers on finite amplitude disturbances of the basic state, one can (as usual) learn from the linear stability problem, in which one writes

$$(5.1) \qquad \mathbf{B} = \mathbf{B}_0 + \mathbf{b}, \qquad \mathbf{V} = \mathbf{V}_0 + \mathbf{v}, \qquad \text{etc.,}$$

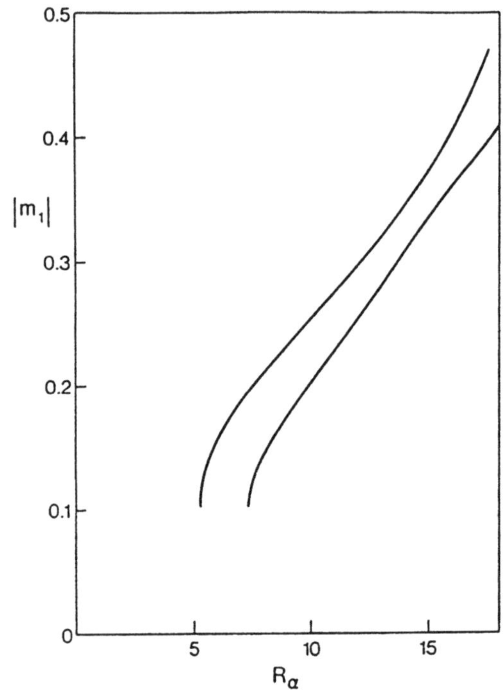

Figure 27. Indications of subcritical bifurcations in two intermediate models integrated by Roberts (1988). Note the vertical tangents where the curves terminate, suggesting bifurcations to unstable branches that link the stable branches to the $m = 0$ axis, i.e. to the marginal state.

and neglects the squares and products of the perturbation fields $\mathbf{b}$, $\mathbf{v}$, etc. It is useful to orientate oneself by, as a first step, ignoring dissipation. So we study "ideal instabilities" by setting $\nu = \kappa = \eta = 0$.

One immediately learns something of geophysical import. Suppose the density stratification is absent ($\nabla T_0 = 0$). The elasticity of the magnetic field lines allows wave propagation, but the waves are drastically dispersed by the Coriolis force. They do not travel with the Alfvén wave speed, V_A. Their characteristic "magneto-Coriolis timescale" is

$$\tau_{\mathrm{MC}} = 2\Omega b^2/V_A^2 \,. \tag{5.2}$$

Since $B_{0T} \gg B_{0P}$, these waves ride on the toroidal field lines, so we can take $B_0 = 0.02\mathrm{T}$, which implies that $V_A = 0.02\mathrm{m/s}$ and $\tau_{\mathrm{MC}} = 10^3\mathrm{y}$, a timescale redolent of the secular variation. It is considerably shorter than the τ_η given by (3.10); $\tau_{\mathrm{MC}} = \tau_\eta$ for $B = B_{\mathrm{ch}}$, but $B_T \sim B_{\mathrm{ch}} R^{1/2}$ by (4.20), so that for $B = B_T$ we have $\tau_{\mathrm{MC}} = \tau_\eta/R$.

If the gradients of $\mathbf{B}_0$ are sufficiently large, $\mathbf{B}_0$ may be prone to the so-called 'Field Gradient Instability', which is fueled by the magnetic energy stored in $\mathbf{B}_0$.

But the instability we seek is to produce the α that indirectly creates that $\mathbf{B}_0$. It is therefore hard to believe that this instability is relevant to the dynamo.

When a top-heavy density gradient, $\beta\rho_0$, exists the ideal system remains stable provided that β lies below a certain threshhold, β_c. The waves are now called "MAC Waves", the 'A' standing for 'Archimedean'. The new timescale that enters,

$$\tau_A = (g\beta)^{-1/2}, \tag{5.3}$$

is the buoyancy or 'Archimedean' timescale. When the density gradient is bottom-heavy, $(-g\beta)^{1/2}$ is known as the 'Brunt frequency'.

If $\beta > \beta_c$, the waves become instabilities that grow on the MAC-wave timescale, i.e. some of the normal modes, which in the present perturbation analysis are proportional to $\exp[i(m\phi + \omega t)]$, amplify with time: $\Im(\omega) < 0$. The most unstable modes are asymmetric ($m \neq 0$) and belong to small m, i.e. they are of "planetary scale". Moreover, there is a term in the governing equations that distinguishes between westward and eastward propagating waves. In general, both types of waves are possible. And one can be turned into the other by the Doppler-shift of an appropriate $\overline{V}_\phi$. If no such zonal flow is present, Acheson (1973) found that unstable waves would propagate westward in an ideal fluid, i.e. $m\Re(\omega) > 0$. Fearn (1979), analyzing a model of thermal convection in a rotating sphere and including the effects of magnetic diffusion (see below), found that, if Λ is large enough, marginal disturbances propagate westward, in agreement with the geomagnetic data.

There are two less encouraging aspects however. The first concerns the timescale. The geodynamo is "slow"; it loses energy at a rate determined by τ_η and requires the instabilities to draw energy from the working of the gravitational force at the same rate. Now the ideal timescales τ_{MC} and τ_A are surely much less than τ_η. Of course, above but sufficiently close to threshhold, the most unstable mode will grow on a timescale that is much greater than τ_{MC} and τ_A, and perhaps this can be made of order τ_η by careful arrangement. But clearly, it is a somewhat delicate balance that might easily be upset.

The second disappointment concerns α. Naturally, α must be quadratically infinitesimal for the perturbations we have so far been considering, but one might reasonably hope that, in the context of small but finite amplitude disturbances, he would obtain a nonzero α from the linear eigenfunctions. Unfortunately however, the phase relationships are such that, in a steady state, $\alpha \equiv 0$.

Both these deficiencies could be removed by restoring the magnetic diffusivity. This would destroy the phase relationships that caused α to vanish. Moreover, even when the system is stable according to ideal theory, it is prone when $\eta \neq 0$ to slow "resistive" instabilities that are the analogues for highly rotating systems of the tearing and g-modes well-known to the plasma physicist; see e.g. Kuang and Roberts (1990, 1991). Moreover, it is slow modes that we are seeking for field generation.

The prospect of generalizing the already challenging MAC-wave theory by adding magnetic diffusion is not exactly enthralling. It is hardly surprising that, despite heroic efforts by Fearn and his collaborators, much remains to be done. We shall here merely discuss one simplistic model that bears on the question, and also questions raised in §§3 and 4. Before doing so, let us take stock.

When one considers the delicate energy balance of the Earth's core, he may feel that our arguments have moved inexorably onto increasingly shakey ground. In §3, we abandoned the weak field dynamo, even though its axisymmetric toroidal field dissipates far less energy than does the $\overline{B}_T$ of a strong field dynamo. In §4, we favored model-Z even though, through its large geostrophic flow, this added a significant viscous energy penalty that the Taylor state was not required to pay. At first sight, our direction seems misguided, to say the least. The dynamics are, however, rather subtle.

We have already remarked in §4 that, through its larger meridional flow, a model-Z may actually operate subcritically. This is just one instance of what seems to be an all-pervading theme: the fluid tries to establish the magnetic environment that best allows the convective motions that cool it most efficiently to take place. And, in a highly rotating system, that optimal environment is *not* $\mathbf{B} = 0$! The fluid creates the $\mathbf{B}$ it requires by dynamo action, and that dynamo is *not* of weak-field type. While it is true that the $\overline{B}_T$ of a weak field dynamo dissipates magnetic energy less rapidly, its asymmetric components, $\mathbf{V}'$ and $\mathbf{B}'$, are potentially so energetically expensive that it is very hard to excite them at all in the face of the Coriolis force, i.e. $\widetilde{Ra}$ is extremely large when $\overline{B}_T$ is small. Similarly, among the strong field dynamos, a subcritical model-Z can, perhaps, set up an environment more conducive to convection than can a Taylor-type model?

The idea that magnetic field can promote convection will now be illustrated by the simplistic example referred to earlier. Most details will be omitted, for a more detailed account see, for example, §8.4 of Gubbins and Roberts (1987). Imagine a horizontal plane layer of electrically conducting Boussinesq fluid, of depth d and infinite horizontal extent, rotating about the vertical, heated from below, cooled from above, and pervaded by a uniform horizontal field, $\mathbf{B}$. We ask when the layer will first become linearly unstable to convective motions as the temperature of the lower boundary is gradually increased, or equivalently as $g\gamma\beta$ approaches its threshhold, $g\gamma\beta_c$. If $\mathbf{B} = 0$ is small or absent,

$$g\gamma\beta_c \sim \Omega^{4/3}(\kappa/d^2)^{2/3}(\kappa/\nu)^{1/3}, \tag{5.4}$$

in the critical state, and this leads to a very large value of β_c, underscoring the potency of Coriolis forces in inhibiting convection.

When $\mathbf{B} \neq 0$ and the layer is highly rotating in the sense that $\Omega \gg V_A/d$, we obtain a much smaller β_c. The optimal field strength, i.e. the one that minimizes β_c is such that $\Lambda = O(1)$. The horizontal scale of the most unstable mode is $O(d)$ and

$$g\gamma\beta_c \sim \Omega\kappa/d^2. \tag{5.5}$$

Even if we take $\nu = \eta$ (see below), this β_c is much smaller than that given by (5.4); taking geophysical scales for Ω, κ and d, the ratio of the two β_c is about 10^{-3}.

To establish (5.5), we report that the dispersion relationship governing the growth rate, $s(= i\omega)$, of a disturbance in our plane layer model is, in the magnetostrophic approximation governed by (4.4), (4.5), (4.7), (4.8) and (4.12),

(5.6)

$$(2\Omega k_z)^2(s + \kappa k^2)(s + \eta k^2)^2 + k^2(V_A k_x)^4(s + \kappa k^2) = g\gamma\beta k_H^2(V_A k_x)^2(s + \eta k^2),$$

where $\mathbf{k} = (k_x, k_y, k_z)$ is the wave vector of the disturbance, $\mathbf{B}$ being parallel to Ox; $k_H = \sqrt{(k_x^2 + k_y^2)}$ is the horizontal wave number and $k = |\mathbf{k}|$ is the total wave number. Boundary conditions at $z = 0$ and d require that k_z is a multiple of π/d, and the relevant choice here is

$$k_z = \pi/d. \tag{5.7}$$

The cubic (5.6) possesses one real positive root if

$$g\gamma\beta > 2\Omega\kappa k_z^2 \frac{k^3}{k_z k_H^2}\left[\frac{(V_A k_x)^2}{2\Omega\eta k k_z} + \frac{2\Omega\eta k k_z}{(V_A k_x)^2}\right]. \tag{5.8}$$

Minimize the right-hand side of (5.8) over k_x and k_y, or rather first over k_y and then over k_H. The minimum of the right-hand side of (5.8) over k_y clearly occurs when

$$\frac{(V_A k_x)^2}{2\Omega\eta k k_z} = 1, \tag{5.9}$$

and (5.8) reduces then to

$$g\gamma\beta > 4\Omega\kappa k_z^2 \frac{k^3}{k_z k_H^2} = 4\Omega\kappa k_z \frac{(k_H^2 + k_z^2)^{3/2}}{k_H^2}. \tag{5.10}$$

The minimum over k_H of the new right-hand side arises when

$$k_H^2 = 2k_z^2, \qquad \text{i.e. when} \qquad k^2 = 3k_z^2, \tag{5.11}$$

and (5.10) then becomes

$$g\gamma\beta > 6\sqrt{3}\,\Omega\kappa k_z^2, \tag{5.12}$$

which is equivalent to (5.5). In order to allow (5.11), we must suppose, *a posteriori*, that $k < 2\Lambda k_z$; modes with larger k are more stable and of less interest here (but see below). According to (5.11), the spatial periodicities of the most unstable mode are of order d in all directions. This stands in sharp contrast to the small scale motions that arise at threshhold when $\mathbf{B}_0$ is small or absent. The enormous dissipation associated with these lies behind the large disparity between (5.4) and (5.5).

As a model of the core, our model is oversimple in many ways; for example, all geometrical effects have been excluded. More serious than this is the supposition that a $\mathbf{B}_0$, supplied by external sources, is available. The core must manufacture its own $\mathbf{B}_0$! The analogue for the core of the large scale instability we have just located must exist, but it is required to create the α-effect and so power the geodynamo. The β given by (5.5) is therefore too small; it must greatly exceed[2] this value, though it is unlikely to be as large as the β_c of (5.4).

Once one admits that β exceeds the β_c given in (5.5), he must also investigate disturbances on much shorter length scales, ones that do not have the responsibility of maintaining $\mathbf{B}_0$. When

$$Ra \equiv \frac{g\gamma\beta d^2}{2\Omega\kappa\pi^2} \gg 1, \tag{5.13}$$

where Ra is the (modified) Rayleigh number, we see from (5.8) that there are now very many unstable modes, namely those for which

$$k_z \geq \pi/d, \tag{5.14}$$

and

$$Q/Ra < Q_c \equiv (\pi/k_z d)^2, \tag{5.15}$$

where

$$Q \equiv \frac{k^3}{k_z k_H^2}\left[\frac{(V_A k_x)^2}{2\Omega\eta k k_z} + \frac{2\Omega\eta k k_z}{(V_A k_x)^2}\right]. \tag{5.16}$$

It is convenient to think of a small-scale mode in terms of its z-periodicity (or equivalently through its k_z), and its overall form (i.e the *ratios* between k_x, k_y and k_z, which are all that affect Q). We clearly no longer need the restriction (5.7); the cells can now have a comparatively small z-scale, in one sense the smaller the better since the small-scale modes are not involved in powering the dynamo. On the other hand, if k is too large, the neglect of ν (and the neglect of κ that comes presently) is unjustified. Moreover, if k_z is too large, the restriction (5.15) on Q may come into force, since it might not be possible to make Q smaller than $Q_c Ra$. Otherwise the requirement (5.15) mainly rules out cells for which k_H/k_z is too large (or too small); Q can be large for the remainder.

Since $\eta \gg \kappa$ in the core, we may seek modes for which

$$Q/Ra \gg \kappa Q_c/\eta. \tag{5.17}$$

Using (5.15) and (5.17), we can then simplify (5.6):

$$\left(\frac{s}{\eta k^2}\right)^3 + 2\left(\frac{s}{\eta k^2}\right)^2 + \frac{V_A^2}{2\Omega}\frac{k_x^2 k_H^2}{k^2}\left[Q\left(\frac{s}{\eta k^2}\right) - \frac{\kappa Q_c Ra}{\eta}\right] = 0. \tag{5.18}$$

[2]If we equate Coriolis and buoyancy forces, we find that $g\gamma\beta L$ must be of order $2\Omega V$, i.e. the actual β in the core is probably of order $\beta_{\text{actual}} = 2\Omega V/g\gamma\beta L$. Taking $d = L$ and comparing this with the β given in (5.5), we see that $\beta_{\text{actual}}/\beta_c \approx (\eta/\kappa)R_\omega$. Both η/κ and R_ω are large.

Since $Q_c Ra\, \kappa k^2 \ll Q\eta k^2$ by (5.17), one root of (5.18) is small and positive; the other two are larger and negative. Only the former is of interest:

$$(5.19) \qquad s = \frac{g\gamma\beta}{2\Omega}\left(\frac{k}{k_z}\right)^2 Q^{-1},$$

which is independent of κ and small compared with ηk^2. Maximizing s with respect to k_x, we again recover (5.9) and reduce (5.19) to

$$(5.20) \qquad s = \frac{g\gamma\beta}{4\Omega}\frac{k_H^2}{kk_z}.$$

The maximum of this s over k_H occurs for $k_H/k_z \to \infty$, so that $k \approx k_H$ and

$$(5.21) \qquad s_{\max} = \frac{g\gamma\beta}{4\Omega}\frac{k_H}{k_z}.$$

Of course, this result is inapplicable when k_H/k_z is too large, because then inequality (5.17) would be violated, and the root (5.19) of (5.18) would no longer be small. The large-scale mode defined by (5.9) and (5.11) which reached threshhold first, also has a growth rate of order (5.19) when (5.13) holds; in fact, it grows slightly more rapidly than the small-scale modes we have been considering. This is, however, irrelevant, because the situation is surely reversed in the geodynamo, for the reasons we have given. According to (5.9), $k_x \sim (k_H k_z/\Lambda)^{1/2}$ so that, if $\Lambda = 1$, we would have $k_z < k_x < k_H \approx k_y$. Actually Λ based on the toroidal field strength is large according to (4.17), so that k_x and k_z are in reality of roughly the same size, while k_y is much greater. Thus the most rapidly growing of the short wavelength modes have "plate-like" forms, with their large dimensions in directions parallel to $\mathbf{\Omega}$ and $\mathbf{B}_0$ and their small dimension in the direction parallel to $\mathbf{\Omega} \times \mathbf{B}_0$, i.e. in the direction away from the rotation axis in the case of the Earth's core.

Braginsky (1964b) and Braginsky and Meytlis (1990) recognized that the instability of the short wavelength modes betrayed the existence of turbulence on similar length scales. They argued that this turbulence would be far more effective in transporting heat and chemical inhomogeneities than the molecular diffusion processes we have so far considered. They estimated in fact that the turbulent diffusion coefficients of these fields in the x- and z-directions would be of order η, but (because of the plate-like form of the turbulent eddies, which they supposed was a property not only of the infinitesimal waves but also of the finite amplitude turbulence) diffusion in the y-direction would not be so greatly enhanced. The turbulent diffusion of heat and chemical composition would thus be highly anisotropic. The diffusion of magnetic field and the α-effect would not be increased by the turbulence, i.e. field generation would still be the task of the large eddies alone. As a result of the turbulent transport, the analysis of the large-scale modes requires reconsideration, as does the entire magnetohydrodynamics of the dynamo. Further investigation of this unusual type of magnetofluid turbulence, and its implications for geomagnetic theory are exciting tasks for future study.

6. Reversals; final discussion. It is unfortunately true that very little is known about why the geodynamo reverses its polarity. It is of course obvious that the full magnetohydrodynamic equations are invariant under the change of $\mathbf{B}$ to $-\mathbf{B}$. Thus, by reversing the initial $\mathbf{B}$, we also reverse $\mathbf{B}$ for all times; this conclusion is unaffected by asymmetries in the Earth's internal structure. It is no longer true if another mechanism, or mechanisms, such as thermoelectric or electrochemical potential differences inside the Earth, assist electromagnetic induction in creating $\mathbf{B}$. The statistics of reversals indicate that the Earth has spent comparable times in each polarity state during geological time. This may indicate that other mechanisms of field production are not effective in the case of the Earth.

It is inconceivable on dynamical grounds that the core would overturn with its field; the field lines must therefore have time to detach themselves from the fluid in which they lie, and take up their new reversed orientation, leaving the fluid core the "same way up". Thus a reversal cannot be completed in a time much less than the electromagnetic time constant of the core [see (3.10) and the discussion below (3.10)]. This seems to be borne out by the observational facts; see §1. But why does the field reverse at all? Probably the most popular view is that the core is an auto-reversing MHD system. Several such systems are known, albeit very simple ones. Figure 28 shows Rikitake's two disk dynamo, a simple generalization of the homopolar dynamo shown in Figure 15. Figure 29 shows the result of integrating the third order differential system governing this device[3] The current in one of the wire loops is shown as a function of time. It clearly reverses continually. The dynamo essentially defines a strange attractor.

Even granted that the geodynamo auto-reverses, it is very hard to understand why the average time that elapses between reversals is so long. At present the rate is about once per million years, but what combination of core properties would yield such a long timescale? And during other geological epochs, the interval between reversals is much greater than 10^6y, maybe once it was even 8.10^7y. That is even harder to relate to the MHD of the core. For this reason extraneous causes have been sought for the reversals. One possibility is that the MHD of the core is controlled by the thermodynamic state of the mantle and that, as the latter changes, the core's MHD state makes a transition from a regime of frequent reversal to one of infrequent reversal, or *vice versa*.[4] It is certainly true that small modifications of the defining parameter of the Rikitake dynamo can create substantial changes in its behavior; see Figure 30.

To conclude this presentation, let us look back at our twenty one questions, and ask how successfully they have been answered. To Q1 and Q2 we would reply, "Because a dynamo is operating inside the core, and has been operating in the core throughout its existence". Q3 has been dealt with in a footnote in §2. We have made progress with Q4. We have understood the order of magnitude of $\mathbf{B}_P$; see (4.20).

---

[3]It is third order only because the disks and driving torques are assumed to be identical. The more general unsymmetric system is governed by four first order equations that have recently been studied by Ershov *et al.* (1990).

[4]Another opinion is that reversals are a bye–product of the impact of bolides with the Earth's surface.

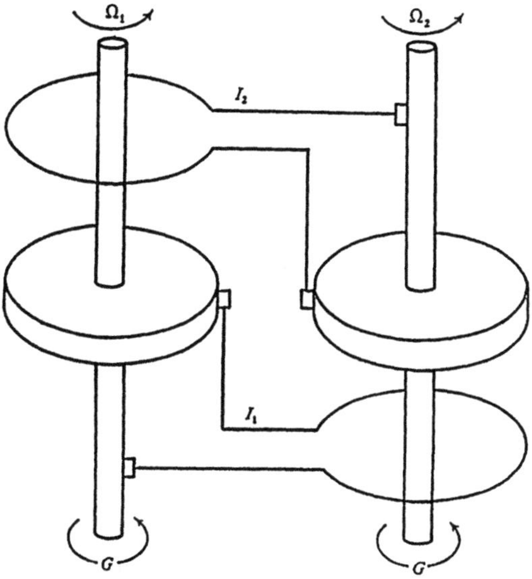

Figure 28. The Rikitake two-disk dynamo.

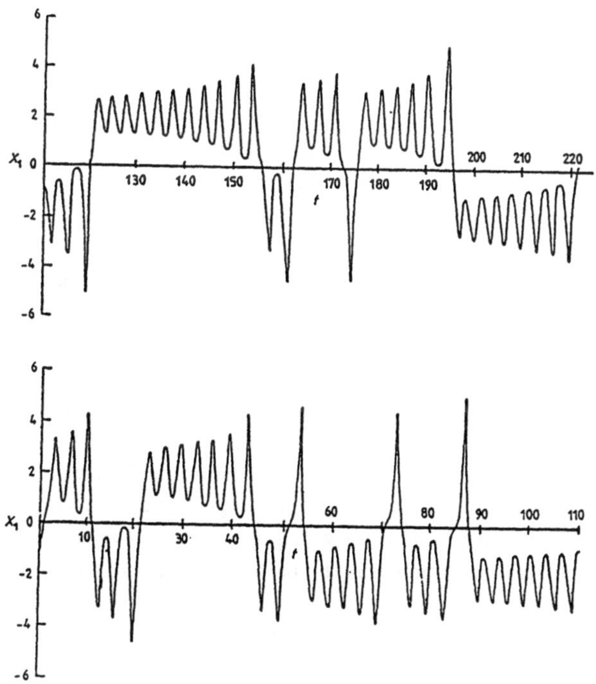

Figure 29. The behavior of the Rikitake dynamo. (After Cook and Roberts, 1970.)

In answer to Q5, we might argue that the rough constancy of the field strength

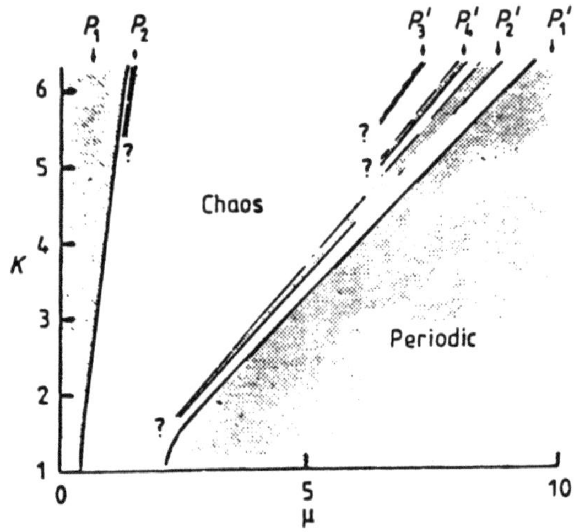

Figure 30. Phase diagram of the Rikitake dynamo. The character of the solution is indicated by the two parameters μ and K defining the model. The solution is periodic in the shaded regions and is non-periodic in the region labelled "chaos". (After Ito, 1980.)

indicates a similar constancy in the sources powering the dynamo. Q6 and Q7 have been explained by the dominance of the Coriolis force on core MHD. Virtually no progress has been made in answering Q8–Q10, but there seems to be nothing about reversals that would rule out the dynamo hypothesis. We have seen MAC-waves as an answer to Q11 and Q12. Concerning Q13, we have identified torsional oscillations of the core as one source of short period field variations, but there may be others, for example waves in a stable layer at the top of the core, a topic that we have not had time to discuss here. We have not attempted to answer Q14; indeed, so far there has been no inkling of how impulses could be so rapid and so worldwide. There seems to be no inherent obstacle in invoking the dynamo mechanism to explain the magnetism of other planets that have fields. Since, however, so many of the physical parameters involved in the dynamo mechanism vary considerably from planet to planet, it is hard to imagine (Q15) that they could be fitted into any kind of "Bode's law of planetary magnetism". Of course, our central theme is that electromagnetic induction is the battery sought in Q16, and Q17 would be answered in the same way that we have answered Q5 above. The last four questions were posed in §3 and have been answered in a qualitative way in our discussion of MAC waves in §5. In short, while many of its details are yet to be worked out, dynamo theory has encountered no fundamental difficulty in answering all but one of the twenty one questions, and it is hard to believe that the totally unexplained impulses will sound the death knell of geodynamo theory.

Much remains to be done, and the resolution of the remaining mysteries promises exciting times ahead.

Acknowledgement I am grateful to Dr. Stainslav Braginsky for criticising this review.

REFERENCES

Acheson, D.J., "Hydromagnetic wavelike instabilities in a rapidly rotating stratified fluid," *J. Fluid Mech.*,**61**, 609 (1973).

Braginsky, S.I. "Self-excitation of a magnetic field during the motion of a highly conducting fluid," *J. Exptl. Theoret. Phys. CCCP*, **47**, 1084 (1964a).Trans. *Soviet Phys. JETP*, **20**, 726 (1965).

Braginsky, S.I. "Magnetohydrodynamics of the Earth's core," *Geomag. & Aeron.*, **4**, 898 (1964b). Trans. *Geomag. & Aeron.*, **4**, 698 (1964).

Braginsky, S.I. "Nearly axially symmetric model of the hydromagnetic dynamo of the Earth," *Geomag. & Aeron.*, **18**, 340 (1978). Trans. *Geomag. & Aeron.*, **18**, 225 (1978).

Braginsky, S.I. and Meytlis, V. P. "Local turbulence in the Earth's core", *Geophys. & Astrophys. Fluid Dynam.*, **55**, 71 (1990).

Braginsky, S.I. and Roberts, P.H. "A model-Z geodynamo", *Geophys. & Astrophys. Fluid Dynam.*, **38**, 327 (1987).

Bucha, V. ed. "Magnetic Field and the Processes in the Earth's Interior", Academia, Czechoslovakian Academy of Sciences, Prague (1983).

Cook, A.E. and Roberts, P.H. "The Rikitake two-disc dynamo system," *Proc. Camb. Phil. Soc.*, **68**, 547 (1970).

Ershov, S.V. Malinetskii, G.G. and Ruzmaikin, A.A. "A generalized two-disc dynamo system", *Geophys. & Astrophys. Fluid Dynam.*, **47**, 251 (1989).

Fearn, D.R., "Thermal and magnetic instabilities in a rapidly rotating sphere," *Geophys. & Astrophys. Fluid Dynam.*, **14**, 103 (1990).

Gubbins, D. and Roberts, P.H. "Magnetohydrodynamics of the Earth's core," in *Geomagnetism*, (Ed. J.A. Jacobs), Academic Press, London UK and New York (1987).

Irving, E. and Pullaiah, G. "Reversals of the geomagnetic field, magnetostratigraphy, and relative magnitude of paleosecular variation in the Phanerozoic," in *Earth Sci. Rev.* **12**, 35 (1976).

Ito, K. "Chaos in the Rikitake two–disc dynamo system", *Earth planet. Sci. Lett.*, **51**, 451 (1980).

Kuang, W. and P. H. Roberts, "Resistive instabilities in rapidly rotating fluids: linear theory of the tearing mode", *Geophys. & Astrophys. Fluid Dynam.*, **55**, 199 (1990).

Kuang, W. and P. H. Roberts, "Resistive instabilities in rapidly rotating fluids: linear theory of the g-mode", *Geophys. & Astrophys. Fluid Dynam.*, Submitted (1991).

Langel, R.A. "The main field," in "Geomagnetism 1" (ed. J. A. Jacobs), Academic Press, London UK and New York (1987).

Le Mouël J.-L. and Courtillot, V. "Core motions, electromagnetic core-mantle coupling and variations in the Earth's rate of rotation: new constraints from secular variation impulses," in *Phys. Earth & planet. Inter.***24**, 236 (1981) .

Madden, T. and Le Mouël, J.-L."The recent secular variation and motions at the core surface," in *Phil. Trans. R. Soc. Lond. A* **306**, 271 (1982).

McElhinny, M.W. "Paleomagnetism and plate tectonics," Cambridge University Press, Cambridge UK (1973).

McElhinny, M.W. and Senenayake, W.E. "Variations in the geomagnetic dipole 1. The past 50,000 years," in *J. Geomag. Geoelectr.* **34**, 39 (1982).

Merrill, R.T. and McElhinny, M.W. "The Earth's Magnetic Field. Its History, Origin and Planetary Perspective." Academic Press, London UK and New York. (1983)

Merrill, R.T. and McFadden, P.L. "Paleomagnetism and the nature of the geodynamo," in *Science* **248**, 345 (1990).

Prévot, M. Mankinen, E.A. Grommé, C.S. and Coe, R.S. "How the geomagnetic field vector reverses polarity", *Nature*, **316**, 230 (1985).

Roberts, P.H. "Kinematic dynamo models", *Phil. Trans. R. Soc. Lond.* **A272**, 663 (1972).

Roberts, P.H. "Future of geodynamo theory", *Geophys.& Astrophys. Fluid Dynam.*, **44**, 3 (1988).

Roberts, P.H. "From Taylor State to Model-Z?", *Geophys. & Astrophys. Fluid Dynam.*, **49**, 143 (1989).

Roberts, P.H. and Stewartson, S. "Double roll convection in a rotating magnetic system," *J. Fluid Mech.*,**68**, 447 (1975).

Taylor, J.B. "The macrodynamics of a rotating fluid and the Earth's dynamo problem, *Proc. R. Soc. London* **A274**, 274 (1963).

THEORY OF THE GEODYNAMO AND CORE-MANTLE COUPLING

F. H. BUSSE*

Abstract. The theoretical problem of the origin of geomagnetism is outlined and the important role played by lateral inhomogeneities of the electrical conductivity in the lower mantle of the Earth is emphasized. The latter may be responsible for the separation between standing and drifting fields found in the analysis of secular variation. It is even possible in principle to find dynamo action based on lateral variations of conductivity.

1. Introduction. To the fluid dynamicist interested in nonlinear processes, the Earth sciences offer a number of challenging fundamental problems. Traditionally, most attention has been focussed on the general circulation of the atmosphere for which the most detailed observational data exist. The general circulation of the oceans and its variability is another topic to which increasing attention is devoted, and since the advent of plate tectonics considerable progress has been made in understanding convection flows in the Earth's mantle. The fundamental process least constrained by observational data appears to be the generation of the geomagnetic field in the Earth's liquid iron core. Although from paleomagnetic evidence and from measurements of the geomagnetic secular variation throughout the past centuries there exists a considerable amount of data, it has not been possible so far to use this data to specify the mechanism of the geodynamo with the exception of a few general properties such as its non-oscillatory nature.

Because of the remoteness of the Earth's core, special significance is attached to the core mantle boundary (CMB) which represents the major interface separating us from the magnetic field producing region. For a number of different scientific disciplines the lowermost mantle and the CMB have become of special interest in recent years. Results of seismic mantle tomography indicate strong variations in properties near the CMB. The D"-layer just above the CMB has received much attention from seismologists for a long time. But recent evidence (Young and Lay, 1987; Doornbos and Hilton, 1989) indicates that it may exist only over parts of the CMB. High pressure experiments suggest that percolation of liquid iron into the lowermost mantle may take place (Jeanloz, 1990) and there is evidence that FeO becomes metallic at the physical conditions of the CMB (Knittle and Jeanloz, 1986). In fact there is an emerging consensus that the lower boundary of the mantle exhibits a similar degree of complexity as the upper boundary does. In this paper we plan to consider some of the implications of lateral heterogeneity for the theory of the geodynamo.

In the following we start with a review of the basic symmetry properties of the problem of the geodynamo in section 2. We then turn to the importance of electromagnetic coupling for the determination of the state of relative rotation between core and mantle. In section 4 we describe a simple model illucidating the effect of a lateral inhomogeneity in electrical conductivity on the geomagnetic field. Finally

*Institute of Physics, University of Bayreuth, 858 Bayreuth, FRG.

we briefly discuss in section 5 the possibility of a dynamo action induced by a differential rotation between core and mantle in the presence of lateral variations of conductivity.

2. Generation of magnetic fields in rotating spherical shells. In order to provide a physically consistent model for a magnetohydrodynamic dynamo resembling the geodynamo, the following system of equations has been proposed (Busse and Zhang, 1988; Zhang and Busse, 1989).
Equations of motions:

$$\nabla \cdot \mathbf{u} = 0 \tag{2.1a}$$

$$\left(\frac{\partial}{\partial t} + \mathbf{u}\cdot\nabla\right)\mathbf{u} + 2\boldsymbol{\Omega}\times\mathbf{u} = -\nabla\pi - \mathbf{g}\beta\Theta + \nu\nabla^2\mathbf{u} + (\nabla\times\mathbf{B})\times\mathbf{B}/\rho\mu \tag{2.1b}$$

Diffusion equation for the buoyancy field Θ:

$$\left(\frac{\partial}{\partial t} + \mathbf{u}\cdot\nabla\right)\Theta + \mathbf{u}\cdot\nabla S = \kappa\nabla^2\Theta \tag{2.1c}$$

Equation of induction for the magnetic flux density $\mathbf{B}$:

$$\left(\frac{\partial}{\partial t} + \mathbf{u}\cdot\nabla\right)\mathbf{B} - \mathbf{B}\cdot\nabla\mathbf{u} = \lambda\nabla^2\mathbf{B} \tag{2.1d}$$

Here the Boussinesq approximation has been assumed in which the density ρ is regarded constant except in the gravity term where a linear dependence of ρ on the temperature or on the concentration of light elements has been taken into account with β as coefficient. In this simple model we do not distinguish between the two buoyancy providing fields. The field Θ may either describe the deviation of temperature from the static basic state of pure conduction or the deviation of the concentration of light elements from the distribution in the static solution of the problem. The quantity κ thus denotes either the thermal diffusivity or an average constant of diffusion of light elements. The basic static state enters through the spherically symmetric field S which represents the unstable buoyancy field of the basic static state. In the case of thermal buoyancy, S stands for the superadiabatic part of the static temperature distribution. The angular velocity of rotation has been denoted by $\boldsymbol{\Omega}$ and ν and λ are the kinematic viscosity and the magnetic diffusivity, respectively.

Since both the velocity field $\mathbf{u}$ and the magnetic field $\mathbf{B}$ are solenoidal, it is convenient to use the general representation

$$\mathbf{u} = \nabla\times(\nabla\times\mathbf{r}\phi) + \nabla\times\mathbf{r}\psi, \quad \mathbf{B} = \nabla\times(\nabla\times\mathbf{r}h) + \nabla\times\mathbf{r}g$$

where the first terms on the right hand sides are called the poloidal parts while the second terms are referred to as the toroidal parts. There is no need here to derive the dimensionless equations for the scalar fields ϕ, ψ, h and g (see, for example,

Zhang & Busse, 1988). We only mention the important dimensionless parameters of those equations,

Rayleigh number	$R \equiv \beta(T_2 - T_1)\gamma(r_2 - r_1)^3/\nu\kappa$
Taylor number	$T \equiv \left[\frac{\Omega(r_2 - r_1)^2}{\nu}\right]^2$
Prandtl number	$P \equiv \nu/\kappa$
magnetic Prandtl number	$P_m \equiv \nu/\lambda$
radius ratio of spherical boundaries	$\eta \equiv r_2/r_1$

where γ is the gravity at the CMB, and where the difference between the radius r_2 of the CMB and the radius r_1 of the inner-outer core boundary enters because it has been introduced as length scale.

The onset of convection in the form of non-axisymmetric motions represents the first bifurcation from the basic spherically symmetric state. The preferred mode of convection is characterized by an azimuthal wavenumber m_0 depending on T and P and described by fields ϕ and Θ that are symmetric with respect to the equatorial plane (Busse, 1970). The corresponding field ψ is antisymmetric. Non-axisymmetric convection is necessarily time-dependent with respect to the rotating frame of reference. But since this time dependence assumes the form of a drift of the convection pattern in the azimuthal direction it is always possible to find a frame of reference with respect to which the motions remain steady.

We now turn to the second bifurcation in the solution space of the problem. This bifurcation could either be a hydrodynamic instability of the finite amplitude convection flow or it could manifest itself as a growing solution of equation (2.1d) in which case we speak of a dynamo. We focus the attention on the latter possibility and deduce from the symmetry of equation (2.1d) that the growing magnetic field could either belong to the dipolar class or to the quadrupolar class of magnetic fields as shown in figure 1. In both cases the azimuthal periodicity of the magnetic field could differ from that of the velocity field. But we shall not consider this possibility since it would not correspond to an axisymmetric component of **B** which only arises if the magnetic field exhibits the same azimuthal wavenumber m_0 as the convection flow. A further distinction can be made in terms of a vanishing or non-vanishing imaginary part σ_i of the growth rate σ of the growing magnetic field. Accordingly we can separate the classes of steady and oscillatory dynamos. Depending on the region in the R-T-P parameter space, one or another of the four different types of dynamos (dipolar–quadrupolar, steady–oscillatory) corresponds to the preferred lowermost value of the magnetic Prandtl number. The reader should be reminded at this point that the relevant parameter for the onset of dynamo action is the magnetic Reynolds number which is represented by the product of the magnetic Prandtl number times a typical velocity of the convection flow. Since a change of the latter flow requires new computations, it is much more easy to determine the onset of a dynamo by changing the magnetic Prandtl number.

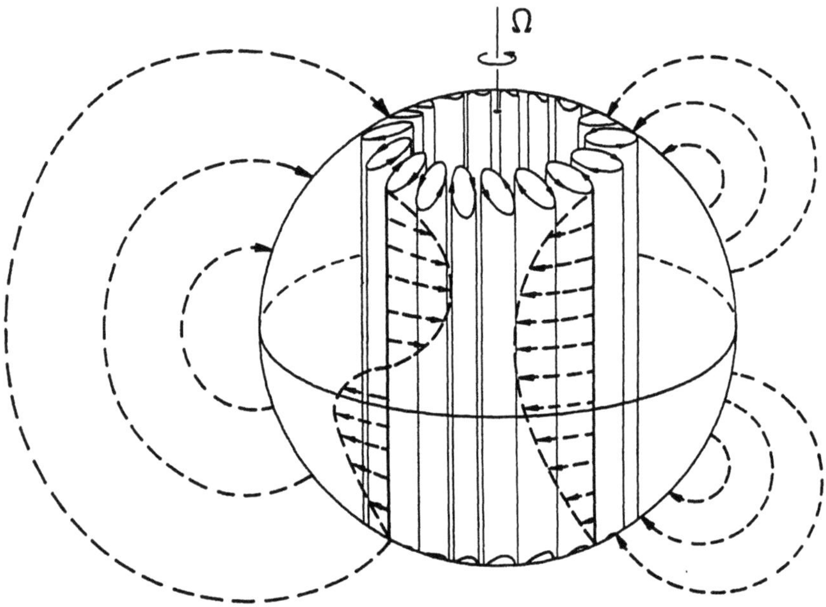

Figure 1: The dipolar class (left side) and the quadrupolar class (right side) of magnetic fields obtained as solutions of equation (2.1d) for velocity fields exhibiting the symmetry of the convection columns obtained as asymptotic solutions in a rotating sphere (Busse, 1970). Only the axisymmetric components of poloidal and corresponding toroidal fields are indicated by the broken field lines.

The fact that dipolar and quadrupolar dynamos seem to be rather competitive raises the question whether quadrupolar dynamos may actually be realized in some planetary cores. In view of the unusual dipole fields observed at Uranus and Neptun one may speculate that quadrupolar dynamos are operating in those planets. A dipolar perturbation of the field of the same relative order as the quadrupolar component of the geomagnetic field easily dominates the observed field at the distance from the planetary core at which the measurements by the Voyager probe have been made.

3. Core-Mantle coupling. Since it is known that in rapidly rotating systems the influence of rigid boundaries on convection is similar to that of stressfree boundaries except for thin Ekman layers which are present in the former case (Chandrasekhar, 1961; Roberts, 1968), the stressfree boundary conditions are usually assumed at the CMB for mathematical convenience. In first approximation the assumption of an electrically insulating mantle is appropriate. For an axisymmetric CMB these assumptions imply that no angular momentum can be transferred across the CMB. Even a small electrical conductivity of the mantle, however, will in general lead to a torque which will induce an adjustment of the rotation rates of core and mantle such that the time averaged torque vanishes if the total angular momentum of the Earth is assumed to be constant. It is important to realize as has been pointed out by Bullard et al. (1950) that a vanishing electromagnetic torque is likely to correspond to a finite differential rotation between mantle and core. Braginsky (1964) has discussed this condition in a kinematic model of the geodynamo. If other torques are present such as those generated by deviations from axisymmetry of the geometrical shape of the CMB, the electromagnetic torque will in general have a finite time averaged component. But we shall not consider this case here.

The electromagnetic boundary conditions are derived from the continuity of the magnetic field and of the tangential component of the electric field since it is assumed that the magnetic permeability does not change across the CMB. It is convenient to convert the condition for the electric field into a condition for the magnetic field since only the latter appears in the equation of induction (2.1d). The continuity of the tangential component of the electric field $\mathbf{E}$ is equivalent to the conditions

$$[\nabla \cdot (\mathbf{n} \times \mathbf{E})] = 0, \qquad [\nabla_2 \cdot \mathbf{E}] = 0 \tag{3.1a,b}$$

where the brackets indicate the jump across the boundary, $\mathbf{n}$ denotes the normal unit vector and ∇_2 indicates the tangential part of divergence operator. The special case of a constant electric field is not included in conditions (3.1), but it is not relevant for the geophysical application and can eventually be added separately. In the case of a spherical or planar boundary condition (3.1a) is equivalent to

$$[\mathbf{n} \cdot \nabla \times \mathbf{E}] = 0 \tag{3.2}$$

Because (3.2) is implied by the continuity of the normal component of the magnetic field according to Maxwell's equations, the electric condition is reduced to condition (3.1b). Using Ohm's law we can write this condition in the form

$$[\nabla_2 \cdot (\mathbf{u} \times \mathbf{B} - \lambda \nabla \times \mathbf{B})] = 0 \tag{3.3}$$

The problem of electromagnetic coupling in the presence of a spherically symmetric distribution of electrical conductivity has been considered extensively in the literature (see, for example, Stix and Roberts, 1984). A major problem in the case of spherically symmetric conductivity in the lower mantle arises from the fact, that the conductance, which is a radial integral over the conductivity distribution, must be sufficiently high in order to permit electromagnetic torques as high as $10^{18}Nm$ required by the observed changes in the length of the day. On the other hand, the observed secular variation provides an upper bound on a spherically symmetric conductance. This problem of possible contradictions disappears, however, when lateral inhomogeneities of the conductivity are admitted.

4. The effect of laterally varying conductivity. In order to understand the effect of lateral variations of the electrical conductivity in the boundary, it is best to consider a simple model with a Cartesian geometry. In figure 2 a layer of highly conducting fluid is shown which is bounded by two plates of low, but laterally varying conductivity. For simplicity we assume that the thickness δd of the plates is small in comparison with the layer thickness d and with the lateral scale of the conductivity variations. Outside the plates electrically insulating material is assumed. For a dimensionless description of the problem we introduce d as length scale and d^2/λ_0 as time scale where λ_0 is the uniform magnetic diffusivity of the fluid. We shall assume that the diffusivity λ of the plates does not depend on $z, \lambda = \lambda_0 \varepsilon^{-1} \xi(x, y)$.

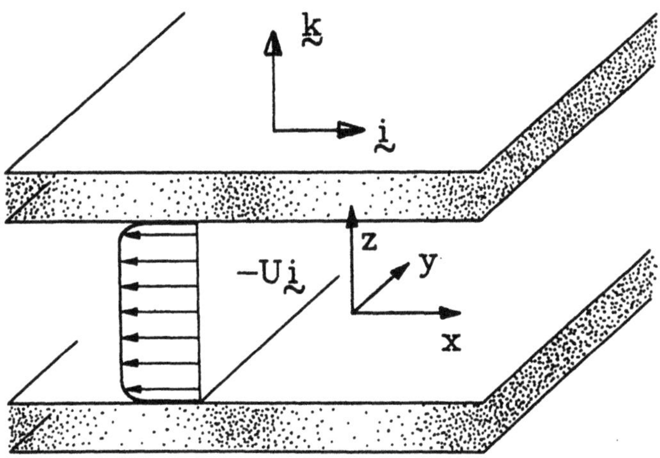

Figure 2. The configuration used for the discussion of the effect of lateral variations of the electrical conductivity in the boundaries of a fluid channel.

Using the general representation for the magnetic field

$$\mathbf{B} = \nabla \times (\nabla \times \mathbf{k}h) + \nabla \times \mathbf{k}g \tag{4.1}$$

where $\mathbf{k}$ is the unit vector normal to the plates we find the following conditions at the inner boundaries of the plates,

(4.2)

$$h = \hat{h}, \quad g = \hat{g}, \quad \frac{\partial}{\partial z} h = \frac{\partial}{\partial z} \hat{h},$$

$$\left[\frac{\partial}{\partial z}\Delta_2 g - (\mathbf{u} \times \nabla \Delta_2 h) \cdot \mathbf{k}\right] = \nabla_2 \cdot \left(\xi \nabla_2 \frac{\partial}{\partial z} \hat{g}\right) + \mathbf{k} \cdot \nabla_2 \xi \times \nabla_2 \nabla^2 \hat{h} \quad \text{at } z = \pm\frac{1}{2}$$

A Cartesian system of coordinates has been introduced with the origin on the mid-plane of the layer and the z-coordinate in the direction of $\mathbf{k}$. The equations for the poloidal and toroidal functions $\hat{h}$ and $\hat{g}$ in the plates are given by

(4.3a)

$$\nabla_2 \cdot (\xi \nabla_2 \nabla^2 \hat{h}) = \mathbf{k} \cdot \nabla_2 \hat{g} \times \nabla_2 \xi + \varepsilon \frac{\partial}{\partial t} \Delta_2 \hat{h}$$

(4.3b)

$$\nabla^2 \Delta_2 \hat{g} + \nabla_2 \cdot (\Delta_2 \hat{g} \xi^{-1} \nabla_2 \xi) = \varepsilon \left[\nabla_2 \cdot \left(\xi^{-1} \nabla_2 \frac{\partial}{\partial t} \hat{g}\right) + \mathbf{k} \cdot \left(\nabla_2 \xi^{-1} \times \nabla_2 \frac{\partial^2}{\partial z \partial t} \hat{h}\right)\right]$$

where ∇_2 represents the x- and y-components of ∇ and Δ_2 is defined by $\Delta_2 = \nabla_2 \cdot \nabla_2$.

Since ε is regarded as a small parameter it is convenient to express solutions of equations (4.3) in terms of perturbation expansion

$$\hat{h} = \hat{h}_0 + \varepsilon \hat{h}_1 + \ldots, \qquad \hat{g} = \hat{g}_0 + \varepsilon \hat{g}_1 + \ldots \tag{4.4}$$

Because the conductivity of the plates vanishes for $\varepsilon = 0$, $\hat{g}_0$ must vanish identically.

To illustrate some typical properties of the problem we consider the special case where the function ξ and the fields g_0, h_0, $\hat{h}_0$ are given by

(4.5a)

$$\xi = 1 + \kappa_0 \sin \hat{\alpha} x, \quad 0 < \kappa_0 < 1, \quad g_0 = \hat{g}_0 \cos \beta \cos \pi z,$$

(4.5b)

$$h_0 = \tilde{h}_0 \cos \alpha(x - ct) \sin \beta y \cos \gamma z, \quad \hat{h}_0 \equiv \tilde{h}_0 a_0 \cos \alpha(x - ct) \sin \beta y \exp\{-a|z|\}$$

The constants γ, a_0, a are determined by the conditions

$$\gamma \tan \gamma \frac{1}{2} = a, \quad a = \sqrt{\alpha^2 + \beta^2}, \quad a_0 \equiv \cos \gamma \frac{1}{2} \exp\left\{a \frac{1}{2}\right\} \tag{4.5c}$$

Even for the simple Ansatz (4.5) the order ε^1 of equations (4.3) can not be solved by explicit expressions. Using the assumption $\delta \ll 1$ we write the solution of (4.3b) in the form

$$\hat{g}_1 = \left[z \mp \left(\frac{1}{2} + \delta\right)\right] \{b_1 + \hat{b}_1 \sin \hat{\alpha} x + \tilde{b}_1 \cos[(\alpha - \hat{\alpha})x - ct] + \ldots\} \cos \beta y \tag{4.6a}$$

where the dots indicate similar terms as those given explicitly, but with more complex combinations of the arguments $\hat{\alpha}x$ and $\alpha(x-ct)$. For h_1 we obtain the equation (4.6b)

$$\nabla^2\hat{h}_1 = \{a_1(z)\sin\alpha(x-ct)+\hat{a}_1(z)\kappa_0\hat{g}_0\cos\hat{\alpha}x+\tilde{a}_1(z)\cos[(\alpha-\hat{\alpha})x-ct]+\ldots\}\sin\beta y$$

where again only the principal terms have been denoted explicitly of an infinite series of analogous terms as indicated by the dots. For the following qualitative discussion there is no need to calculate the constants $b_1, \hat{b}_1, \tilde{b}_1$ and functions $a_1(z), \hat{a}_1(z), \tilde{a}_1(z)$ explicitly. The main conclusion drawn from expression (4.6b) is that the field observed from the outside in the case $\varepsilon \neq 0$ differs qualitatively from the case $\varepsilon = 0$. If we identify $\hat{g}_0$ with the amplitude of the axisymmetric component of the toroidal field in the Earth's core we may conclude that a strong "standing" poloidal field can be produced by variations of the conductivity in the lowermost mantle.

Additional conclusions can be drawn if the spatial average of the force F exerted on the plates in the x-direction is considered. Indicating the average over a plate by angular brackets we find in the order ε

$$\langle F\rangle = -\varepsilon\langle\left(\frac{\partial^2}{\partial y\partial z}\hat{g}_1 + \frac{\partial}{\partial x}\nabla^2\hat{h}_1\right)\Delta_2\hat{h}_0\rangle \tag{4.7}$$

As expected a time independent contribution will be made by expressions (4.6). Since only discrete values of α, $\hat{\alpha}$ are possible when we interprete x as the azimuthal coordinate in the case of the Earth, it is appropriate to consider the special case $\alpha = \hat{\alpha}$. In this case oscillatory components appear in the coupling force (4.7). Since in general a variety of scales of the core magnetic field may interact with a spectrum of spatial variations of the conductivity in the lower mantle, it becomes obvious that the coupling torque at the CMB may exhibit a complex history in time, as has indeed been observed in the decade fluctuations of the length of the day (Morrison, 1979).

5. Dynamo action caused by lateral variations of conductivity. A well known theorem of dynamo theory originally mentioned by Elsasser (1946) and proved by Bullard and Gellman (1954) states that magnetic fields in the presence of a purely toroidal velocity must decay if the magnetic diffusivity is a function of the radial coordinate only. This result is due to the fact that a toroidal velocity field may produce a toroidal magnetic field from a poloidal one, but this process does not work the other way around if the conductivity is laterally uniform. As soon as this latter condition is dropped, new possibilities for dynamo action become available, however. In the following we intend to show that even the most simple toroidal motion can give rise to dynamos when the conductivity of the boundary varies laterally.

We consider the configuration shown in figure 3. A plate of thickness d separates an electrically conducting fluid of constant magnetic diffusivity λ_0 with constant velocity $U\mathbf{i}$ on the one side from an electrical insulator on the other side. The electrical conductivity in the plate does not change in the direction normal to the plate but exhibits a long wavelength variation in the direction of the unit vector $\mathbf{i}$

parallel to the plate. Using d as length scale and d^2/λ_0 as time scale we can formulate the problem of magnetic induction based on the representation (4.1) for the magnetic field in the three regions indicated in figure 3.

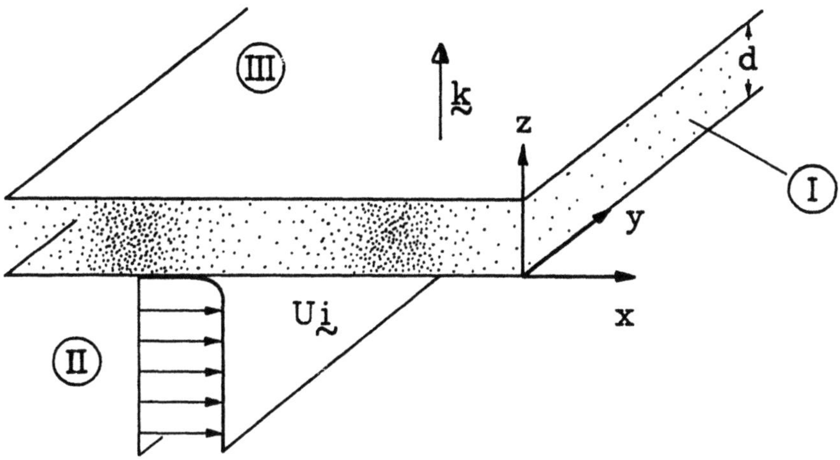

Figure 3. The three regions of the configuration that produces a dynamo caused by conductivity variations in the plate of thickness d.

Region I:

(5.1a) $$\lambda = \lambda_0 \xi(x, y),$$

(5.1b) $$\nabla_2 \cdot (\xi \nabla_2 \nabla^2 h) = -\mathbf{k} \cdot \nabla_2 \xi \times \nabla_2 \frac{\partial}{\partial z} g$$

(5.1c) $$\nabla^2 \Delta_2 g + \nabla_2 \cdot (\Delta_2 g \xi^{-1} \nabla_2 \xi) = 0$$

Region II: $\lambda = \lambda_0$

(5.2a) $$\left(\nabla^2 - U \frac{\partial}{\partial x}\right) \Delta_2 h^{II} = 0$$

(5.2b) $$\left(\nabla^2 - U \frac{\partial}{\partial x}\right) \Delta_2 g^{II} = 0$$

Region III:

(5.3) $$g^{III} \equiv 0, \qquad \nabla^2 h^{III} = 0$$

where a Cartesian system of coordinates with the x- and z-coordinates in the directions of the unit vectors $\mathbf{i}$ and $\mathbf{k}$ has been introduced and the previous definitions for ∇_2, Δ_2 hold. Matching conditions at $z=0$ and $z=1$ are given by

$$\text{(5.4a)} \quad h = h^{II}, \quad g = g^{II}, \quad \frac{\partial}{\partial z}h = \frac{\partial}{\partial z}h^{II}, \qquad \text{at } z=0$$

$$\text{(5.4b)} \quad \nabla_2 \cdot \left(\xi \nabla_2 \frac{\partial}{\partial z} g\right) - \mathbf{k}\cdot\nabla_2\xi \times \nabla_2\nabla^2 h = \frac{\partial}{\partial z}\Delta_2 g^{II} - U\frac{\partial}{\partial y}\Delta_2 h, \quad \text{at } z=0$$

$$\text{(5.4c)} \quad h = h^{III}, \quad g = 0, \quad \frac{\partial}{\partial z}h = \frac{\partial}{\partial z}h^{III} \qquad \text{at } z=1$$

Assuming that x, y-derivatives of g are small compared to the z-derivative we can simplify equation (5.1a) to $\partial^2\Delta_2 g/\partial z^2 \approx 0$. Accordingly we write the solution of equations (5.1), (5.2) for the special case

$$\text{(5.5)} \qquad \xi(x,y) = \xi_0 + \kappa_0 \cos \alpha x$$

in the form

$$\text{(5.6a)} \qquad h(x,y,z) = \cos\beta y \sum_n H_n(z)\exp\{in\alpha x\}$$

$$\text{(5.6b)} \qquad g(x,y,z) = (z-1)\sin\beta y \sum_n G_n \exp\{in\alpha x\}$$

$$\text{(5.6c)} \qquad g^{II}(x,y,z) = -\sin\beta y \sum_n G_n \exp\{\mu_n z + in\alpha x\}$$

where the definitions

$$\mu_n^2 = in\alpha U + \gamma_n^2, \quad \gamma_n^2 = n^2\alpha^2 + \beta^2$$

$$H_n(z) = H_{on}\gamma_n^{-2}\Big[-1 + \frac{1}{2}\exp\{\gamma_n(z-1)\} + \frac{1}{2}\exp\{-\gamma_n z\}(\exp\{-\gamma_n\}(\gamma_n - \mu_n) + 2\mu_n)(\gamma_n + \mu_n)^{-1}\Big]$$

have been used.

The full set of linear algebraic equations for the complex coefficients H_{on}, G_n which are obtained after expansions (5.6) have been introduced into equations (5.1b) and (5.4b) can be solved only numerically (Busse and Wicht, 1991). An idea of the interaction of the different fields can be obtained, however, when we introduce a most severe truncation by setting all coefficients equal to zero except for $H_{00}, G_0, H_{1i}, G_{1i}$ where the subscript i refers to the imaginary part. From the resulting equations

$$\text{(5.7a)} \qquad \beta^2\xi_0 H_{00} = \kappa_0\alpha\beta G_{1i}/2$$

$$\text{(5.7b)} \qquad -\beta^2\xi_0 G_0 = \mu_0\beta^2 G_0 + U\beta H_{00}(1-e^{-\beta})/2 - \kappa_0\alpha\beta H_{1i}/2$$

(5.7c)

$$(\alpha^2+\beta^2)G_{1i}(\xi_0 + \mu_{1r}) = \kappa_0\alpha\beta H_{00} - U\beta H_{1i}(1-e^{-\gamma_1})\left(1 - \frac{\gamma_1\mu_{1r} + |\mu_1|^2}{|\gamma_1+\mu_1|^2}\right)$$

(5.7d)

$$(\alpha^2+\beta^2)H_{1i}\xi_0 = \kappa_0\beta\alpha G_0$$

one obtains the relationship

(5.8)

$$(U_c\kappa_0)^2(1-e^{-\beta})(1-e^{-\gamma_1})\left(1-\frac{|\mu_1|^2+\gamma_1\mu_{1r}}{|\gamma_1+\mu_1|^2}\right)=[2\alpha^{-2}(\alpha^2+\beta^2)\xi_0(\xi_0+\mu_{1r})-\kappa_0^2]$$
$$[2\xi_0(\xi_0+\mu_0)(\alpha^2+\beta^2)-\kappa_0^2\alpha^2]$$

for the critical magnetic Reynolds number U_C. Obviously the truncation (5.7) of the complete system of equations is much too severe to yield a reliable expression for U_c. Expression (5.8) just indicates the possibility of a balance required for dynamo action. The detailed numerical investigation of the problem confirms this expectation (Busse and Wicht, 1991).

6. Concluding remarks. The example of dynamo action caused by lateral conductivity variations presented in the preceeding section is not likely to be of geophysical relevance. The magnetic Reynolds numbers of any reasonable steady relative motion between core and mantle are much too low. The predominant flows at the CMB exhibit an oscillatory time dependence and are caused by the effects of tides, nutations, and precession of the Earth. It remains yet to be seen how effective the dynamo process turns to be if the flow shown in figure 3 oscillates in the time. As the period of oscillation becomes small in comparison with the time scale of magnetic diffusion, the magnetic Reynolds number for dynamo action will increase and thus again the process is not likely to be geophysically relevant. In the context of this paper the analysis of the preceeding section just demonstrates the complexities and the potential for qualitatively new effects that are introduced by the phenomenon of lateral variation of conductivity.

The possibility of lateral variations of conductivity in the lower mantle with maximal values of the order of the core conductivity (Jeanloz, 1990) also adds to the difficulties of the interpretation of the secular variation in terms of fluid motions near the CMB. The traditional assumption of an insulating mantle may have to be dropped and processes such as those indicated in section 3 may have to be considered. Clearly, more detailed studies of the consequences of lateral conductivity variations are needed for a better understanding of the origin of the geomagnetic field.

REFERENCES

BRAGINSKY, S.I., *Kinematic Models of the Earth's Hydromagnetic Dynamo*, Geomagn. Aeron., 4 (1964), pp. 572–583.

BULLARD, E.C., AND GELLMAN, H., *Homogeneous Dynamos and Terrestrial Magnetism*, Phil. Trans. Roy. Soc. London, Ser. A, 247 (1954), pp. 213–278.

BULLARD, E.C., FREEDMAN, C., GELLMAN, H., AND NIXON, J, *The Westward Drift of the Earth's Magnetic Field, Phil. Trans. Roy. Soc., London, Ser A.*

BUSSE, F.H., *Thermal Instabilities in Rapidly Rotating Systems*, J. Fluid Mech., 44 (1970), pp. 441–460.

BUSSE, F.H., AND WICHT J, *A Simple Dynamo Caused by Conductivity Variations*, to be published, (1991).

BUSSE, F. H. AND ZHANG, K.-K., *Dynamics of the Earth's Core and the Geodynamo*, Structures and Dynamics of the Earth's Deep interior, Geophys. Monograph 46, AGU/IUGG, D.E. Smylie and R. Hide, eds. (1988).

CHANDRASEKHAR, S., *Hydrodynamic and Hydromagnetic Stability*, Oxford: Clarendon Press (1961).

DOORNBOS, D.J., AND HILTON,T., *Models of the Core-Mantle Boundary and the Travel Times of Internally Reflected Core Phases*, J. Geophys. Res., 24 (1989), pp. 15741-51.

ELSASSER, W.M., *Induction Effects in Terrestrial Magnetism*, Phys. Rev., 69 (1946), pp. 106–116.

JEANLOZ, R., *The Nature of the Earth's Core*, Annual Rev. Earth Planet. Sci., 18 (1990), pp. 357–386.

KNITTLE, E. AND JEANLOZ, R., *High Pressure Metallization and FeO and Implications for the Earth's Core*, Geophys, Res. Lett, 13 (1986), pp. 1541–44.

MORRISON, L.V., *Re-Determination of the Decade Fluctuations in the Rotation in the period 1861-1978*, Geophys. J.R. Astr. Soc., 58 (1979), pp. 349–360.

ROBERTS, P.H., *On the thermal instability of a Rotating-Fluid Sphere Containing Heat Sources*, Phil. Trans. Roy. Soc., London, A (1968), pp. 93–117.

STIX, M., AND ROBERTS, P.H., *Time-Dependent Electromagnetic Core-Mantle Coupling*, Phys. Earth Plan. Int., 36 (1984), pp. 49–60.

YOUNG, C.J., AND LAY, T., *The Core-Mantle Boundary*, Annual Rev. Earth Planet. Sci., 15 (1987), pp. 25–46.

ZHANG, K.-K. AND BUSSE, F.H., *Finite Amplitude Convection and Magnetic Field Generation in a Rotating Spherical Shell*, Geophys. Astrophys. Fluid Dyn., 44 (1988), pp. 33–53.

ZHANG, K.-K., AND BUSSE, F.H., *Convection Driven Magnetohydrodynamic Dynamos in Rotating Spherical Shells*, Geophys. Astrophys. Fluid Dyn., 49 (1989), pp. 97–116.

NONLINEAR DYNAMICAL MODELS FOR EARTHQUAKES AND FRICTIONAL SLIDING

JOHN B. RUNDLE*

Abstract. This paper describes two new approaches to understanding the nonlinear dynamics of earthquakes. These models are motivated by the inability to observe in nature effects associated with classical growth mechanisms for shear fractures. Moreover, friction experiments in the laboratory display a rate dependence through a series of state variables whose physical interpretation is unclear. Of the two nonclassical approaches examined here, the first is a field-theoretic extension of the classical Griffith–Gibbs nucleation approach. The second is a lattice automaton in which each site is characterized by threshold force below which the site is stable. Above the threshold, the site is in a state of metastable equilibrium, and will decay to a more stable state over a characteristic relaxation time. Qualitatively, the decay of each site is analogous to local barrier penetration, as understood from nucleation problems involving a Landau "double well" potential. However, the mathematical isomorphism for this analogy has not yet been made rigorous.

Key words. Nucleation, lattice automata, nonclassical fracture, friction

1. Introduction. A primary goal of earthquake research has been to understand how these massive shear fractures nucleate and grow. However, despite the careful efforts of a number of investigators, no unambiguous precursors to major earthquakes have been observed [1,8,15,16,21].

There are *only three* possible alternatives that can explain these negative results. The *first* is there exist no precursors to observe at the precision we have available to us with our current measurement techniques. If this is the case, one would like to prove it theoretically, so that no more money will be spent on fruitless searches. The *second* is that the space-time coverage of the affected region is inadequate. If this is the case, one would like to determine, using theoretical approaches, what constitutes adequate coverage. And the *third* alternative is that the precursory signals are presently there in the data, but we do not yet know how to recognize them. This alternative implies that theoretical models for the earthquake nucleation process must be constructed to provide predictions of the corresponding geodetic signature. Observational techniques must then be devised to optimize chances for detection of these signals.

For the most part, existing theoretical models of the earthquake rupture process rely either on approaches growing out of conventional engineering fracture mechanics (see, e.g., [12,13] for reviews), or on the use of rate and state dependent friction laws (see, e.g. [3,22]). These models correspond to *classical nucleation*, a process in which earthquake slip nucleates as a small patch, followed by rapid growth of the slipped area and the amplitude of slip. The important condition in these models is that the stress change over a bounded slipped region is uniform. Changes in surface deformation predicted by these models would be gradual, monotonic acceleration of horizontal displacement, uplift, and strain. As emphasized in the preceding paragraph, changes of this type have not been observed in field data.

*Earth Science Department, L-202, Lawrence Livermore National Laboratory, P.O. Box 808, Livermore, CA 94550

However, classical nucleation is not the only possibility. There are a variety of *nonclassical* mechanisms for nucleation that have recently been proposed (e.g., [2,5,9,10]). Imagine a volume of space in which there exists a distribution of interacting particles, for example, molecules or perhaps spin dipoles. Furthermore, suppose that the force $f(\mathbf{x})$ at any site due to all the other particles is given by an expression like $\int d\boldsymbol{x}'T(\boldsymbol{x}-\boldsymbol{x}')\rho(\boldsymbol{x}')$, where $T(\boldsymbol{x}-x')$ represents an interaction potential, and $\rho(x)$ represents the density of particles, or spin dipoles. An important quantity that determines the nature of the nucleation process in these systems is the range of interaction, R, defined by the expression:

$$R^2 = \frac{1}{2} \int d\boldsymbol{x}/\boldsymbol{x}/^2 \, T(\boldsymbol{x}) \tag{1.1}$$

In thermal and magnetic systems, for example, nonclassical mechanisms become important when R becomes large, $R \to \infty$. For example, a particular kind of nonclassical nucleation, referred to as "spinodal nucleation", differs from classical nucleation in that droplets of spins or monomers nucleate as fractals rather than compact objects (nuclei). The nonclassical droplets are highly *ramified* (i.e., diffuse) collections of particles whose dynamics are correlated over large ranges. Early stage growth of the nonclassical droplets consists of a period of "filling in", in which the droplet is altered from a diffuse to a more compact object. Subsequently, a period of rapid growth follows in which the droplet grows more or less classically at its boundaries. However, the ramified nonclassical part of the droplet remains as a "halo" surrounding the compact classical region.

2. Classical nucleation. The classical theory of nucleation is due to Gibbs [4]. For classical nucleation in thermal systems, one thinks, for example, of liquid droplets surrounded by vapor. The free energy F_l for the nucleating liquid droplet of size r in spatial dimension $d = 3$ is then [7]:

$$F_l = \frac{4}{3}\,\pi r^3 f_b + 4\pi\sigma r^2 \tag{2.1}$$

The quantity $\sigma > 0$ is the well known surface tension, whereas $f_b < 0$, a function of temperature and volume, is the free energy density of the bulk liquid phase.

For classical nucleation in magnetic systems, one thinks of a droplet of "down" spins surrounded by "up" spins. The free energy F_m for a droplet of down spins of size r in d-dimensional space is then [5]:

$$F_m = 2Hr^3 + \sigma r^{d-1} \tag{2.2}$$

H is the magnetic field strength in the up direction, and $\sigma > 0$ is the "surface tension" of the droplet. The dynamics of nucleation in this system are generated by applying a down magnetic field $H < 0$ to a lattice of up spins, causing the nucleation and growth of droplets of down spins.

In both (2.1) and (2.2), the first terms have opposing signs to the second terms, thus the terms compete. Whereas the bulk free energy is lowered by expansion of the droplet, there is a free energy cost associated with creating more surface area. It can easily be shown that there exists one unstable maximum to the curve of F against r. That maximum, located at r_c, represents the critical size. As droplets of size r form through fluctuations, droplets with $r < r_c$ lower their free energy by disappearing (evaporating). However, droplets formed with $r > r_c$ lower their free energy by growing in size, causing nucleation of the entire system.

Now consider fracture. Following Griffith [6], the Helmholtz free energy F_c for a tensile crack of semilength r under an applied tensile stress p is written as:

$$F_c = -B^2 p^2 r^2 + 2\gamma r \tag{2.3}$$

The quantity $-B^2$ depends only on the elastic constants and the boundary conditions of the elastic solid. The first term is an elastic term, giving the decrease in free energy due to the decrease in stored elastic strain energy as the crack extends. The second term is the surface energy term, describing the energy needed to break atomic bonds at the crack tip so that the crack may lengthen. The quantity 2γ is the binding energy needed to separate atomic bonds at the crack tip.

3. Nonclassical nucleation. The Griffith theory, usually thought to describe the result of applying a variable tensile stress p to a preexisting crack of fixed semilength r, could equally well be a description of the growth of cracks of varying length r in an applied stress field of fixed strength p [19]. In this view, the Griffith theory is essentially a nucleation theory that can be obtained from a "free energy functional" $F/\zeta(\boldsymbol{x})]$ written in terms of the tensile offset field $\zeta(\boldsymbol{x})$ across a lattice plane in the solid [19]:

$$F[\zeta(\boldsymbol{x})] = \int d\boldsymbol{x}\Big[-p\zeta(\boldsymbol{x}) + \int d\boldsymbol{x}' T(\boldsymbol{x}-\boldsymbol{x}')\zeta(\boldsymbol{x})\zeta(\boldsymbol{x}') + f_s[\zeta(\boldsymbol{x})]\Big] \tag{3.1}$$

As before, $T(\boldsymbol{x}-\boldsymbol{x}')$ in (3.1) is the interaction potential, also called "stress Green's function". The field $\zeta(\boldsymbol{x})$ is called an *order parameter* field. For thermal phase transitions, the corresponding order parameter field is the mass density field; for magnetic phase transitions, it is the spin density field.

The first two terms under the integral in (3.1) originate from the elasticity of the material surrounding the lattice plane. The third term, $f_s[\zeta(\boldsymbol{x})]$, is the surface energy potential function [19]. A particular choice of $f_s[\zeta(\boldsymbol{x})]$ appropriate to tensile (Griffith) fracture is:

$$\begin{aligned} &\zeta(\boldsymbol{x}) \geq 0 \ : \ f_s[\zeta(x)]2\gamma\{1-\exp[-\zeta^2(\boldsymbol{x})/a^2]\} \\ &\zeta(x) < 0 \ : \ f_s[\zeta(\boldsymbol{x})] \rightarrow \infty \end{aligned} \tag{3.2}$$

Again, 2γ represents the binding energy across the lattice plane whose lattice constant is a. The shape, or profile of the critical crack $\zeta_c(\boldsymbol{x})$ just prior to failure of the lattice is given by the solution of the Euler–Lagrange equation obtained from $\delta F[\zeta(\boldsymbol{x})] = 0$. Using the surface potential (3.2), we obtain:

$$(3.3) \qquad -p + \int d\boldsymbol{x}'\ T(\boldsymbol{x} - \boldsymbol{x}')\zeta_c(\boldsymbol{x}') + \frac{4\gamma}{a^2}\ \zeta_c(\boldsymbol{x})\ \exp[-\zeta_c^2(\boldsymbol{x})/a^2] = 0$$

The dynamics of crack growth, i.e., evolution of the field $\zeta(\boldsymbol{x}, t)$, can be obtained by appealing to kinetic arguments familiar from nonequilibrium thermodynamics and statistical mechanics. Details can be found in [18,19].

For nonzero but sufficiently small p, the solution ζ_c corresponds to a state of metastable equilibrium. The value $p = p_0$ above which no stable solution ζ_c exists corresponds to the *classical spinodal* in thermal and magnetic systems [19]. It has been shown that the decay of the system from the state ζ_c is isomorphic to the problem of quantum tunneling through a potential barrier [14]. Moreover, the lifetime Γ of the system in the metastable state can be calculated by path integral methods. Thus, the system has multiple states ζ_n, some of which may be metastable, and at least one of which represents a stable, minimal free energy state. The probability with which the system decays from the metastable state to a minimal free energy state depends on the magnitude of p, and the type of nucleation mechanism exhibited by the system depends on the value of $(p - p_0)$. Again, details may be found in [19].

These arguments demonstrate that field theories for fracture can plausibly be written in analogy to modern field theories of nucleation [19]. In this case, the mathematics of fracture is isomorphic to nucleation, and it follows that one can make predictions about fracture based upon results obtained from nucleation models. For example, it is known that tensile cracks can be described by a suitable distribution of dislocations. Furthermore, ideal linear elasticity in solids leads to an interaction $T(\boldsymbol{x} - \boldsymbol{x}')$ between elementary dislocations that falls off as $T(\boldsymbol{x} - \boldsymbol{x}') \sim /\boldsymbol{x} - \boldsymbol{x}'/^{-3}$. Using (1.1), it is easy to see that the range of interaction R is infinite. One should therefore expect to find examples of nonclassical effects in the nucleation of cracks. In particular, nucleation of fractures may not occur classically, but may instead nucleate by the ramified droplet mechanism [19]. This may explain why precursors to earthquakes have been difficult to observe, i.e., that slip would appear to nucleate over a very large area "spontaneously", followed by explosive growth of the amplitude of slip, with little growth of the slipping area. Moreover, the fracture process zone, a region of partial separation often seen in laboratory experiments in advance of the obvious crack tip in many materials, may correspond to the ramified halo surrounding the classical part of the nucleating droplet [19]. A simple application of these ideas is shown in Figure 1, taken from [19]. In (a), a lattice plane in an otherwise linear elastic solid is subjected to an applied tensile stress whose magnitude is near the breaking strength across the lattice plane. After 5 time steps, the lattice is in metastable equilibrium with the applied stress. A small spatial fluctuation (random number) is then introduced to nucleate the lattice, and the subsequent development is a complex evolution of the offset field over time. The final result is an open crack with the classical offset profile (the lattice ends at 800Å, thereby cutting off the nonclassical region).

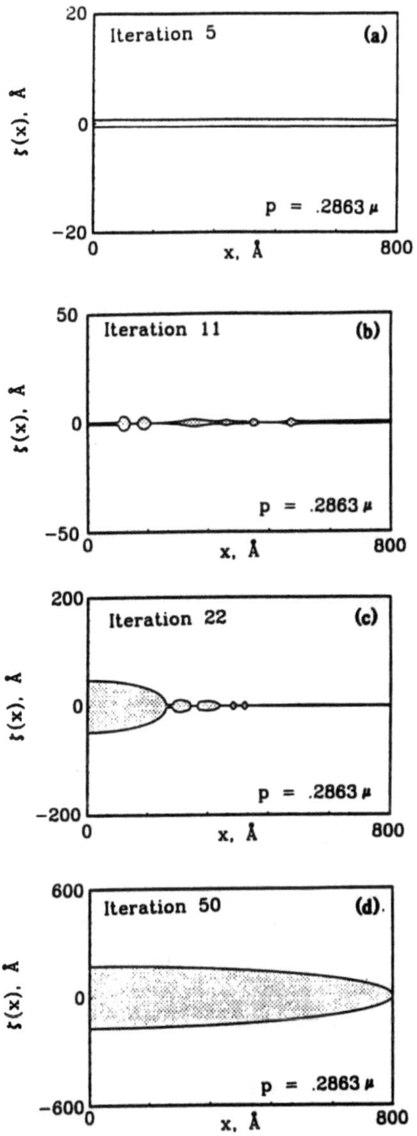

Figure 1. Plot of the offset field $\zeta(x, t)$ as a function of position x for four separate times of the model crack described in the text (after [19]).

4. Lattice automata. So far, we have examined the problem of tensile fracture. Shear fracturing has the additional complication of friction. During the shear sliding that accompanies shear fracture, there is a process of bonds initially breaking, then reforming at points of contact across the surface, then breaking again, and so forth. In our simple treatment, we neglect the processes of spallation, in which pieces of the solid break off entirely, and comminution, in which grains of material caught between the surfaces are ground up by the rolling and shearing action. Microscopic examination of laboratory specimens, or field examination of joints, fractures, and faults reveals that cohesion across the surfaces occurs primarily at the bumps at which the contact is maintained. It is the continual breaking and reforming of the cohesive bonds at the points of contact or "asperities" that give rise to frictional forces. The lattice automaton approach therefore attempts to describe a fairly limited and specific set of effects involved in the generation of frictional forces.

To motivate our automation model, consider two surfaces in contact. When a macroscopic, tangential traction force $p(t)$ is applied across the surfaces, the microscopic contact points experience increased stress. Depending on the existing level of total stress σ_i (externally applied plus local effects), some of these cohesive points break loose, jumping a distance s_i apart. This jump clearly results in a lower local stress on the points that moved, as well as a transfer of stress to neighboring points. Thus our model leads to a picture in which cascades of failed points progressively spread out from initial failure loci. Moreover, we have a *multiple state* model, just as in the nucleation approach. As the stress σ_i increases, the slip variables s_i remain fixed, or nearly so, until a threshold value $\sigma_i = \sigma_i^F$ is reached. At this point, site i decays, allowing s_i to jump ahead to a new larger value. Here, the threshold σ_i^F plays a role similar to p_0, the spinodal stress in the nucleation picture.

To understand more quantitatively the dynamics of these processes, we need to construct a mathematical model corresponding to the physical picture outlined above. Returning to the surfaces in contact, imagine a spatial averaging (coarse-graining) operation that divides the contact surface into a regular gridwork (lattice), each block having a side length a. Across each block of the lattice there are cohesive forces of a specified maximum strength σ_i^F. In general, application of external force $p(t)$ raises the level of local stress everywhere on the lattice.

Suppose the experiment is conducted so that one surface is dragged across another at an instantaneous velocity V. A force arises at all points on the lattice because the slip state on the surfaces falls behind the slip measured at the reference point. When the local stress exceeds the local failure strength σ_i^F, the cohesive bond is broken, and the surface within the lattice block jumps ahead by the amount of slip s_i. Cohesion is then reestablished. Transfer of local stress is described by an interaction T_{ij}, which expresses the change in local stress for a given jump in slip.

The mathematical description therefore begins with the construction of a spatial lattice, at each site of which a slip variable $s_i(t)$ is defined for each time t. The local stress $\sigma_i(t)$ at each point is given by the expression:

$$\sigma_i(t) = p(t) + \Sigma_j T_{ij} \sigma_j(t) \tag{4.1}$$

where $p(t)$ is the externally applied stress field, T_{ij} is the interaction, $\phi_j(t) = sj - Vt$ is the deficit in slip, and V is the externally imposed velocity. In the case of interest for frictional sliding, we set $p(t) = 0$, and (4.1) becomes:

$$\sigma_i(t) = \Sigma j T_{ij} sj(t) - V\ t\ T^{(0)} \tag{4.2}$$

where the first moment $T^{(0)}$ of T_{ij} is $T^{(0)} = \Sigma j T_{ij}$.

In simulating the laboratory experiments on sliding friction, two important assumptions are made, based upon a physical picture of the sliding process. First, an increase in the local stress σ_i above the threshold σ_i^F puts the lattice site i into an excited, metastable state, whose average lifetime for decay is Γ. The existence of a nonzero Γ is consistent with the picture of nonclassical nucleation described above. Second, it should be recalled that friction experiments are controlled by an automated system that keeps the slip surfaces moving across each other at a desired rate V [17]. The output of a servo controller measuring the state of slip is fed to the ram driving the surfaces. Clearly, this electromechanical system must have a response time Δ between when the deformation is measured, and when the ram force increment is applied. In this physical picture, the surfaces are driven by a series of force increments $-V\Delta T^{(0)}$ separated by time increments $\delta t = \Delta$. The major assumptions in the present model are that Γ and Δ are constant, independent of V, and that $\Delta \gg \Gamma$ (see, e.g., [20]). For convenience, Δ is set equal to 1.

Now consider a T_{ij} that is nonzero for self- and nearest neighbor-interactions only. This T_{ij} represents all interactions in the limit that only spatial variations in slip containing small wavenumber modes are kept. A nearest neighbor T_{ij} also has an appealing physical interpretation, the Burridge-Knopoff (BK) spring-and-block model for earthquakes. In the BK model, the blocks are linked together by coupling springs with constant K_C, and connected to a translating rigid plate by loading springs with constant K_L. Denoting the self interaction at the ith location by $T_{\langle s\rangle}$, and the nearest neighbor interaction by $T_{\langle nn\rangle}$, one can write the spring constants as $K_L = -T^{(0)}, K_C = T_{\langle nn\rangle}$. In the BK model described here, it is assumed that $T^{(0)} < 0$.

The dynamics in the model are generated by the use of a threshold force σ_i^F defined at each lattice location, and by a jump rule. These define a nonlinear map [11] taking the slip variables s_i at time $t-1$ into the slip at time t:

$$s_i(t) = s_i(t-1) + \{\Omega_i\}\Theta[\sigma_i(t) - \sigma_i F] \tag{4.3}$$

In (4.3), Θ is the Heaviside step function, $\sigma_i(t)$ is defined in (4.2), and $\Omega_i = \Omega(\boldsymbol{x}_i)$ is the amplitude of the jump in slip. A variety of distributions Ω were tried, and all yielded the same fundamental results (statistics, exponents, etc.). In the results described below, Ω is taken to be fractal, with Hausdorff dimension $D = 2.2$.

The map (4.3) proceeds as follows. Starting from random initial conditions, time is incremented by $\delta t = 1$, increasing $\sigma_i(t)$ by $-VT^{(0)}$. The argument of Θ is sequentially evaluated at each site i. If the argument is positive, $s_i(t)$ increases by Ω_i. Sequential passes through the lattice continue until the argument of Θ is negative everywhere. At this point, t again increases by 1 and the process repeats.

Figure 2. Plot of clusters of failed lattice sites on a square lattice of size 100 × 100. Each dot represents an individual lattice site that failed on time step 20101. After [20].

For all classes of models described by (4.3) in which $T^{(0)} \leq 0$, $T_{\langle nn \rangle} > 0$, evolution of the order parameter field occurs by the formation and growth of clusters of failed lattice points. In the calculations, random initial conditions $s_i(0)$ are used, together with boundary conditions corresponding to $K_L = -T^{(0)} = 0$ on the edge blocks. V was chosen to be 8, and a value for Ω_i was assigned to each location according to $\Omega_i = (\sigma^D - \sigma_i^F)/T_{\langle s \rangle}$, where σ_i^F is a self-affine (fractal) surface bounded by the upper limit 350, and the lower limit 50, and $\sigma^D = 10$.

This BK model was run for several tens of thousands of time steps. Shown in Figure 2 are all of the lattice sites that failed on time step 20101. For the most part, the clusters of failed points are compact rather than ramified (they have few holes in their interiors), although one relatively ramified cluster can be seen in the lower left hand corner of the figure. It can be shown that the compactness of the clusters arises in part from the local nature of the interaction [9].

One intuitively feels that the clusters of failed sites in Figure 2 are analogous to the clusters of occupied lattice sites appearing in percolation problems. This is in fact the case [11]. Because the physics of thermal and magnetic systems near the critical point can be mapped onto percolation problems near the percolation threshold, our model for frictional sliding can therefore be understood using concepts from critical phenomena.

Perhaps the most important concept is the correlation length ξ. Two lattice sites are correlated if failure at one site depends on failure at the other. Quantity ξ is essentially the maximum distance over which the dynamics at the sites are correlated. Thus, the physics of the system will be independent of scale up to the length ξ. Alternately, one typically finds that the probability of simultaneous failure for two sites separated by a distance r is proportional to $e^{-r/\xi}$.

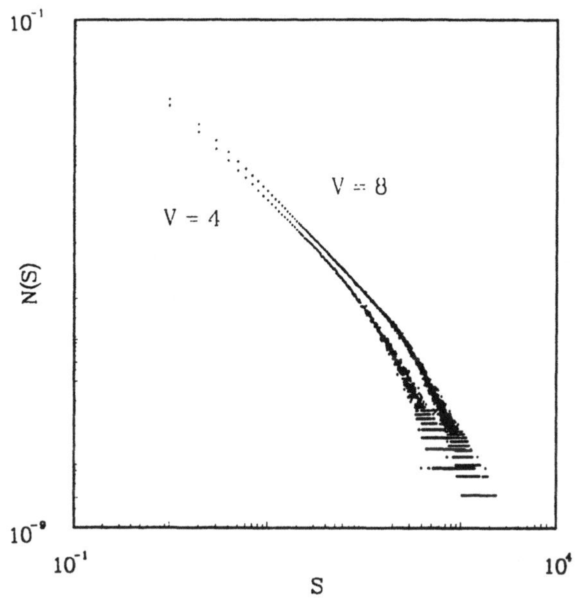

Figure 3. Plot of frequency of appearance $N(s)$ of clusters of size s for two different sliding velocities $V = 4$ and $V = 8$. A scaling region is apparent for each value of V.

The ***scaling hypothesis*** asserts that near a critical point, all important quantities are proportional to a power of the correlation length. This hypothesis, which has been verified for a variety of systems, implies that the physically meaningful observables are the scaling functions and their exponents. An an example, Figure 3 is a log-log plot illustrating the frequency of appearance of clusters $N(s)$ as a function of cluster size s, for two values of V, 4 and 8. All other parameter values are the same as for Figure 2. It can be seen that a scaling region exists in which $N(s) \propto s^{-\tau}$ and that $\tau \approx 1.5$ for both values of V. τ is called the Fisher exponent. The break in slope apparent on both curves in Figure 3 indicates that there is a critical cluster size s_c beyond which scaling no longer holds, the value of s_c clearly increasing with

V. According to the scaling hypothesis, one expects that the mean cluster size $\langle s \rangle$, and therefore also the critical cluster size, depend on a power of ξ. We therefore expect that $s_c \propto \xi^{D_c}$ where D_c is *average* fractal dimension. Clearly, ξ increases with V, and we infer that a critical point exists at which $\xi \to \infty$ for sufficiently large V.

5. Conclusion. Both the field theoretic nonclassical nucleation models and the lattice automata represent complementary ways of viewing the earthquake source process. The traditional picture of an earthquake as a compact area of slip on a fault plane has no counterpart in the physical processes predicted by these nonclassical approaches. While it is not yet clear exactly what deformation or other signals will characterize the failure process, these models predict that failure will be associated with the growth of the most unstable Fourier modes. A primary goal of future research will be to identify these modes by the use of models like those above.

Acknowledgements. This research has been conducted under the auspices of the US Department of Energy by Lawrence Livermore National Laboratory under contract W-7405-ENG-48. The author would also like to thank S.R. Brown and W. Klein for many stimulating and helpful conversations.

REFERENCES

[1] W.H. BAKUN, M.M. CLARK, R.S. COCKERHAM, W.L. ELLSWORTH, A.G. LINDH, W.H. PRESCOTT, A.F. SHAKAL, and P. SPUDICH, *The 1984 Morgan Hill, California earthquake*, Science, 225 (1984), pp. 288–291.

[2] K. BINDER, *Decay of metastable and unstable states: mechanisms, concepts and open problems*, in Statistical Physics, Invited Lectures from STATPHYS 16, ed. H.E. Stanley, North-Holland, Amsterdam (1986), pp. 35–43.

[3] J.H. DIETERICH, *Mechanical behavior of crustal rocks (the Handin volume)*, ed. N.L. Carter, J.M. Logan, and D.W. Stearns, Geophys. Monograph Ser. 24, Amer Geophys. Un. (1981), pp. 103–120.

[4] J.W. GIBBS, *On the Equilibrium of Heterogeneous Substances, 1*, Trans. Conn. Acad. Arts Sci (1876).

[5] J.D. GUNTON and M. DROZ, *Introduction to the Theory of Metastable and Unstable States*, Springer–Verlag, Berlin (1983).

[6] A.A. GRIFFITH, *The phenomena of rupture and flow in solids*, Phil. Trans. Roy. Soc. Lon. A, 221 (1920), pp. 163–198.

[7] K. HUANG, *Statistical Mechanics*, John Wiley & Sons, New York (1963).

[8] M.J.S. JOHNSTON, A.T. LINDE, M.T. GLADWIN, AND R.D. BORCHERDT, *Fault failure with moderate earthquakes*, Tectonophysics, 144 (1987), pp. 189–206.

[9] W. KLEIN and C. UNGER, *Pseudospinodals, spinodals, and nucleation*, Phys. Rev. B, 28 91983), pp. 445–448.

[10] W. KLEIN and F. LEYVRAZ, *Crystalline nucleation in deeply quenched systems*, Phys. Rev. Lett., 57 (1986).

[11] W. KLEIN and RUNDLE, to be published (1991).

[12] B.V. KOSTROV and S. DAS, *Principles of Earthquake Source Mechanics*, Cambridge University Press, Cambridge (1988).

[13] V.C. LI, *Mechanics of shear rupture applied to earthquake zones*, in Fracture Mechanics of Rock, ed. B.K. Atkinson, Academic Press, London (1987), pp. 351–428.

[14] J.S. LANGER, *Theory of the condensation point*, Ann. Phys. N.Y., 41 (1967), pp. 108–157.

[15] M. Lisowski, W.H. Prescott, J.S. Savage, and J.L. Svarc, *A possible gedletic anomaly observed prior to the Loma Prieta, California, earthquake*, Geophys. Res. Lett., 17 (1990), pp. 1211–1214.

[16] National Research, *Geodesy, A Look to the Future*, National Academy Press, Washington, DC, (1985).

[17] A.L. Ruina, *Slip instability and state variable friction laws*, J. Geophys. Res., 88 (1983), pp. 10359–10370.

[18] J.B. Rundle, *A physical model for earthquakes, 3, Thermodynamical approach and its relation to nonclassical theories of nucleation*, J. Geophys. Res., 94 (1989), pp. 2839–2855.

[19] J.B. Rundle and W. Klein, *Nonclassical nucleation and growth of cohesive tensile cracks*, Phys. Rev. Lett., 63 (1989), pp. 441–444.

[20] J.B. Rundle and S.R. Brown, *Origin of rate dependence in frictional sliding*, J. Stat. Phys., in press, 1991.

[21] J.S. Savage, W.H. Prescott, and M. Lisowski, *Deformation along the San Andreas Fault 1982–1986 as indicated by frequent geodolite measurements*, J. Geophys. Res., 92 (1987), pp. 4785–4797.

[22] S. Tse and J.R. Rice, *Crustal earthquake instability in relation to the depth variation of frictional slip properties*, J. Geophys. Res., 91 (1986), 9452–9472.

THE SPATIAL STRUCTURE OF ISOTROPIC TURBULENCE AND THE RELATED THREE-DIMENSIONAL GRAPHICS PROBLEMS

A. VINCENT*, M. MENEGUZZI*AND J.D. VILLASENOR*

Vorticity tubes are observed with great details in data processing from direct numerical simulations of homogeneous turbulence at resolution 240^3 ($R_\lambda \approx 140$ with an inertial subrange extended over more than one decade). These tubes correspond to the regions of strongest vorticity as well as regions of lowest pressure. Their width is of the order of a few dissipation scales, while their length can reach the integral scale of the flow. The large velocity regions are sheets located near and parallel to the axis of the tubes but directed perpendicularly to it. The maxima of both the compressive (negative eigenvalue) or extensive (positive eigenvalue) part of the rate of strain tensor are located near the tubes and oriented perpendicularly to their axis. Further, the eigenvector associated with the intermediate eigenvalue is most of time a stretching (positive eigenvalue) and oriented along the axis of the tube. These results are in excellent agreement with recent laboratory measurements (Tsinober et al. 1990).

The advantage of direct numerical simulations is to provide a complete description in space and time of the various quantities involved in the physical process. Up to now, direct simulations of the three-dimensional Navier-Stokes equations were restricted to modest Reynolds numbers (Orszag and Patterson (1972), Siggia and Patterson (1978), Siggia (1981), Kerr 1985, Yamamoto and Hosokawa 1988, She et al. 1988, 1990). The reasons are not only the limited power of the available computers but also the problems resulting of management, access and processing of the resulting data. The object of this communication is to present some new results in the physical domain.

We solve the Navier-Stokes equation for incompressible fluids in rotational form

$$\frac{\partial \mathbf{v}}{\partial t} = \mathbf{v} \times \omega - \nabla(p + v^2/2) + \nu \nabla^2 \mathbf{v} + \mathbf{f} \tag{1}$$

with the continuity equation:

$$\nabla \cdot \mathbf{v} = 0$$

$\mathbf{v}$ is the velocity field, $\omega = \nabla \times \mathbf{v}$ the vorticity, p the pressure, ν the kinematic viscosity and $\mathbf{f}$ a force field. Since we are interested in (statistically) homogeneous turbulent flows, we take periodic boundary conditions in all directions, with period 2π. A pseudospectral method is used to compute the right-hand side of this equation (see Gottlieb and Orszag 1977). The time marching is done using a second order finite-difference scheme. We chose to force the field in a deterministic way, as was

*CERFACS, 42 av. G. Coriolis, 31000 Toulouse, France

done for instance by Kerr (1985). All Fourier modes with wavenumber components equal to 1 or 0 are forced with a constant amplitude $\mathbf{f}$.

The calculation presented here was done with 240^3 Fourier components, and a viscosity of 10^{-3}. This gives $R \approx 1000$ ($R_\lambda \approx 150$).

After integration over a few tens of turnover times we obtain a statistically stationary regime in the sense that the mean kinetic energy density fluctuates by a few percent in time around a constant value. The velocity derivative distribution, known to be strongly non gaussian, is found to be close to but not exponential (Vincent and Meneguzzi 1990). A method for estimating the degree of isotropy of the flow in the spectral space, is to project the velocity $\mathbf{V}(\mathbf{k})$ along two unit vectors $\mathbf{e}_1(\mathbf{k})$ and $\mathbf{e}_2(\mathbf{k})$ which form with $\mathbf{k}$ an orthogonal frame (Curry et al. 1984). The isotropy I is then defined as:

$$I = \left[\frac{\langle |v_1|^2 \rangle}{\langle |v_2|^2 \rangle} \right]^{1/2}$$

In the calculations presented here, the isotropy I fluctuates by a few percent around 0.95. Therefore, our flow is close to be statistically isotropic. Our time unit is such that the large eddy turnover time is 0.35. The energy spectrum $E(k)$ at a given time is shown in Fig. 1(a) where it has been multiplied by $k^{5/3}$. A power raw range can be seen for $k < 25$, with an exponent a little larger than $-5/3$. The small increase in the energy spectrum at the highest wavenumbers is a numerical effect due to a lack of resolution of the very small scales in the dissipative range. It is not due to aliasing and does not affect the inertial range. In Fig. 1(b), we show a power spectrum in time at a given point inside the box. Although the time series is not long enough, an inertial subrange is clearly seen. We have redone a part of our runs with aliasing removed by the method of shifted grids described in Patterson and Orszag (1971), and we did not find any significant difference in the results. We have also made some comparisons with a better resolution 300^3 without any noticeable differences. The fact that this is an inertial range is confirmed by inspection of the energy flux spectrum, shown in Fig. 2(a). The energy flux $\Phi(k)$, which is the rate of energy transferred per unit mass by all modes with $k' < k$ to modes with $k' > k$, is constant in the inertial range. In Fig. 2(b), we show the energy transfer $T(k)$. The inertial range is seen in Fig. 2 to occupy a little more than one decade in wavenumber space.

A 3D view of the vorticity field is given in Fig. 3 (see Plates section). The vorticity at each grid point is represented by a vector, here so small that individual vectors can hardly be seen. Vectors are only plotted if their modulus is larger than a given threshold. One can see that the vorticity is organized in thin elongated tubes, as previously reported by Siggia (1981), Kerr (1985), She et al. (1990). These tubes also correspond to the regions of lowest pressure. The length of these tubes seem to be of order 1, i.e. of the same order as the integral scale (the cube size is 2π). Their thickness is of the order of a few dissipation scales, here a few grid points. This is confirmed by a more detailed analysis. e.g. cut through a typical vorticity tube, and shows clearly that its characteristic thickness is a few grid points. The

(a)

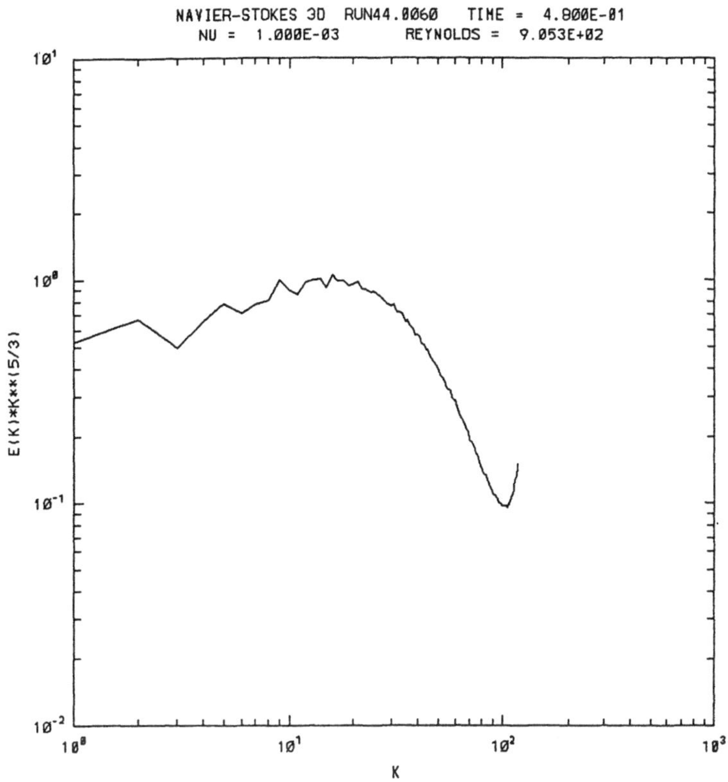

Figure 1a: Energy spectrum at a given time and multiplied by $k^{5/3}$

dissipation scale l_D, in our simulation, is of the order of the mesh size, while the Taylor microscale is approximately ten mesh sizes. Therefore, the characteristic tube thickness seems to seems to be intermediate between these two lengths.

We have done a filter of the vorticity field after removing all dissipation range scales and also the large scales to which the forcing is applied. The external regions of the tubes are still present. Hence, one is lead to the conclusion that these vorticity tubes, which seem to be the basic structure of three-dimensional homogeneous turbulence, involve all the scales of the flow.

Extensive use of NCAR-gks has been done for the graphics. For exploring cubes we have written "flight simulator"-type graphic packages running on an IBM5080 and on a Silicone Graphics 4D220. However, due to the size of the field to explore (240**3 vectors) and also to the limited power of the line that we used for transfering data (48Kbit/sec), we have isolated particular regions of space, like tubes for example, easier to manage on our graphics workstation. A single tube along with its environment has been extracted from the vorticity field by a pattern recognition algorithm. The underlying principle of this algorithm involves a search using a "probe" which provides a measure of the location of the structure in question. This is working because a vorticity tube can be regarded as an elongated, connected

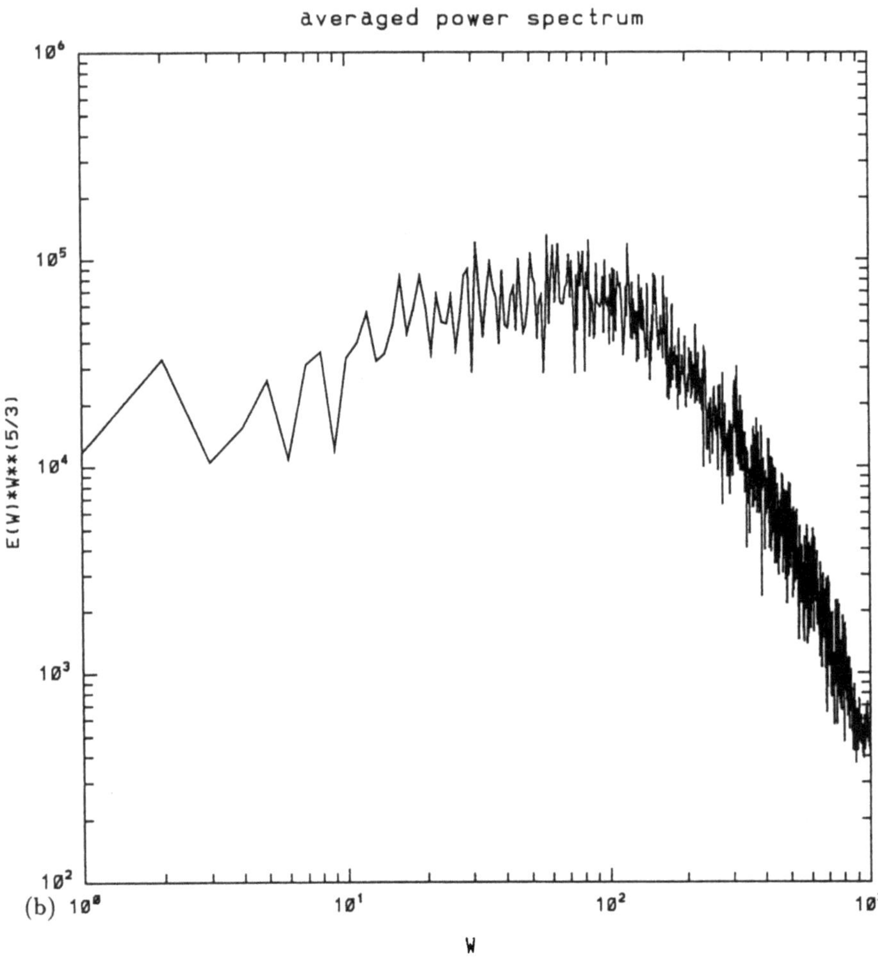

(b)

Figure 1b: Energy spectrum in time at a given point in space multiplied by $k^{5/3}$.

structure which is characterized by a "skeleton" (e.g. Serra 1990). The first step is to extract the skeleton of a given tube by following the central region of the tube where the vorticity is maximal. When the skeleton has been found, selection of the actual vectors forming part of the tube is accomplished by choosing all vectors lying within a given distance of the skeleton. This is accomplished using overlapping volumes described topologically as balls. The second step is to find the next position of the tube at the next instant in time. A progressive and exhaustive exploration of the space around the central region of the tube is done by a search in spiral. The algorithm is described in details in Villasenor and Vincent (1990). Some helical structure can be seen, in the external part of the tube, as noted by She et al. (1990). This might correspond to secondary instabilities closely related to the dynamic of the tube itself.

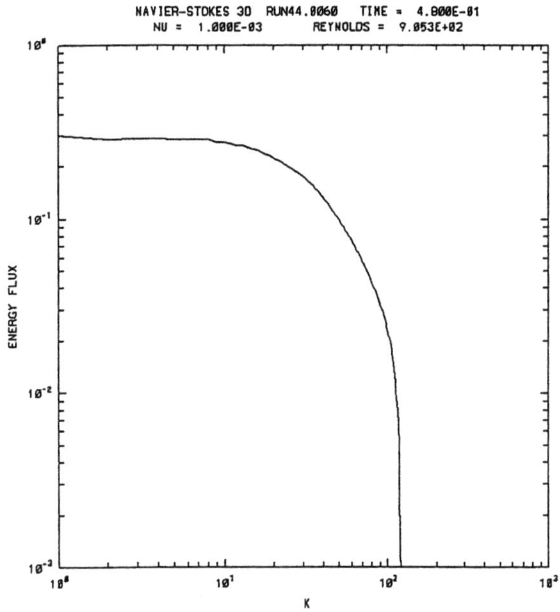

Figure 2 a

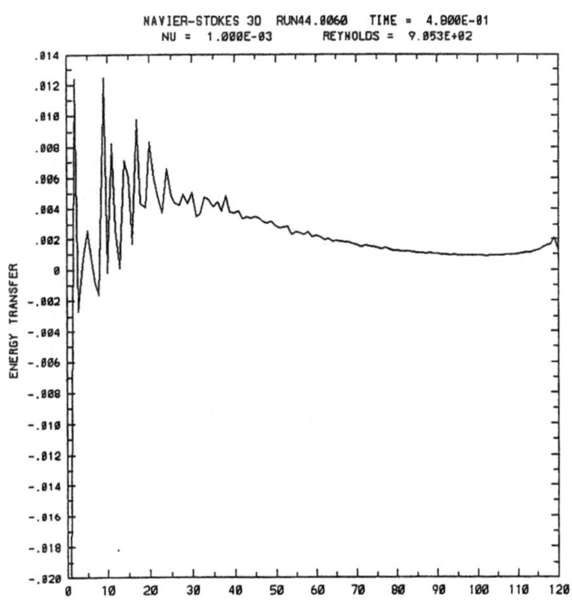

Figure 2 b

Figure 2: Energy flux corresponding to the spectrum of Fig. 1.

Space correlations between vorticity and rate of strain were reported by Kerr (1985). At each grid point, we compute the eigenvalues $\lambda_1, \lambda_2, \lambda_3$ of the rate of strain tensor $S_{ij} = \frac{1}{2}(\partial_i v_j + \partial_j v_i)$, with λ_1 always negative and λ_3 always positive ($\nabla \cdot \mathbf{v} = 0$ implies $\lambda_1 + \lambda_2 + \lambda_3 = 0$). The intermediate eigenvalue λ_2 is found to be positive in approximately two thirds of the cases, in agreement with the results of the analysis of Kerr (1985) and Ashurst et al. (1987). Three-dimensional plots of the eigenvector fields, are used to study the correlations with vorticity. Let us call $\mathbf{e}_1, \mathbf{e}_2, \mathbf{e}_3$ the eigenvectors associated with $\lambda_1, \lambda_2, \lambda_3$ respectively, each eigenvector being given the amplitude and sign of its associated eigenvalue. The highest rate of strain is seen to be generally in the vicinity of vorticity tubes. The eigenvector $\mathbf{e}_3$, associated with the positive eigenvalue, i.e. the stretching direction, is perpendicular to the vorticity (Fig. 6(a)) (see Plates section). The eigenvector $\mathbf{e}_1$, associated with the negative eigenvalue of the strain, is also perpendicular to the vorticity (Fig. 6(b)) (see Plates section). The intermediate eigenvector $\mathbf{e}_2$ is the one which is generally aligned with vorticity (Fig. 6(c)) (see Plates section). Our conclusions are drawn from fully three-dimensional examination of the flow using rotation, zoom and clipping. We could not find an example of $\mathbf{e}_3$ aligned with vorticity.

A more quantitative picture of the alignment between the strain eigenvectors and vorticity is given by the distribution of the cosine of the angles between ω and $\mathbf{e}_1, \mathbf{e}_2, \mathbf{e}_3$. In the case of $\mathbf{e}_2$, we computed separately the distribution for $\lambda_2 < 0$, $\lambda_2 > 0$, and the complete distribution. We find $\lambda_2 > 0$ in 75% of the cases. We obtained the same results than Tsinober et al. (1990) from experimental data.

The general alignment of vorticity with $\mathbf{e}_2$ reflects the quasi two-dimensionality of the vortex tubes. Indeed, in two dimensions, $\lambda_2 = 0$, $\lambda_3 = -\lambda_1$ and ω would be perpendicular to the plane $(\mathbf{e}_1, \mathbf{e}_3)$. A discussion of this correlation of vorticity with the intermediate rate of strain direction is presented in Ashurst et al. (1987), where an explanation is proposed using a model due to Vieillefosse (1982, 1984). According to calculations using this model, a vorticity tube is created along the direction $\mathbf{e}_3$ of maximum stretching, but very soon becomes aligned with $\mathbf{e}_2$. However, we have followed vorticity tubes back in time until the time of their formation without noticing any change in this configuration. Moreover, we observe that the maxima of velocity form a sheet perpendicular to the axis of the tube. In Fig. 5 is shown the projection of the velocity field on a plane perpendicular to a vorticity tube. Our preliminary impression is that the tubes are generated by an instability due to the shear associated with the velocity sheets. If this is the case, the above picture would not be correct. The results of this investigation will be presented in a forthcoming paper.

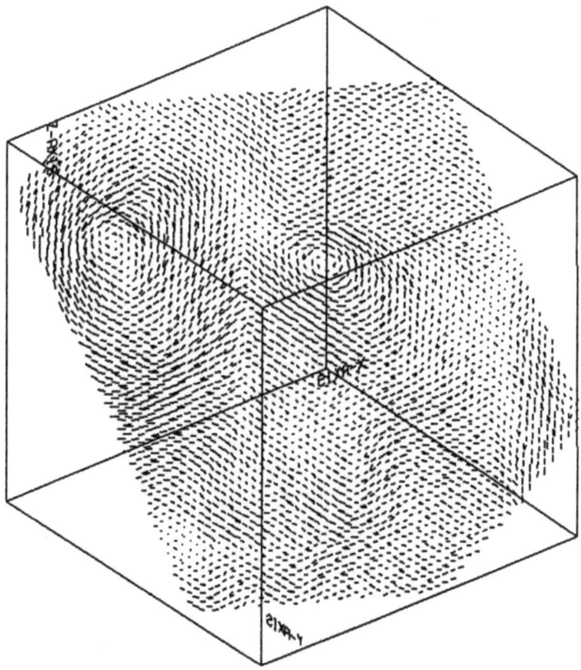

Figure 5: Projection of the velocity field on a plane perpendicular to a particular vorticity tube.

Acknowledgements. The calculations presented were done on the Cray-2 of CCVR (Centre de Calcul Vectoriel pour la Recherche), Palaiseau, France.

REFERENCES

ASHURST W.T., KERSTEIN A.R., KERR R.M., GIBSON C.H., Phys. Fluids, 30, 2343 (1987).

CURRY J.H., HERRING J.R., LONCARIC J., ORSZAG S.A., J. Fluid Mech., 147, 1 (1984).

GOTTLIEB D., ORSZAG S.A., *Numerical Analysis of Spectral Methods*, SIAM CBMS-NSF Regional Conference Series in Applied Math (1977).

KERR R., J. Fluid Mech, 153, 31 (1985).

ORSZAG S.A., PATTERSON G.S., Phys. Rev. Lett, 28, 76 (1972).

PATTERSON G.S., ORSZAG S.A., Phys. Fluids, 14, 2538 (1971).

SHE Z.S., JACKSON E., ORSZAG S.A., J. Sci. Computing, 3, 407 (1988).

SHE Z.S, JACKSON E., ORSZAG S.A., Nature, 344 , 226 (1990 1988).

SIGGIA E.D., PATTERSON G.S., J. Fluid Mech, 86, 567 (1978).

SIGGIA E.D., J. Fluid Mech, 107, 375 (1981).

TSINOBER A., KIT E., DRACOS T., submitted to J. Fluid Mech. (1990).

YAMAMOTO K., HOSOKAWA I., J. Phys. Soc. Japan, 57, 1532 (1988).

VIEILLEFOSSE P., J. de Physique, 43, 837 (1982).

VIEILLEFOSSE P., Physica, 125A, 150 (1984).

VINCENT, A.P., MENEGUZZI, M., to appear in the journal of Fluid Mechanics, 1990.

VILLASENOR, J.D., VINCENT, A.P., submitted to Computer Vision, Graphics, and Image Processing (1990).

OTHER TITLES AND ABSTRACTS

The following are titles and abstracts of those workshop talks for which the speakers did not prepare an article for the proceedings volume.

U.R. Christensen

Max-Planck Inst. für Geochemie,
Mainz, Germany

Three-dimensional mantle convection with variable viscosity: Mantle plumes and toroidal excitation

Abstract: We report numerical calculations for three-dimensional convection with variable viscosity. A hybrid spectral and finite difference method is used. The coupling of modes in the equation of motion, which is caused by lateral viscosity variations, is treated iteratively. Solutions for bimodal, hexagonal, square, triangular and spoke patterns are reported for bottom heated convection at infinite Prandtl number. The Rayleigh number is between critical and 10^5 and temperature-induced viscosity contrasts up to 100 are considered. In agreement with results from laboratory experiments we find that temperature-dependent viscosity favours at low Rayleigh number flow patterns like squares or hexagons, where a columnar rising current is surrounded by sheet like descending flow. The dichotomy in geometry between upwelling and sinking flow becomes more pronounced with increasing viscosity contrast. The temperature-dependence of viscosity gives rise to a toroidal velocity component. However, it amounts only to a few percent of the total velocity. In contrast, at the earth's surface an approximate equipartitioning of poloidal and toroidal energy is found. We show that with non-Newtonian and depth-dependent rheology the toroidal component at the free surface can become significant, and a pattern reminiscent of plate motion can arise in a free convection model. Although these results are obtained in a parameter range which is not directly applicable to the earth, they support the conclusions that (i) upwelling flow in the mantle is unlikely to be sheet-like and will probably be in the form of columnar plumes, and that (ii) the toroidal motion found at the earth's surface is due to the highly non-linear rheology which leads to the existence of mobile surface plates and is not caused by viscosity variations related to lateral temperature contrasts deeper in the mantle.

Michael C. Gurnis

University of Michigan,
Ann Arbor, Michigan

Large-scale mantle convection and dynamic interaction with plates

Abstract: The dynamic coupling of tectonic plates (including non-subducting continental lithosphere and subducting oceanic lithosphere) with thermal convection is being investigated with finite element models. In a range of cases, the plates are viscous entities (like the background fluid) and are introduced into the system with various rheologies and boundary conditions. The models are meant to explore rather specific parts of the over-all coupling between tectonic plates and convection and to help us understand the relationship between various quantities measured on the Earth's surface and the dynamic processes in the interior.

Particular attention will be given to the interaction between large non-subducting continental lithosphere and the deep mantle and how these interactions effect seismic structure, dynamic topography, the geoid, and sea-level (measured as continental flooding). I will demonstrate that such interactions have a first order influence on the spatial and temporal pattern of continental flooding; I will suggest how the pattern of observed marine deposits on continental platforms can be used to constrain aspects of mantle rheology and mantle flow.

U. Hansen Univ. zu Köln, Germany

Chaotic thermal convection at infinite Prandtl number

Abstract: Numerical simulations of two-dimensional high Rayleigh (Ra) number, base heated convection in large aspect ratio boxes are presented for infinite Prandtl number fluids as applied to the earth's mantle. Typically, flows at $Ra > 10^6$ consist of large-scale cells with intermittent boundary layer instabilities. Due to the instabilities the flow displays strong temporal variations while the existence of the large-scale flow leads to spatial coherent structures. Changes in the large-scale component (appearance or disappearance of convection cells) are found to operate on a time scale much longer than the one associated with the boundary layer instabilities. For Ra exceeding 10^7 we find the heat transfer mechanism changing from one, characterized by mushroom-like plumes to one consisting of disconnected instabilities. The evolution and the fate of the boundary-layer instabilities is determined by the kind of interaction between those instabilities and a 'wind', generated and maintained by the large scale circulation. Collisions of instabilities within the thermal boundary layers lead to pronounced plumes having the tendency to develop 'pulse-like behavior'. Tilting of the instabilities, resulting from the shear which is exterted by the wind is another typical feature of the multiscale flow. The richness of the dynamical phenomena resulting from the interaction of the different scales is potentially interesting with respect to the variety of geodynamical features in the Earth's mantle, as observed from seismic tomography.

George W. Bergantz University of Washington,
Seattle, Washington

Conjugate solidification and melting in open and closed multi-component systems

Abstract: Underplating and intrusion of continental crust by basaltic magma has repeatedly been invoked as a mechanism to provide the needed heat and mass transfer to initiate and sustain magmatism in a variety of tectonic settings. This superposition yields a strongly time dependent and non-linear feedback system with both conductive and convective components. Theoretical and experimental studies have been undertaken which focus on the manner in which transport properties and thermal history are buffered by enthalpy changes associated with simultaneous solidification and melting as a function of the change of melt fraction for naturally ocurring compositions. This is particularly important as the crystallization of basalts and the melting interval for common crustal rocks overlap; this yields strong changes in material properties which govern the heat transfer systematics at the solidification-melting interface.

Experiments: The underplating process was modeled by superposing mixtures of fluid and solid paraffins representing "basalt" and "country rock" in a vareity of geometries. The paraffins have similar Stefan numbers, and relationships between density, viscosity, enthalpy, and most importantly, melting interval, as magmas. These paraffins exhibit no nucleation delay. Turbulent convection in the underplated material was not observed: the system exhibted a dynamical and thermal history largely consistent with models based on a coupled conductive solidification of the "basalt" and partial melting of the "country rock".

Numerical modeling: of this conjugate system in the presence of both forced and fully coupled free convection yields a variety of time dependent interactions from melting back of the wall in the case of turbulent forced flow to a short-lived period of natural convection associated with the fully coupled and more geologically realistic case. The governing dynamical quantities are the slope of the solid fraction distribution $1/e$, and the background temperatures of the country rock. The Peclet number associated with interdendritic flow is demonstrably small, and can be neglected. The formulation of the problem in terms of a general dependent variable, the chemical potential, will be discussed.

David J. Stevenson — Division of Geol. and Plant. Sci., Caltech, Pasadena, California

Aspects of the dynamics in two-phase media

Abstract: A large fraction of the Earth is either solid nor liquid but a combination of these phases. I will concentrate on partial melts, the precursor to all igneous activity. After a brief introduction to the underlying theory and a summary of the properties of magmons (non-linear solitary waves) I will shift to recent work concerning the possible instability of partial melts under shear, especially the fascinating and non-linear phenomenon of the possible development of a dendritic drainage network as the origin of macrosegregation.

Phillippe Machetel — French Space Agency of Toulouse, France

Short-wavelength lower mantle seismic velocity anomalies

Abstract: The lower mantle seismic velocity anomalies display very smooth shapes and low amplitudes in apparent contradiction with the thermal anomalies obtained from chaotic convection models. An attempt to reconcile theoretical convection models with tomographic results is presented in this talk. Along with a coherent approach of lower mantle convection, the influences of radial and angular filtering on the shapes and on the amplitudes of anomalies have been monitored. A tremendous spreading of the radial anomalies into horizontal structures and a significant loss of amplitude of the anomalies result from this exercise. The level of the Root Mean Square of the thermal anomalies falls from several tens of degrees to values in close agreement with tomographic results. This study shows that the short-wavelength of lower mantle velocity anomalies are dramatically hidden by the truncation level of seismic tomography.

Gary A. Glatzmaier — Earth Sciences Division,
Los Alamos National Lab., New Mexico

Three-dimensional numerical simulations of chaotic convection in the Earth's mantle

Abstract: Three-dimensional numerical simulations of highly viscous thermal convection in a spherical fluid shell used to try to understand the basic properties of convection in the Earth's mantle. The anelastic approximation filters out sound waves while accounting for a density stratification. All dependent variables are expanded in spherical harmonics to describe their latitudinal and longitudinal structures and in Chebyshev polynomials to describe their radial structures. A semi-implicit time integration scheme that use a spectral transform method to compute the nonlinear terms evolves the time-dependent solution. Two cases are presented. One is totally heated from within the model mantle and has a Rayleigh number of 120,000. It initially has a tetrahedral spatial pattern that is periodic in time. After several periods this spatial and temporal pattern evolves into a chaotic time-dependent structure. The second case is 80% heated from within the mantle and 20% heated from below and has a Rayleigh number of 1,600,000. This case, which is more representative of the Earth's mantle, is strongly and chaotic and characterized by long narrow downflows near the outer boundary and cylindrical upflows near the inner boundary. 16 mm. movies of both cases are shown.

Stephen Morris — Dept. of Mech. Engineering,
UC Berkeley, California

Rheological control of solid-state transformations occurring by nucleation and growth

Abstract: A solid transforming by nucleation and growth consists of grains of new phase growing in a matrix of parent phase. The grain radius is seen to increase by four or five orders of magnitude as neighbouring grains grow and merge. It is argued here that growth therefore involves flow in the matrix unless the two phases have very nearly the same density. What follows is a theoretical study of the coupling of flow and growth-rate when the parent and product phases have the same composition. The standard metallurgical treatments of this problem assume growth is controlled completely by interfacial kinetics. The object of the present work is to show why the coupling of flow and growth-rate is important and to show how it can be observed in experiments.

The coupling of flow and growth-rate is studied initially for an isolated spherical grain by taking the interfacial kinetics to be infinitely fast. The timescale for growth is then determined by the rheology of the matrix and by the amount $(P - P')$ by which the applied pressure P exceeds the equilibrium pressure P', i.e., the pressure P' at which the two phases are in equilibrium across a plane interface. Detailed analysis shows that rheologically-controlled growth has a characterisitc signature: when the excess pressure $(P - P')$ is independent of time, the volume of an isolated spherical grain grows exponentially with time. This result is general and holds for any matrix in which the deviatoric stress depends on the strain-history and not simply on the instantaneous strain. (For example, it applies when the deviatoric stress is proportional to the strain-rate.)

By contrast, previous studies of kinetic control have shown the volume of an isolated spherical grain to grow algebraically with time. It should therefore be possible to distinguish the two types of control in high-pressure experiments. An unambiguous test is possible by experiments involving single grains, so that the nucleation-rate is not a factor in interpreting measurements of the volume fraction.

Recent experiments by Rubie et al (1990) are compatible with the existence of rheological control, but the results are slightly ambiguous: the authors inferred that only a single grain was present, but were not sure. This work is motivated by a problem in geophysics. The pressure in the earth is nearly hydrostatic, and phase transitions therefore occur on level surfaces if the temperature is horizontally-uniform and the transition is in thermodynamic equilibrium. However the conditions for thermodynamic equilibrium are poorly understood. The essential problem is one of scaling: for application to the mantle, measurements taken in the laboratory must be extrapolated to dilatation-rates 10-12 orders of magnitude slower. If kinetics alone control the growth, this extrapolation leads to the well-known conclusion that significant vertical deflexions of the transition-layer are unlikely at the high temperatures typical of the deep mantle.

By contrast, it is shown here that the existence of rheological control means that laboratory experiments are compatible with significant vertical deflexions of the transition-layer if the strain-rate in the matrix depends strongly on the deviatoric stress. The problem is analysed for two particular choices of matrix. The first example is a power-law matrix in which the strain-rate varies as the n-th power of the deviatoric stress. It is shown that if the matrix is a linear solid (i.e. $n = 1$), rheological control will cause no major deflexion of the transition-layer. In contrast, major deflexions are possible if the strain-rate depends strongly on the deviatoric stress (i.e., $n >> 1$). This behaviour is clarified by the second example which studies the limiting case of an elastic, perfectly-plastic solid with yield stress $2m$ and elastic rigidity G. This example allows us to find the additional dissipation produced if mantle convection occurs through the transition-layer. The volume-integrated caused by phase changes is found to exceed that caused by deformation outside the transition-layer if the yield stress of the matrix satisfies $2m > G/3000$. (The corresponding excess pressure is about 0.2 GPa.)

The additional dissipation may have two effects on the large-scale flow. First, it reduces the turnover speed of the cell and is therfore likely to destabilise the horizontal boundary-layers, and enhance mixing. Secondly, layering may result if sufficient dissipation is added.

FSC
www.fsc.org
MIX
Papier aus verantwortungsvollen Quellen
Paper from responsible sources
FSC® C105338

If you have any concerns about our products,
you can contact us on
ProductSafety@springernature.com

In case Publisher is established outside the EU,
the EU authorized representative is:
Springer Nature Customer Service Center GmbH
Europaplatz 3, 69115 Heidelberg, Germany

Printed by Libri Plureos GmbH
in Hamburg, Germany